Lecture Notes in Mathematics   1609

Editors:
A. Dold, Heidelberg
F. Takens, Groningen

Subseries: Fondazione C.I.M.E., Firenze

Advisor: Roberto Conti

T0185156

Springer
Berlin
Heidelberg
New York
Barcelona
Budapest
Hong Kong
London
Milan
Paris
Tokyo

L. Arnold   C. Jones
K. Mischaikow   G. Raugel

Dynamical Systems

Lectures given at the 2nd Session of the
Centro Internazionale Matematico Estivo
(C.I.M.E.)
held in Montecatini Terme, Italy,
June 13-22, 1994

Editor: R. Johnson

Fondazione
C.I.M.E.

 Springer

Authors

Ludwig Arnold
Institut für Dynamische Systeme
Universität Bremen
D-28334 Bremen, Germany

Christopher K. R. T. Jones
Division of Applied Mathematics
Brown University
Providence, RI 02912, USA

Konstantin Mischaikow
Department of Mathematics
Georgia Institute of Technology
Atlanta, GA 30332, USA

Geneviève Raugel
Laboratoire d'Analyse Numérique
Université de Paris-Sud
F-91405 Orsay, France

Editor

Russell Johnson
Dipartimento di Sistemi e Informatica
Università di Firenze
Via di S. Marta 3
I-50139 Firenze, Italy

Mathematics Subject Classification (1991): 34C35, 34C37, 34E15, 35B40, 35K55, 36Q30, 54H20, 58F12, 60H10, 60H25

ISBN 3-540-60047-7 Springer-Verlag Berlin Heidelberg New York

Dynamical systems : held in Montecatini Terme, Italy, June 13-22, 1994 / L. Arnold ... Ed.: R. Johnson. – Berlin ; Heidelberg ; New York ; Paris ; Tokyo ; Hong Kong : Springer, 1995
(Lectures given at the ... session of the Centro Internationale Matematico Estivo (CIME) ... ; 1994,2) (Lecture notes in mathematics ; Vol. 1609 : Subseries: Fondazione CIME)
ISBN 3-540-60047-7 (Berlin ...)
ISBN 0-387-60047-7 (New York ...)
NE: Arnold, Ludwig; Johnson, Russell [Hrsg.]; 2. GT

© Springer-Verlag Berlin Heidelberg 1995
Printed in Germany

Typesetting: Camera-ready by authors
SPIN: 10130360   46/3142-543210 - Printed on acid-free paper

Preface

This volume contains the lecture notes written by the four principal speakers at the C.I.M.E. session on Dynamical Systems held at Montecatini in June 1994. A goal of the session was to illustrate how methods of dynamical systems can be applied to the study of differential equations. And indeed the reader will find discussed in these notes a broad range of techniques recently developed in dynamical systems theory together with numerous applications to problems, both conceptual and specific, in ordinary and partial differential equations.

The lectures were delivered in the format of mini-courses of six hours each by Prof. Geneviève Raugel of the Université de Paris-Sud, Prof. Konstantin Mischaikow of Georgia Tech University, Prof. Christopher Jones of Brown University, and Prof. Ludwig Arnold of the Universität Bremen. They were attended by scholars from Italy and several other countries; their good-natured and active participation was essential to the success of the session.

As the reader will discover, the content of each of the courses is very rich. The remarks which follow are for purposes of general orientation only. For an adequate outline of the material in the courses the reader is referred to the statements written by the speakers themselves.

Prof. Arnold outlined a theory of random dynamical systems, discussing foundational points and applications to ordinary differential equations with random coefficients. He gave results on invariant manifolds and normal forms, and among other applications considered the breakdown of stability of a fixed point in certain differential equations with stochastic coefficients.

Prof. Jones reviewed the approach to singular perturbation theory recently developed by him and his collaborators. This approach is based on certain fundamental structure results of Fenichel together with the "exchange lemma". These matters and other were thoroughly discussed and numerous applications were given.

Prof. Mischaikow defined the Conley index of an isolated invariant set in a flow, then went on to consider the nature of the corresponding Morse sets and the set of connecting orbits. He reviewed a recent application of the Conley index theory: a proof of the existence of chaos in the Lorenz equations for certain parameter values.

Prof. Raugel considered recent work in the theory of partial differential equations and their singular perturbations. In particular she discussed partial differential equations in thin domains and the semi-continuity properties of attractors for such equations. She illustrated the theory by applications to the damped wave equation, the Navier-Stokes equations, and others.

Thanks are due to the main lecturers for the care they took in preparing their courses and for the speed with which they readied their notes for publication.

Thanks are also due to the six other participants of the session who delivered single lectures. Unfortunately it was not possible to publish their contributions in this volume. However a list of the titles of their lectures is given in the succeeding pages.

Finally, all participants, and especially the organizer, are indebted to the Director of the C.I.M.E., Prof. Roberto Conti, the Secretary Prof. Pietro Zecca, and to the C.I.M.E. staff for cooperation manifested in many ways.

Russell Johnson

TABLE OF CONTENTS

Random Dynamical Systems

Ludwig Arnold
Institut für Dynamische Systeme
Universität, Postfach 33 04 40
D-28334 Bremen, Germany
email:arnold@mathematik.uni-bremen.de

Contents

Introduction

The theory of random dynamical systems continues, extends, and unites various developments in probability theory and dynamical systems.

Roughly speaking, a *random dynamical system* is a combination of

- a measure-preserving dynamical system $(\Omega, \mathcal{F}, \mathbb{P}, (\theta(t))_{t \in \mathbb{T}})$ in the sense of ergodic theory, and

- a smooth (or topological) dynamical system, typically generated by a differential or difference equation $\dot{x} = f(x)$ or $x_{n+1} = \varphi(x_n)$.

Both parts have been very well investigated, but the symbiosis of them leads to a new research program which is the subject of these lectures (and of a forthcoming monograph, see Arnold [4]). It has significant applications in sciences and engineering.

From an abstract dynamical systems point of view, we 'just' deal with dynamical systems which possess a measure-preserving factor.

From the point of view of differential equations, we go beyond periodic forcing and consider non-autonomous equations $\dot{x} = f(\theta(t)\omega, x)$ coupled to a dynamical 'bath'. In this sense our theory extends the book by Guckenheimer and Holmes [44].

If the flow $\theta(t)$ is a flow of homeomorphisms of a compact space we are in the realm of *skew-product flows* in the sense of Sacker, Sell and Johnson (see e.g. [48], [72], [73], [75]). We go beyond this by stripping off all the topology, but adding an invariant measure, i.e. we go from 'almost periodic' to 'random'.

We also generalize *smooth ergodic theory*, as e.g. treated by Mañé [61]. There one has a deterministic diffeomorphism or vector field with an invariant measure on a manifold. We consider a random diffeomorphism or vector field, and a measure which is equivariant.

From a probabilistic point of view our approach offers a new look at the now quite classical subject of stochastic differential equations (stochastic analysis). The door between stochastic analysis and dynamical systems was opened around 1980 when several people (see e.g. Baxendale [21], Elworthy [39], Kunita [54]) realized that a stochastic differential equation generates a *flow* of random diffeomorphisms. This makes a *dynamical* theory of these equations possible. In particular, we can dramatically improve many classical results on stochastic stability (see Khasminskii [50]) and stochastic bifurcation theory (see Horsthemke and Lefever [46]).

It will turn out to be a characteristic feature of the theory of random dynamical systems that every problem involves ergodic theory. The *multiplicative ergodic theorem* which applies to the linearization of a smooth random system is of fundamental importance: It provides a random substitute of linear algebra, making it possible to develop a local nonlinear theory.

We will also describe a circle of problems non-existent in the nonrandom case: the interplay of measurability and dynamics.

Lecture 1: The concept of a random dynamical system

1.1 Metric, topological, and smooth dynamics

The theory of dynamical systems studies those properties of a family of self-mappings of a space which unfold asymptotically through iterations. This family is (algebraically) always a semigroup (often a group) $\mathbb{T}$ which we call *time*. In these lectures we exclusively consider the cases

$\mathbb{T} = \mathbb{Z}^+$ (or $\mathbb{N}$) respectively. $\mathbb{Z}$: *one-sided* respectively *two-sided discrete time*, or

$\mathbb{T} = \mathbb{R}^+$ respectively $\mathbb{R}$: *one-sided* respectively *two-sided continuous time*.

A *dynamical system*[1] on the set $X \neq \emptyset$ is a mapping

$$\varphi : \mathbb{T} \times X \to X, \quad (t, x) \mapsto \varphi(t)x,$$

for which the family of mappings $\varphi(t) : X \to X$, $x \mapsto \varphi(t)x$, forms a *flow*, i.e. satisfies

  (i) $\varphi(0) = \mathrm{id}_X$,
  (ii) $\varphi(t + s) = \varphi(t) \circ \varphi(s)$.

Here $\circ$ denotes composition. For two-sided time, (i) and (ii) imply that $\varphi(t)$ is bijective with $\varphi(t)^{-1} = \varphi(-t)$, and for discrete time $\varphi(n) = \varphi(1)^n$, $n \in \mathbb{T}$.

According to the choice of the category of the space X and its self-mappings, we have three giant sub-disciplines: metric dynamics or ergodic theory, topological dynamics, and differentiable or smooth dynamics.

Metric dynamics/ergodic theory

A *metric* DS $(\Omega, \mathcal{F}, \mathbb{P}, (\theta(t))_{t \in \mathbb{T}})$ is a flow of mappings of a probability space $(\Omega, \mathcal{F}, \mathbb{P})$, such that

  (i) $(t, \omega) \mapsto \theta(t)\omega$ is $\mathcal{B} \otimes \mathcal{F}, \mathcal{F}$ measurable, where $\mathcal{B}$ is the Borel σ-algebra of $\mathbb{T}$ (such a flow is called *measurable*),
  (ii) $\theta(t) : \Omega \to \Omega$ is measure preserving, i.e. $\theta(t)\mathbb{P} = \mathbb{P}$, where

$$\theta(t)\mathbb{P}(B) := \mathbb{P}\{\omega : \theta(t)\omega \in B\}, \quad B \in \mathcal{F}.$$

The study of metric DS is traditionally called *ergodic theory*. Of the many excellent textbooks we just mention Cornfeld, Fomin and Sinai [31] and Walters [79], and the survey volume of Sinai [77].

We will later come back to the following models of 'noise':

1.1. Example ('real noise'). *Stationary stochastic process as metric DS*: Every stationary stochastic process $(\xi_t)_{t \in \mathbb{T}}$ with state space $(E, \mathcal{E})$ generates a probability measure $\mathbb{P}$ on the product space $(\Omega, \mathcal{F}) = (E^{\mathbb{T}}, \mathcal{E}^{\mathbb{T}})$ which is invariant with respect to the shift $\theta(t)\omega := \omega(t + \cdot)$. ∎

---

[1]'Dynamical system' is henceforth abbreviated as 'DS'

1.2. Example ('white noise'). *Brownian motion/Wiener process as a metric DS*: Let $(W_t)_{t \in \mathbb{R}}$ be standard Brownian motion/Wiener process in $\mathbb{R}^m$, i.e. a stochastic process which has stationary independent increments, $W_{t+h} - W_t \sim \mathcal{N}(0, |h|I)$, $W_0 = 0$, and continuous trajectories. Put $\Omega = \{\omega \in \mathcal{C}(\mathbb{R}, \mathbb{R}^m) : \omega(0) = 0\}$, $\mathcal{F}$ the Borel σ-algebra of Ω, $\mathbb{P}$ the measure on $\mathcal{F}$ generated by W (so-called *Wiener measure*). The shift

$$\theta\omega(\cdot) := \omega(t + \cdot) - \omega(t)$$

is measure preserving and ergodic (with infinite entropy), and $W_t(\omega) = \omega(t)$ is Brownian motion. The generalized derivative of W_t is called *white noise*. ∎

Topological dynamics

Here X is a topological space, and $(t, x) \mapsto \varphi(t)x$ is assumed to be continuous. For two-sided time, $(\varphi(t))_{t \in \mathbb{T}}$ is then a flow of homeomorphisms.

Important notions of topological dynamics are: recurrence, limit sets, transitivity, topological equivalence, and structural stability, see Irwin [47] and Shub [76].

The 'intersection' of metric and topological dynamics contains *topological ergodic theory* (Brown [29]) and the *thermodynamic formalism* with its *variational principles* (Ruelle [69]).

Differentiable/smooth dynamics

Here X is a smooth manifold, and $(t, x) \mapsto \varphi(t)x$ is assumed to be continuous and k times $(1 \leq k \leq \infty)$ continuously differentiable with respect to x. For two-sided time, $(\varphi(t))_{t \in \mathbb{T}}$ is then a flow of $\mathcal{C}^k$-diffeomorphisms.

For $\mathbb{T} = \mathbb{R}$, every differential equation $\dot{x} = f(x)$, f a $\mathcal{C}^k$ vector field, generates a (local) $\mathcal{C}^k$ DS via its solution flow $x \mapsto \varphi(t)x$. Conversely, every smooth DS for which $t \mapsto \varphi(t)x$ is absolutely continuous with respect to t is generated by an autonomous differential equation. Hence we have a basically one-to-one correspondence between smooth vector fields and smooth DS which can be symbolically expressed as "smooth DS = exp(vector field)".

A prototypical problem of the *local* theory of smooth DS is to study the solution flow φ of $\dot{x} = f(x)$ with $f(0) = 0$ in a neighborhood of $x = 0$ by means of the linearization $\dot{v} = Av$, $A := \frac{\partial f}{\partial x}(0)$. It is a simple fact with far-reaching consequences that the behavior of a linear DS is completely described by linear algebra (eigenvalues, eigenspaces of A). The problem is now to turn this into statements about the *nonlinear DS* φ, e.g.

- to prove the *stability* of the fixed point $x = 0$ of φ in case all eigenvalues of A have negative real part,

- to construct *invariant manifolds* of φ tangent to the eigenspaces of A,

- to simplify (if possible: to linearize) φ by means of a nonlinear coordinate transformation (*normal form theory, Hartman-Grobman theorem*),

- to study local 'qualitative changes' (*bifurcations*) in parametrized families $\dot{x} = f^\alpha(x)$, where 'qualitative change' has been very successfully formalized by the topological concept of *structural stability*.

From the immense literature we only quote the books by V. I. Arnold [18], Guckenheimer and Holmes [44], and Ruelle [71].

The technique of drawing conclusions on φ from its linearization $T\varphi$ also works in a neighborhood of a periodic orbit (thanks to Floquet theory). For the almost-periodic case see Johnson [48] and Sell [75]. It might come as a surprise that local theory is possible in full generality for *random* DS due to the availability of the multiplicative ergodic theorem which replaces deterministic linear algebra.

In the enormously fertile intersection of topological and smooth dynamics we find the theories of *structural stability, hyperbolicity* and *Markov partitions*, leading to *symbolic dynamics*. See Irwin [47], Shub [76], and Szlenk [78].

The intersection of metric and smooth dynamics contains *smooth ergodic theory* (see Mañé [61]) with its fundamental *Oseledets' multiplicative ergodic theorem* and its celebrated *Pesin's formula*. This is now a particular (deterministic) case of the theory of random DS.

1.2 The notion of a random dynamical system

Intuitive description

Imagine a mechanism which for each discrete time n throws a (possibly complicated, many-sided) coin in order to select a mapping φ_n by which a point x_n is moved to $x_{n+1} = \varphi_n(x_n)$. The selection mechanism is permitted to remember all past (and even to foresee future) decisions. The only thing of importance is that at each step the *same* mechanism is applied. This scenario, called *product of random mappings*, is one of the prototypes of a random DS.

For continuous time the random selection at t would be e.g. from a set of differential equations $\dot{x} = f(t, x)$, again with the only condition that the *statistics* of $f(t, \cdot)$ be independent of t.

The formal definition of a random dynamical system[2] is tailor-made to cover all systems with randomness which are presently of interest, in particular random and stochastic difference and differential equations.

According to the above description an RDS consists of two ingredients:
(i) a topological/smooth DS which is perturbed by 'noise',
(ii) a metric DS which models the 'noise'.

The theory of such systems will be a symbiosis of ergodic theory and topological/smooth dynamics.

---

[2]'Random dynamical system' is henceforth abbreviated as 'RDS'

Formal definition of an RDS. Skew product flows

1.3. Definition. Let $(\Omega, \mathcal{F}, \mathbb{P}, (\theta(t))_{t\in\mathbb{T}})$ be a metric DS, and $(X, \mathcal{B})$ a measurable space[3]. Let

$$\varphi : \mathbb{T} \times \Omega \times X \to X, \quad (t, \omega, x) \mapsto \varphi(t, \omega)x,$$

be a mapping with the following properties:

(i) $\varphi(0, \omega) = \mathrm{id}_X$,

(ii) *Cocycle property:* For all $s, t \in \mathbb{T}$ and all $\omega \in \Omega$

$$\varphi(t + s, \omega) = \varphi(t, \theta(s)\omega) \circ \varphi(s, \omega).$$

1. If φ is measurable, then it is called a *measurable* RDS over θ.

2. If, in addition, X is a topological space, and the measurable RDS φ satisfies

$$(t, x) \mapsto \varphi(t, \omega)x \quad \text{continuous for all } \omega \in \Omega,$$

then φ is called a *continuous* or *topological* RDS over θ.

3. If, in addition, X is a smooth manifold, and the continuous RDS φ satisfies for some k, $1 \le k \le \infty$,

$$x \mapsto \varphi(t, \omega)x \quad \text{is } \mathcal{C}^k \quad \text{for all } (t, \omega) \in \mathbb{T} \times \Omega,$$

(i.e. it is k times differentiable with respect to x, and the derivatives are continuous with respect to (t, x)), then φ is called *smooth*, more precisely $\mathcal{C}^k$-RDS over θ. ∎

1.4. Remark. (i) Let φ be a (e.g. measurable) RDS over θ. Then

$$(\omega, x) \mapsto (\theta(t)\omega, \varphi(t, \omega)x) =: \Theta(t)(\omega, x)$$

is a (measurable) flow on $\Omega \times X$, the so-called *skew-product* of θ and φ. Conversely, each skew-product flow defines a cocycle φ on X. We will thus use "RDS φ", "cocycle φ" and "skew-product flow Θ" synonymously. For skew-product flows in a topological setting see Sacker and Sell [72], [73].

(ii) It will turn out to be very useful to imagine φ to act on the trivial bundle $\Omega \times X$: While ω is moved in time t to $\theta(t)\omega$ on the base space Ω, φ moves the point x in the fiber $\{\omega\} \times X$ to $\varphi(t, \omega)x$ in $\{\theta(t)\omega\} \times X$. Nontrivial bundles will appear later. ∎

---

[3]If X is a topological space, then $\mathcal{B}$ is always taken to be its Borel σ-algebra generated by the open sets.

Consequences of the cocycle property

1.5. THEOREM. *Suppose $\mathbb{T}$ is a group (i.e. $\mathbb{T} = \mathbb{Z}$ or $\mathbb{R}$).*
  (i) Let φ be a measurable RDS. Then for all (t, ω), $\varphi(t, \omega)$ is measurably invertible, with
$$\varphi(t, \omega)^{-1} = \varphi(-t, \theta(t)\omega).$$
Moreover, $(t, \omega, x) \mapsto \varphi(t, \omega)^{-1}x$ is measurable.
  (ii) Let φ be a topological RDS. Then for all (t, ω), $\varphi(t, \omega) \in \mathrm{Homeo}\,(X)$. Moreover, if
  1. $\mathbb{T} = \mathbb{Z}$, or
  2. $\mathbb{T} = \mathbb{R}$ and X is a topological manifold, or
  3. $\mathbb{T} = \mathbb{R}$ and X is a compact Hausdorff space
then
$$(t, x) \mapsto \varphi(t, \omega)^{-1}x = \varphi(-t, \theta(t)\omega)x$$
is continuous.
  (iii) Let φ be a $\mathcal{C}^k$-RDS. Then for all (t, ω), $\varphi(t, \omega) \in \mathrm{Diff}^{\,k}(X)$. Moreover, $(t, x) \mapsto \varphi(t, \omega)^{-1}x$ is $\mathcal{C}^k$.

Proof. For a detailed proof see Arnold [4], Theorem 1.6.
  (i) is an easy exercise.
  (ii) For case 2 observe that $(t, x) \mapsto (t, \varphi(t, \omega)x)$ is a continuous (by Definition 1.3) and bijective (by part (i) of this theorem) mapping of $\mathbb{R} \times X$ onto itself. This mapping is thus a homeomorphism by Brouwer's theorem (cf. Dieudonné [37], page 52), so the inverse $(t, x) \mapsto (t, \varphi(t, \omega)^{-1}x)$ is continuous, and so is $(t, x) \mapsto \varphi(t, \omega)^{-1}x$. The proof in case 3 is based on the fact that a continuous bijection of a compact space into a Hausdorff space is a homeomorphism.
  (iii) Use the formula for the derivative of the inverse of a diffeomorphism. □

Lecture 2: Generation of random dynamical systems

We will now associate (infinitesimal) generators to all reasonably regular RDS. The task is trivial in the discrete time case where all RDS have generators (their time one maps) (see section 2.1).
  In the continuous time case $\mathbb{T} = \mathbb{R}$, we ask for stochastic equivalents of the symbolic relation "flow = exp(vector field)". There are two basically different answers:
  (i) RDS φ for which $t \mapsto \varphi(t, \omega)x$ is absolutely continuous are solution flows of (pathwise, i.e. ω-wise) *random* differential equations of the form $\dot{x} = f(\theta(t)\omega, x)$, and conversely (see section 2.2).
  (ii) It seems surprising that there is a large and very important class of RDS for which $t \mapsto \varphi(t, \omega)x$ is not even of bounded variation, but which have generators. These generators are *stochastic* differential equations (defined by stochastic analysis).

2.1 Discrete time: Products of random mappings

Let φ be a measurable/continuous/$\mathcal{C}^k$ RDS on X over θ with time $\mathbb{T} = \mathbb{Z}^+$. Put $\psi(\omega) := \varphi(1, \omega) : X \to X$, the *time one mapping* of φ. Repeated application of the cocycle property gives

$$\varphi(n, \omega) = \begin{cases} \psi(\theta^{n-1}\omega) \circ \cdots \circ \psi(\omega), & n \geq 1, \\ \mathrm{id}_X, & n = 0. \end{cases} \tag{2.1}$$

The RDS φ is measurable if and only if $(\omega, x) \mapsto \psi(\omega)x$ is measurable. It is continuous/$\mathcal{C}^k$ if and only if $x \mapsto \psi(\omega)x$ is, in addition, continuous/$\mathcal{C}^k$.

Conversely, given a random mapping $\psi(\omega) : X \to X$ such that $(\omega, x) \mapsto \psi(\omega)x$ is measurable, in addition $x \mapsto \psi(\omega)x$ is continuous/$\mathcal{C}^k$. Then φ defined by (2.1) is a measurable/continuous/$\mathcal{C}^k$ RDS. We say that φ is *generated* by ψ.

For two-sided discrete time $\mathbb{T} = \mathbb{Z}$, the time one mapping $\psi(\omega) = \varphi(1, \omega)$ and the time minus one mapping $\varphi(-1, \omega)$ are related by

$$\varphi(-1, \omega) = \varphi(1, \theta^{-1}\omega)^{-1} = \psi(\theta^{-1}\omega)^{-1},$$

and the cocycle property gives

$$\varphi(n, \omega) = \begin{cases} \psi(\theta^{n-1}\omega) \circ \cdots \circ \psi(\omega), & n \geq 1, \\ \mathrm{id}_X, & n = 0, \\ \psi(\theta^n\omega)^{-1} \circ \cdots \circ \psi(\theta^{-1}\omega)^{-1}, & n \leq -1. \end{cases} \tag{2.2}$$

The RDS φ is measurable if and only if

$$(\omega, x) \mapsto \psi(\omega)x \quad \text{and} \quad (\omega, x) \mapsto \psi(\omega)^{-1}x \tag{2.3}$$

are measurable. φ is, moreover, continuous or $\mathcal{C}^k$ if and only if $\psi(\omega) \in \mathrm{Homeo}(X)$ or $\mathrm{Diff}^k(X)$, respectively.

Consequently, every discrete time RDS is a *product of* (a stationary sequence of) *random mappings*. Equivalently, we can think of it as the solution of the *random difference equation*

$$x_{n+1} = \psi(\theta^n\omega)x_n, \quad n \in \mathbb{Z}^{(+)}, \quad x_0 \in X. \tag{2.4}$$

The case $X = \mathbb{R}^d$ and $\psi(\omega) \in Gl(d, \mathbb{R})$ is the case of *products of random matrices* with many fundamental papers, e.g. Furstenberg [41], Oseledets [64], Ruelle [70], Guivarc'h and Raugi [45], and the book by Bougerol and Lacroix [27]. For a random Perron-Frobenius theory for products of positive random matrices see Arnold, Demetrius and Gundlach [8].

The case $X = \mathbb{R}^d$, $\psi(\omega)$ an affine mapping, is the case of *iterated function systems*, important for the possibility of encoding and visualizing fractals, see Barnsley [20] and Arnold and Crauel [6].

Products of iid random mappings and Markov chains

A measurable cocycle φ with time $\mathbb{T} = \mathbb{Z}^+$ which is a product of independent and identically distributed (iid) random mappings $\psi_n = \psi(\theta^n \cdot)$ generates a homogeneous Markov chain (x_n), $x_{n+1} = \psi_n x_n$, in the state space X with transition probability

$$P(x, B) = \mathbb{P}\{\omega : \psi(\omega)x \in B\}.$$

For a proof of this fact and a systematic study of the iid case see Kifer [51]. If X is a metric space and $\psi(\omega) : X \to X$ is continuous, then $P(x, B)$ is Feller, i.e. $P(f)(x) = \int_X f(y)P(x, dy) = \mathbb{E}f(\psi(\omega)x)$ is in $\mathcal{C}_b(X)$ if f is.

The inverse problem of constructing a measurable/continuous/$\mathcal{C}^k$ cocycle of iid mappings with a prescribed transition probability is largely unsolved. A general abstract answer is the following: If $(X, \mathcal{B})$ is a standard measurable space[4] and $P(x, B)$ a transition probability on it then there is a metric DS θ and a function $\psi(\omega) : X \to X$ for which $(\omega, x) \mapsto \psi(\omega)x$ is measurable, $\psi(\theta^n \cdot)$ are iid and $\mathbb{P}\{\omega : \psi(\omega)x \in B\} = P(x, B)$.

2.2 Continuous time 1: Random differential equations

Let $\mathbb{T} = \mathbb{R}$ and $X = \mathbb{R}^d$. We will now establish a basically one-to-one correspondence between continuous/$\mathcal{C}^k$ RDS φ over θ which are *absolutely* continuous with respect to t, and random differential equations coupled to θ. The correspondence is given by

$$\varphi(t, \omega)x = x + \int_0^t f(\theta(t)\omega, \varphi(s, \omega)x)\, ds. \tag{2.5}$$

If (2.5) holds, we say that $t \mapsto \varphi(t, \omega)x$ *solves* the random differential equation $\dot{x} = f(\theta(t)\omega, x)$, or that the random differential equation *generates* φ.

2.1. Definition. (i) Let $k \in \mathbb{Z}^+$ and $0 \le \delta \le 1$. Denote by $\mathcal{C}_b^{k,\delta}$ the Banach space of functions $f : \mathbb{R}^d \to \mathbb{R}^d$ which are k times continuously differentiable and (for $\delta > 0$) whose k-th derivative is δ-Hölder continuous (for $\delta = 1$: Lipschitz continuous) for which the norm

$$\|f\|_{k,0} := \sup_{x \in \mathbb{R}^d} \frac{|f(x)|}{1 + |x|} + \sum_{1 \le |\alpha| \le k} \sup_{x \in \mathbb{R}^d} |D^\alpha f(x)|,$$

$$\|f\|_{k,\delta} := \|f\|_{k,0} + \sum_{|\alpha| = k} \sup_{x \ne y} \frac{|D^\alpha f(x) - D^\alpha f(y)|}{|x - y|^\delta}, \quad 0 < \delta \le 1,$$

is finite. Here $\alpha = (\alpha_1, \ldots, \alpha_d) \in (\mathbb{Z}^+)^d$ is a multi-index, $|\alpha| = \alpha_1 + \cdots + \alpha_d$, and

$$D^\alpha f(x) = \frac{\partial^{|\alpha|}}{(\partial x_1)^{\alpha_1} \ldots (\partial x_d)^{\alpha_d}} f(x).$$

---

[4] A measurable space is called *standard* if there exists a bimeasurable bijection to a Borel subset of a Polish space.

(ii) Denote by $L_{loc}(\mathbb{R}, \mathcal{C}_b^{k,\delta})$ the Fréchet space of measurable functions $f : \mathbb{R} \times \mathbb{R}^d \to \mathbb{R}^d$ for which $f(t, \cdot) \in \mathcal{C}_b^{k,\delta}$ for all $t \in \mathbb{R}$ and

$$\int_a^b \|f(t, \cdot)\|_{k,\delta}\, dt < \infty \quad \text{for all } a < b.$$

■

2.2. THEOREM (RDS THROUGH RANDOM DIFFERENTIAL EQUATION).
Let $f : \Omega \times \mathbb{R}^d \to \mathbb{R}^d$ be measurable and consider the pathwise random differential equation

$$\dot{x} = f(\theta(t)\omega, x). \tag{2.6}$$

Let for fixed ω, $f_\omega(t, x) := f(\theta(t)\omega, x)$.
(i) If $f_\omega \in L_{loc}(\mathbb{R}, \mathcal{C}_b^{0,1})$ then (2.6) uniquely generates a continuous RDS φ over θ.
(ii) If $f_\omega \in L_{loc}(\mathbb{R}, \mathcal{C}_b^{k,0})$ for $k \geq 1$ then (2.6) uniquely generates a $\mathcal{C}^k$ RDS φ over θ. The Jacobian of $\varphi(t, \omega)$ at x,

$$D\varphi(t, \omega, x) := \left(\frac{\partial(\varphi(t, \omega)x)_i}{\partial x_j} \right)$$

is a matrix cocycle over the skew product $\Theta(t)(\omega, x) = (\theta(t)\omega, \varphi(t, \omega)x)$ and uniquely solves the variational equation

$$D\varphi(t, \omega, x) = I + \int_0^t Df(\theta(s)\omega, \varphi(s, \omega)x) D\varphi(s, \omega, x)\, ds.$$

Finally, we have Liouville's equation

$$\det D\varphi(t, \omega, x) = \exp \int_0^t (\text{trace } Df)(\theta(s)\omega, \varphi(s, \omega)x)\, ds.$$

Proof. The existence and uniqueness proof is an ω-wise adaptation of the deterministic proof, see e.g. Amann [1], pages 100 ff. The cocycle property of the solution is a consequence of its uniqueness. □

Let $g : \Omega \to \mathbb{R}$ be measurable, and let $g \in L^1(\mathbb{P})$. By Fubini's theorem, there is an ω set of full $\mathbb{P}$ measure such that $t \mapsto g(\theta(t)\omega)$ is locally Lebesgue integrable with

$$\int_a^b g(\theta(t)\omega)\, dt \in L^1(\mathbb{P}) \quad \text{for all } a < b.$$

This fact and the definition of $L_{loc}(\mathbb{R}, \mathcal{C}_b^{k,\delta})$ give the following 'static' sufficient criteria.

2.3. THEOREM. *Assume the situation of Theorem 2.2. If $\omega \mapsto \|f(\omega, \cdot)\|_{0,1} \in L^1(\mathbb{P})$ respectively $\omega \mapsto \|f(\omega, \cdot)\|_{k,0} \in L^1(\mathbb{P})$ then (2.6) uniquely generates a continuous respectively $\mathcal{C}^k$ RDS φ.*

2.4. Remark. We have $\omega \mapsto \|f(\omega, \cdot)\|_{0,1} \in L^1(\mathbb{P})$ if and only if
(i) For some $x_0 \in \mathbb{R}^d$, $f(\cdot, x_0) \in L^1(\mathbb{P})$,
(ii) $|f(\omega, x) - f(\omega, x)| \le R(\omega)|x - y|$ for all $x, y \in \mathbb{R}^d$, and $R \in L^1(\mathbb{P})$. ∎

2.5. Example. (i) Linear random differential equation: Let the measurable function $A : \Omega \to \mathbb{R}^{d \times d}$ satisfy $A \in L^1(\mathbb{P})$ then $\dot{x} = A(\theta(t)\omega)x$ uniquely generates a $\mathcal{C}^\infty$ (even linear) RDS.

(ii) Affine random differential equation: Let the measurable functions $A : \Omega \to \mathbb{R}^{d \times d}$ and $b : \Omega \to \mathbb{R}^d$ satisfy $A \in L^1(\mathbb{P})$ and $b \in L^1(\mathbb{P})$. Then $\dot{x} = A(\theta(t)\omega)x + b(\theta(t)\omega)$ uniquely generates a $\mathcal{C}^\infty$ (even affine) RDS. ∎

The inverse problem of assigning a generator $\dot{x} = f(\theta(t)\omega, x)$ to an RDS φ is easy to solve.

2.6. THEOREM (RANDOM DIFFERENTIAL EQUATION FROM RDS). *Let φ be a continuous/$\mathcal{C}^k$ RDS over θ such that $t \mapsto \varphi(t, \omega)x$ is absolutely continuous for all (ω, x). Then there exists a measurable function $f(\omega, x)$ such that (2.5) holds.*

Proof. By assumption there is a g such that

$$\varphi(t, \omega)x - x = \int_0^t g(s, \omega, x)\, ds =: G(t, \omega, x).$$

g can be chosen to be measurable, in fact

$$g(s, \omega, x) := \limsup_{n \to \infty} \frac{G(s + \frac{1}{n}, \omega, x) - G(s, \omega, x)}{\frac{1}{n}} \tag{2.7}$$

is measurable, is a version of the Radon-Nikodym derivative of G for all (ω, x), and the *limit* exists for Lebesgue-almost all s.

The cocycle property of φ translates into the *helix property* of G over $\Theta = (\theta, \varphi)$,

$$G(t + s, \omega, x) = G(t, \Theta(s)(\omega, x)) + G(s, \omega, x).$$

Inserting this into (2.7) yields

$$g(s, \omega, x) = \limsup_{n \to \infty} \frac{G(\frac{1}{n}, \Theta(s)(\omega, x))}{\frac{1}{n}} = g(0, \Theta(s)(\omega, x)).$$

Now put $f(\omega, x) := g(0, \omega, x)$. □

As a result, we have the basically one-to-one relation between absolutely continuous RDS φ and vector field valued stationary stochastic processes $f(\theta(t)\omega, \cdot)$.

2.3 Continuous time 2: Stochastic differential equations

This subsection is written in a more narrative style than others as a rigorous treatment is beyond the scope of these notes. For the full picture see Arnold and Scheutzow [12] and the forthcoming monograph [4].

The theory of RDS would be much less interesting without the following fact: There is a huge class of important RDS which can principally *not* be obtained from a *pathwise* differential equation, but have *stochastic differential equations* as their generators.

The simplest stochastic differential equation is

$$dx = dW, \quad W = \text{Brownian motion (see Example 1.2)},$$

which generates the $\mathcal{C}^\infty$-RDS $x \mapsto \varphi(t, \omega)x = x + W_t(\omega)$ of random translations, but $t \mapsto \varphi(t, \omega)x$ is known to be nowhere differentiable and of unbounded variation.

The classical case is a (Stratonovich) stochastic differential equation in $\mathbb{R}^d$ with time $\mathbb{T} = \mathbb{R}^+$ of the form

$$dx = f_0(x)dt + \sum_{j=1}^m f_j(x) \circ dW_t^j, \tag{2.8}$$

where $f_0, \ldots, f_m$ are smooth vector fields, and W is standard Brownian motion on $\mathbb{R}^m$. Until around 1980 such an equation was mainly considered as a 'machine' for generating a Markov process on $\mathbb{T} = \mathbb{R}^+$. The door from stochastic analysis to dynamical systems was opened by the discovery (which requires very hard work) that equation (2.8) generates a two-parameter flow $\varphi_{s,t}(\omega)$ of diffeomorphisms (see Baxendale [21], Elworthy [39], Kunita [54]) in the sense that for all $s, t \in \mathbb{R}^+$, $s \leq t$, and all $x \in \mathbb{R}^d$

$$\varphi_{s,t}(\cdot)x = x + \int_s^t f_0(\varphi_{s,u}(\cdot)x)du + \sum_{j=1}^m \int_s^t f_j(\varphi_{s,u}(\cdot)x) \circ dW_u^j.$$

The second integral on the right-hand side is a stochastic Stratonovich integral and can in our situation be defined as

$$\int_s^t f(u) \circ dW_u := \lim \text{ in prob} \sum_{k=0}^{n-1} \frac{f(u_k) + f(u_{k+1})}{2} (W_{u_{k+1}} - W_{u_k}),$$

where the limit in probability runs through a sequence of partitions of $[s, t]$ for which $\max_{0 \leq k \leq n-1}(u_{k+1} - u_k) \to 0$.

In order to obtain an RDS over the DS θ describing Brownian motion with time $\mathbb{T} = \mathbb{R}$ from (2.8) we have to extend stochastic analysis from the traditional $\mathbb{T} = \mathbb{R}^+$ to $\mathbb{T} = \mathbb{R}$ by introducing a *backward* calculus. Unfortunately, uniqueness of solution of (2.8) only yields the *crude* cocycle property for $\varphi(t, \omega) := \varphi_{0,t}(\omega)$ in the sense that

$$\varphi(t + s, \omega) = \varphi(t, \theta(s)\omega) \circ \varphi(s, \omega)$$

is only valid for fixed s, t and $\omega \notin N_{s,t}$, where $N_{s,t}$ is a $\mathbb{P}$-null set depending on s, t. Only very recently it was shown that this cocycle can be perfected, i.e. there exists a solution of (2.8) for which the cocycle property holds identically (cf. Arnold and Scheutzow [12]). Hence we have the following theorem.

2.7. THEOREM (RDS THROUGH STOCHASTIC DIFFERENTIAL EQUATION).
Let for some $k \geq 1$ and $\delta > 0$, $f_0 \in \mathcal{C}_b^{k,\delta}$ and $f_j \in \mathcal{C}_b^{k+1,\delta}$, $j = 1, \ldots, m$. Then there exists a unique $\mathcal{C}^k$ RDS φ over the DS θ describing Brownian motion which solves the stochastic differential equation (2.8) on $\mathbb{T} = \mathbb{R}$.
The Jacobian $D\varphi(t, \omega, x)$ of $\varphi(t, \omega)$ at x satisfies the variational equation

$$dv = Df_0(\varphi(t, \cdot)x)v\,dt + \sum_{j=1}^{m} Df_j(\varphi(t, \cdot)x)v \circ dW_t^j,$$

and we have Liouville's equation

$$\det D\varphi(t, \omega, x) = \exp \int_0^t \operatorname{trace} Df_0(\varphi(s, \cdot)x)\,ds + \sum_{j=1}^{m} \int_0^t \operatorname{trace} Df_j(\varphi(s, \cdot)x) \circ dW_s^j.$$

The inverse problem goes beyond the classical model (2.8). The crucial additional statistical property which a φ should have to be generated by a general stochastic differential equation is that for each x, the stochastic process $t \mapsto \varphi(t, \omega)x$ is a *semimartingale*. To obtain such a φ from a stochastic differential equation, Brownian motion has to be replaced by a vector field valued semimartingale with stationary increments (helix). The result is succinctly written as

semimartingale cocycle = exp (semimartingale helix).

This is more or less the final answer to the generation problem for RDS, as semimartingales are known to be the most general reasonable integrators.

2.8. Example. Let $A_j \in \mathbb{R}^{d \times d}$, $b_j \in \mathbb{R}^d$, $j = 0, \ldots, m$. Then

$$dx = (A_0 x + b_0)dt + \sum_{j=1}^{m}(A_j x + b_j) \circ dW_t^j$$

generates a $\mathcal{C}^\infty$ RDS of affine mappings (linear mappings in case all $b_j = 0$). ∎

Lecture 3: Invariant measures

3.1 Invariant measures for measurable RDS

For all further steps of the theory of RDS we need *invariant measures*. Since our philosophy is that the metric DS $(\Omega, \mathcal{F}, \mathbb{P}, (\theta(t))_{t \in \mathbb{T}})$ is 'given' to us from outside and cannot be manipulated we arrive at the following definition.

3.1. Definition. Let φ be a measurable RDS on $(X, \mathcal{B})$ over θ. A probability measure μ on $(\Omega \times X, \mathcal{F} \otimes \mathcal{B})$ is said to be *invariant* for the RDS φ, or φ-*invariant*, if it satisfies

1. $\Theta(t)\mu = \mu$ for all $t \in \mathbb{T}$, where $\Theta(t)(\omega, x) = (\theta(t)\omega, \varphi(t, \omega)x)$ is the corresponding skew-product flow,

2. $\pi_\Omega \mu = \mathbb{P}$, where π_Ω is the projection onto Ω.

∎

Define

$$\mathcal{P}_\mathbb{P}(\Omega \times X) := \{\mu \text{ probability on } (\Omega \times X, \mathcal{F} \otimes \mathcal{B}) \text{ with marginal } \mathbb{P} \text{ on } (\Omega, \mathcal{F})\},$$

and

$$\mathcal{I}_\mathbb{P}(\varphi) := \{\mu \in \mathcal{P}_\mathbb{P}(\Omega \times X) : \mu \ \varphi\text{-invariant}\},$$

which are both convex sets.

Note that an RDS does in general *not* come equipped with an invariant measure, and that an invariant measure is not a product measure in general.

Suppose $\mu \in \mathcal{P}_\mathbb{P}(\Omega \times X)$. We call a function

$$\mu.(\cdot) : \Omega \times \mathcal{B} \to [0, 1], \quad (\omega, B) \mapsto \mu_\omega(B),$$

a *factorization* (or *disintegration*) of μ with respect to $\mathbb{P}$, if

1. for all $B \in \mathcal{B}$, $\omega \mapsto \mu_\omega(B)$ is $\mathcal{F}$-measurable,

2. for $\mathbb{P}$-almost all $\omega \in \Omega$, $B \mapsto \mu_\omega(B)$ is a probability measure on $(X, \mathcal{B})$,

3. for all $A \in \mathcal{F} \otimes \mathcal{B}$

$$\mu(A) = \int_\Omega \int_X 1_A(\omega, x)\mu_\omega(dx)\mathbb{P}(d\omega). \tag{3.1}$$

We write symbolically

$$\mu(d\omega, dx) = \mu_\omega(dx)\mathbb{P}(d\omega).$$

Introducing the sections $A_\omega := \{x : (\omega, x) \in A\}$, (3.1) can be written as

$$\mu(A) = \int_\Omega \mu_\omega(A_\omega) \mathbb{P}(d\omega).$$

It is a well-known fact that every $\mu \in \mathcal{P}_\mathbb{P}(\Omega \times X)$ has a ($\mathbb{P}$-a.s. unique) factorization provided $(X, \mathcal{B})$ is a standard measurable space (which we assume from now on), in particular if X is a Polish space with its Borel σ-algebra.

We rewrite invariance of μ in terms of its factorization.

3.2. Theorem. *Let φ be a measurable RDS on a standard measurable space $(X, \mathcal{B})$, and let $\mu \in \mathcal{P}_\mathbb{P}(\Omega \times X)$. Then*

(i) $\mu \in \mathcal{I}_\mathbb{P}(\varphi)$ if and only if for all $t \in \mathbb{T}$

$$\mathbb{E}(\varphi(t, \cdot)\mu. | \theta(t)^{-1}\mathcal{F})(\omega) = \mu_{\theta(t)\omega} \quad \mathbb{P}\text{-a. s.} \tag{3.2}$$

(ii) If $\mathbb{T}$ is two-sided then $\theta(t)^{-1}\mathcal{F} = \mathcal{F}$ for all $t \in \mathbb{T}$, and $\mu \in \mathcal{I}_\mathbb{P}(\varphi)$ if and only if for all $t \in \mathbb{T}$

$$\varphi(t, \omega)\mu_\omega = \mu_{\theta(t)\omega} \quad \mathbb{P}\text{-a. s.}$$

Proof. (i) It suffices to check (3.2) for product sets $F \times B \in \mathcal{F} \otimes \mathcal{B}$, for which we have

$$(\Theta(t)\mu)(F \times B) = \mu(\Theta(t)^{-1}(F \times B)) = \int_{\theta(t)^{-1}F} \mu_\omega(\varphi(t, \omega)^{-1}B)\, \mathbb{P}(d\omega)$$

$$= \int_{\theta(t)^{-1}F} (\varphi(t, \omega)\mu_\omega)(B)\, \mathbb{P}(d\omega)$$

and

$$\mu(F \times B) = \int_F \mu_\omega(B)\, \mathbb{P}(d\omega) = \int_F \mu_\omega(B)(\theta(t)\mathbb{P})(d\omega)$$

$$= \int_{\theta(t)^{-1}F} \mu_{\theta(t)\omega}(B)\, \mathbb{P}(d\omega).$$

Thus if μ is invariant then for each fixed B and t, the $\theta(t)^{-1}\mathcal{F}$-measurable function $\omega \mapsto \mu_{\theta(t)\omega}(B)$ is a version of the conditional expectation $\mathbb{E}(\varphi(t, \cdot)\mu.(B)|\theta(t)^{-1}\mathcal{F})$ of the $\mathcal{F}$-measurable function $\omega \mapsto \varphi(t, \omega)\mu_\omega(B)$ (remember that $\theta(t)^{-1}\mathcal{F} \subset \mathcal{F}$).

The exceptional set can depend on B and t. Since $\mathcal{B}$ is countably generated we can find a universal exceptional set (in continuous time still depending on t) outside of which (3.2) holds.

Conversely, if we start with (3.2) we arrive at the invariance of μ.

(ii) follows from (i). $\square$

3.3. Example. (i) When is a product measure $\mu = \mathbb{P} \times \rho$ φ-invariant? For two-sided time $\mu_\omega = \rho$ $\mathbb{P}$-a. s. if and only if $\varphi(t, \omega)\rho = \rho$ $\mathbb{P}$-a. s. for all $t \in \mathbb{T}$, i.e. almost all mappings $\varphi(t, \omega)$ leave the fixed measure ρ invariant — which will be a rare case.

For one-sided time, $\mu = \mathbb{P} \times \rho$ is invariant if and only if

$$\mathbb{E}(\varphi(t, \cdot)\rho|\theta(t)^{-1}\mathcal{F}) = \rho \quad \mathbb{P}\text{-a. s.}$$

for all $t \in \mathbb{T}$. Suppose $\varphi(t, \cdot)$ and $\theta(t)^{-1}\mathcal{F}$ are independent. Then the last condition becomes

$$\mathbb{E}(\varphi(t, \cdot)\rho)(B) = \int_X \mathbb{P}\{\omega : \varphi(t, \omega)x \in B\}\, \rho(dx) = \rho(B), \tag{3.3}$$

meaning that ρ is invariant with respect to φ *on the average*.

For RDS with one-sided time which are either products of iid mappings (see subsection 2.1, $\mathbb{T} = \mathbb{Z}^+$) or solutions of classical stochastic differential equations (see subsection 2.3, $\mathbb{T} = \mathbb{R}^+$) the one-point motions are Markov processes with transition probability

$$P(t, x, B) = \mathbb{P}\{\omega : \varphi(t, \omega)x \in B\}.$$

Then (3.3) reads

$$\int_X P(t, x, B)\rho(dx) = \rho(B),$$

i.e. $\mu = \mathbb{P} \times \rho$ is φ-invariant if and only if ρ is a *stationary* measure of the Markov transition probability P. For continuous time let L be the generator of P. Then ρ is stationary if and only if it solves the *Fokker-Planck equation* $L^*\rho = 0$.

(ii) A φ-invariant measure μ is called a *random Dirac measure* (or *random fixed point*) if there exists a random variable $x_0 : \Omega \to X$ with $\mu_\omega = \delta_{x_0(\omega)}$. Invariance then reads (e.g. for two-sided time)

$$\varphi(t, \omega)x_0(\omega) = x_0(\theta(t)\omega) \quad \mathbb{P}\text{-a.\,s.} \quad \text{for all } t \in \mathbb{T},$$

i.e. the orbit of the cocycle starting at the random initial value $x_0(\omega)$ is a stationary stochastic process in X. This situation, in which μ is supported by the graph of a random variable, is encountered quite frequently. ∎

3.4. Exercise (Affine equation with hyperbolic linear part).

(i) For $\mathbb{T} = \mathbb{Z}$ and $X = \mathbb{R}^d$ let $\psi(\omega)x = Ax + b(\omega)$ be the time one mapping of the affine cocycle φ, where $A \in Gl(d, \mathbb{R})$ has the form

$$A = \begin{pmatrix} A^- & 0 \\ 0 & A^+ \end{pmatrix}$$

with A^- having its spectrum inside and A^+ outside the unit disc in $\mathbb{C}$, and $\log^+ |b| \in L^1(\mathbb{P})$. Put $b = \binom{b^-}{b^+}$. Then the unique φ-invariant measure is the random Dirac measure $\delta_{x_0(\omega)}$ with foot point

$$x_0(\omega) = \begin{pmatrix} x_0(\omega)^- \\ x_0(\omega)^+ \end{pmatrix} = \begin{pmatrix} \sum_{n=1}^{\infty}(A^-)^{n-1}b^-(\theta^{-n}\omega) \\ \sum_{n=0}^{\infty}(A^+)^{-n-1}b^+(\theta^n\omega) \end{pmatrix}.$$

(ii) For $\mathbb{T} = \mathbb{R}$ and $X = \mathbb{R}^d$ the stochastic differential equation

$$dx = \begin{pmatrix} A^- & 0 \\ 0 & A^+ \end{pmatrix} x\,dt + \begin{pmatrix} dW^- \\ dW^+ \end{pmatrix},$$

where A^- has its spectrum in the left-hand side and A^+ in the right-hand side of $\mathbb{C}$, models additive white noise on a hyperbolic fixed point. It generates the affine RDS $x \mapsto \varphi(t,\omega)x = e^{tA}x + \int_0^t e^{(t-s)A} dW_s(\omega)$. It has the unique invariant measure $\mu_\omega = \delta_{x_0(\omega)}$, where

$$x_0 = \begin{pmatrix} \int_{-\infty}^0 e^{-tA^-} dW_t^- \\ -\int_0^\infty e^{-tA^+} dW_t^+ \end{pmatrix}.$$

The first component of x_0 is measurable with respect to the past $\mathfrak{F}_{-\infty}^0 = \sigma(W_t, t \leq 0)$ of the noise, while the second component is measurable with respect to the future $\mathfrak{F}_0^\infty = \sigma(W_t, t \geq 0)$. For white noise, past and future are independent.

It is known that there is a one-to-one correspondence between stationary ρ's of the corresponding Markov transition probability and those invariant μ_ω's which are measurable with respect to the past $\mathfrak{F}_{-\infty}^0$ via

$$\rho \mapsto \mu_\omega = \lim_{t \to \infty} \varphi(-t,\omega)^{-1}\rho, \quad \mu_\omega \mapsto \rho = \mathbb{E}\mu.,$$

see Le Jan [55] and Crauel [32], [33]. Hence, if A^+ is present, the marginal $\rho(\cdot) = \mathbb{E}\mu. = \mathbb{P}\{x_0 \in \cdot\}$ does *not* solve the Fokker-Planck equation since μ_ω anticipates the future.

Observe that the random fixed point x_0 is stable if and only if it is measurable with respect to the past. This foreshadows deep connections between measurability and dynamics, see Crauel [34]. ∎

3.2 Invariant measures for continuous RDS

We sometimes can construct or at least assure the existence of invariant measures for *continuous* RDS.

Let in this whole subsection X be a Polish space. Let $\mathcal{C}_b(X)$ be the Banach space of real-valued bounded continuous functions on X, with sup-norm $\|f\|_b := \sup_{x \in X} |f(x)|$.

Call a function $f : \Omega \to \mathcal{C}_b(X)$ *measurable* if $(\omega, x) \mapsto f(\omega, x)$ is measurable, and define

$$L_\mathbb{P}^1(\Omega, \mathcal{C}_b(X)) = \{f : \Omega \to \mathcal{C}_b(X) \text{ measurable}, \|f\| := \int_\Omega \|f(\omega, \cdot)\|_b \, d\mathbb{P} < \infty\}.$$

For $f \in L_\mathbb{P}^1(\Omega, \mathcal{C}_b(X))$ and $\mu \in \mathcal{P}_\mathbb{P}(\Omega \times X)$ we have $f \in L^1(\mu)$ and, putting as usual $\mu(f) = \int f \, d\mu$,

$$|\mu(f)| \leq \|f\|.$$

3.5. Definition (Weak topology in $\mathcal{P}_\mathbb{P}(\Omega \times X)$). We call the smallest topology in $\mathcal{P}_\mathbb{P}(\Omega \times X)$ which makes $\mu \mapsto \mu(f)$ continuous for each $f \in L_\mathbb{P}^1(\Omega, \mathcal{C}_b(X))$ the *weak topology* in $\mathcal{P}_\mathbb{P}(\Omega \times X)$. ∎

This weak topology is Hausdorff since

$$\mu = \nu \iff \int f d\mu = \int f d\nu \quad \text{for all } f \in L_\mathbb{P}^1(\Omega, \mathcal{C}_b(X)). \tag{3.4}$$

For a systematic study see Crauel [35].

3.6. LEMMA. *Let φ be a continuous RDS on a Polish space X, denote by Θ the corresponding skew-product flow.*

(i) The mappings $f \mapsto \Theta(t)f$, $t \in \mathbb{T}$, defined by $\Theta(t)f(\omega, x) = f(\Theta(t)(\omega, x))$, are a commuting family of continuous linear mappings of $L_{\mathbb{P}}^1(\Omega, \mathcal{C}_b(X))$ to itself (isometries if $\mathbb{T}$ is two-sided).

(ii) The mappings $\mu \mapsto \Theta(t)\mu$, $t \in \mathbb{T}$, defined by $(\Theta(t)\mu)(f) = \mu(\Theta(t)f)$, are a commuting family of affine and weakly continuous mappings of $\mathcal{P}_{\mathbb{P}}(\Omega \times X)$ to itself.

Proof. We prove (i). For fixed t, $(\Theta(t)f)(\omega, x) = f(\theta(t)\omega, \varphi(t, \omega)x)$ is measurable, and for fixed ω it is continuous with respect to x. Since $\|(\Theta(t)f)(\omega, \cdot)\|_b \leq \|f(\theta(t)\omega, \cdot)\|_b$ (this inequality is an equality for two-sided time since $\varphi(t, \omega)$ is bijective),

$$\|\Theta(t)f\| \leq \|f\|.$$

Linearity is clear from the definition. □

3.7. COROLLARY. *Let φ be a continuous RDS on a Polish space. Then $\mathcal{I}_{\mathbb{P}}(\varphi)$ is a (possibly empty) convex and weakly closed subset of $\mathcal{P}_{\mathbb{P}}(\Omega \times X)$.*

3.8. THEOREM (KRYLOV-BOGOLYUBOV PROCEDURE FOR RDS).
Let φ be a continuous RDS on a Polish space X. Define for arbitrary $\nu \in \mathcal{P}_{\mathbb{P}}(\Omega \times X)$ and $N \in \mathbb{T}$, $N > 0$,

$$\mu_N(\cdot) = \begin{cases} \frac{1}{N} \sum_{n=0}^{N-1} \Theta(n)\nu(\cdot), & \mathbb{T} \quad discrete, \\ \frac{1}{N} \int_0^N \Theta(t)\nu(\cdot)\,dt, & \mathbb{T} \quad continuous, \end{cases} \tag{3.5}$$

(similarly for $N < 0$ if $\mathbb{T}$ is two-sided). Then every weak limit point of μ_N for $N \to \infty$ (or for $N \to -\infty$ for two-sided time) is in $\mathcal{I}_{\mathbb{P}}(\varphi)$, and any $\mu \in \mathcal{I}_{\mathbb{P}}(\varphi)$ arises in this way.

Proof. For the last statement put $\nu = \mu$ which produces the constant sequence $\mu_N = \mu$.

We treat only the continuous time case. Assume $\mu_{N_k} \to \mu$. This implies that for each fixed $t_0 \in \mathbb{T}$, $\Theta(t_0)\mu_{N_k} \to \Theta(t_0)\mu$ since $\mu \mapsto \Theta(t_0)\mu$ is weakly continuous by Lemma 3.6(ii).

Let now $f \in L_{\mathbb{P}}^1(\Omega, \mathcal{C}_b(X))$, $t_0 > 0$ fixed and $N_k > t_0$. Since $t \mapsto \Theta(t)\nu(f)$ is measurable and bounded (by $\|f\|$), hence locally integrable, the integral in (3.5) makes sense. We have

$$|\Theta(t_0)\mu_{N_k}(f) - \mu_{N_k}(f)| = \frac{1}{N_k} \left| \left(\int_{N_k}^{N_k+t_0} - \int_0^{t_0} \right) \Theta(t)\nu(f)dt \right|$$

$$\leq \frac{2t_0}{N_k} \|f\| \to 0 \, (k \to \infty),$$

hence $\Theta(t_0)\mu(f) = \mu(f)$. By (3.4), $\Theta(t_0)\mu = \mu$. □

Duality theory for a compact metric space X

If the Polish space X is compact metric we can use classical duality theory. We have $\mathcal{C}_b(X)^* = \mathcal{M}(X)$, where $\mathcal{M}(X)$ is the Banach space of signed measures on $(X, \mathcal{B})$ of finite total variation. As a consequence,

$$L^1_{\mathbb{P}}(\Omega, \mathcal{C}_b(X))^* = L^\infty_{\mathbb{P}}(\Omega, \mathcal{M}(X)),$$

with duality given by

$$\langle f, \mu \rangle = \int_\Omega \int_X f(\omega, x) \mu_\omega(dx)\, \mathbb{P}(d\omega) \tag{3.6}$$

(see Ledrappier [56], page 329).

We identify $\mathcal{P}_{\mathbb{P}}(\Omega \times X)$ with $L^\infty_{\mathbb{P}}(\Omega, \mathcal{P}(X))$ by identifying μ with its factorization (μ_ω),

$$\mathcal{P}_{\mathbb{P}}(\Omega \times X) \ni \mu \cong (\mu_\omega) \in L^\infty_{\mathbb{P}}(\Omega, \mathcal{P}(X)). \tag{3.7}$$

By duality (3.6) and the identification (3.7) the weak topology of $\mathcal{P}_{\mathbb{P}}(\Omega \times X)$ coincides with the weak* topology of $L^\infty_{\mathbb{P}}(\Omega, \mathcal{P}(X))$.

3.9. THEOREM. *Let X be a compact metric space. Then*

(i) $\mathcal{P}_{\mathbb{P}}(\Omega \times X) \cong L^\infty_{\mathbb{P}}(\Omega, \mathcal{P}(X))$ is a convex and weakly (or weak*) compact subset of $L^\infty_{\mathbb{P}}(\Omega, \mathcal{M}(X))$.

(ii) If φ is a continuous RDS on X then the convex and weakly compact set $\mathcal{I}_{\mathbb{P}}(\varphi)$ of its invariant measures is non-void.

Proof. Since by Alaoglu's theorem for each Banach space B the closed unit ball of B^* is compact in the B topology of B^*, the closed sets $L^\infty_{\mathbb{P}}(\Omega, \mathcal{P}(X))$ and $\mathcal{I}_{\mathbb{P}}(X)$ are weak* compact in $B^* = L^\infty_{\mathbb{P}}(\Omega, \mathcal{M}(X))$.

By Lemma 3.6(ii), the mappings $\mu \mapsto \Theta(t)\mu$, $t \in \mathbb{T}$, are a commuting family of affine weak* continuous mappings of $L^\infty_{\mathbb{P}}(\Omega, \mathcal{P}(X)) \cong \mathcal{P}_{\mathbb{P}}(\Omega \times X)$ to itself. By the Markov-Kakutani fixed point theorem there is an element μ for which $\Theta(t)\mu = \mu$ for all $t \in \mathbb{T}$, i.e. $\mathcal{I}_{\mathbb{P}}(\varphi) \neq \emptyset$. □

Lecture 4: The multiplicative ergodic theorem

4.1 Introduction. The concept of Lyapunov exponents

We will now present the most fundamental theorem about RDS, the celebrated *multiplicative ergodic theorem*[5] of Oseledets [64]. The MET provides us with exactly the right substitute of deterministic linear algebra (invariant subspaces, exponential growth rates = Lyapunov exponents) which allow a lift to nonlinear RDS, hence a 'local theory'.

Since the apearance of Oseledets' paper [64] in 1968 there have been numerous attempts to find alternative proofs of the MET[6]. The original proof relies on the

---

[5]'Multiplicative ergodic theorem' is henceforth abbreviated as 'MET'.
[6]In June 1994 I counted fifteen published proofs besides Oseledets'.

triangularization of a linear cocycle and the use of the *classical* ergodic theorem for the triangular cocycle. This technique was taken up again by Johnson, Palmer and Sell [49] (assuming a topological setting for the metric DS) and others.

Another class of proofs uses the singular value decomposition of matrices in combination with Kingman's *subadditive ergodic theorem* (see Ruelle [70], Ledrappier [56], Goldshcid and Margulis [43] and others). The subadditive ergodic theorem allows a proof of the *Furstenberg-Kesten theorem* in a few lines, and the latter enters as an ingredient into the proof of the MET. Here we will sketch this second class of proofs.

For an autonomous linear differential equation $\dot{x} = Ax$ or difference equation $x_{n+1} = Ax_n$ the origin $0 \in \mathbb{R}^d$ (and thus every point) is asymptotically stable if and only if all eigenvalues of A have negative real parts or absolute value less than one, respectively. For stability theory of *non-autonomous* systems $\dot{x} = A(t)x$ ($A(\cdot)$ locally integrable) or $x_{n+1} = A_n x_n$ it turns out *not* to be the right thing to study the eigenvalues of $A(t)$ or A_n, as they have little or nothing to do with the asymptotic properties of solutions, as the example $\dot{x} = A(t)x$ with

$$A(t) = \begin{pmatrix} -1 - 2\cos 4t & 2 + 2\sin 4t \\ -2 + 2\sin 4t & -1 + 2\cos 4t \end{pmatrix}$$

shows (calculate the eigenvalues of $A(t)$, and note that $e^t(\sin 2t, \cos 2t)$ is a solution which grows exponentially).

We should first find a *dynamical* formulation of spectral theory of A, i.e. a formulation which describes spectral objects in terms of the long-term behavior of solutions, and then try to generalize this to the non-autonomous case.

Lyapunov [59] was aware of this problem when he introduced his *characteristic exponents* which today bear his name: Let $\Phi(t)x$ be the solution of $\dot{x} = A(t)x$ or $x_{n+1} = A_n x_n$ starting at $t = 0$ at the state $x \in \mathbb{R}^d$. Then

$$\lambda^+(x) = \lambda(x) := \limsup_{t \to \infty} \frac{1}{t} \log \|\Phi(t)x\|$$

is called the *(forward) Lyapunov exponent* of $\Phi(t)x$. For two-sided time

$$\lambda^-(x) := \limsup_{t \to \infty} \frac{1}{t} \log \|\Phi(-t)x\|$$

is its *backward Lyapunov exponent*. It turns out that this definition is the right generalization of the autonomous 'real part of eigenvalue'.

4.2   The Furstenberg-Kesten theorem

Let $A \in \mathbb{R}^{d \times d}$ and

$$A = VDU, \quad U, V \in O(d, \mathbb{R}), \quad D = \text{diag}(\delta_1, \ldots, \delta_d), \quad \delta_1 \geq \ldots \geq \delta_d \geq 0,$$

be a *singular value decomposition* of A. The singular values $\delta_k = \delta_k(A)$ of A are the eigenvalues of $\sqrt{A^*A}$ and hence uniquely defined. We have

$$\| \wedge^k A \| = \delta_1(A) \cdots \delta_k(A), \quad k = 1, \ldots, d,$$

in particular $\| \wedge^d A \| = \delta_1(A) \cdots \delta_d(A) = | \det A |$, where $\wedge^k A$ is the canonical lift of A to the k-fold exterior power $\wedge^k \mathbb{R}^d$ of $\mathbb{R}^d$, defined by

$$\wedge^k A(u_1 \wedge \ldots \wedge u_k) = A u_1 \wedge \ldots \wedge A u_k$$

(for details see e.g. Arnold [4]).

We now want to study the asymptotic behavior of the singular values of a linear RDS Φ. To simplify the situation, we assume that our metric DS is ergodic, time is $\mathbb{T} = \mathbb{N}$, and our time-one mapping is invertible.

4.1. THEOREM (FURSTENBERG-KESTEN THEOREM [42], TIME $\mathbb{T} = \mathbb{N}$).
Let $\Phi(n, \omega) = A(\theta^{n-1}\omega) \cdots A(\omega)$ be a linear RDS with time $\mathbb{T} = \mathbb{N}$ over the ergodic metric DS $(\Omega, \mathcal{F}, \mathbb{P}, (\theta^n)_{n \in \mathbb{N}})$, where $A : \Omega \to Gl(d, \mathbb{R})$. Assume

$$\log^+ \|A(\cdot)\| \in L^1(\mathbb{P}) \quad and \quad \log^+ \|A(\cdot)^{-1}\| \in L^1(\mathbb{P}). \tag{4.1}$$

Then there exists a θ-invariant set $\tilde{\Omega} \in \mathcal{F}$ of full measure and constants $\gamma^{(k)} \in \mathbb{R}$, $k = 1, \ldots, d$, such that for each $\omega \in \tilde{\Omega}$ the following holds:
(i) For each $k = 1, \ldots, d$

$$\lim_{n \to \infty} \frac{1}{n} \log \| \wedge^k \Phi(n, \omega) \| = \gamma^{(k)}.$$

(ii) The constants Λ_k successively defined by

$$\Lambda_1 + \cdots + \Lambda_k = \gamma^{(k)}, \quad k = 1, \ldots, d,$$

satisfy $\Lambda_1 \geq \ldots \geq \Lambda_d$ and

$$\lim_{n \to \infty} \frac{1}{n} \log \delta_k(\Phi(n, \omega)) = \Lambda_k,$$

where $\delta_k(\Phi(n, \omega))$ is the k-th singular value of $\Phi(n, \omega)$.
(iii) Convergence in (i) and (ii) also holds in L^1.

Proof. The main statement, the $\mathbb{P}$-a.s. convergence in (i), immediately follows from Kingman's subadditive ergodic theorem by noting that $f_n^{(k)} := \log \| \wedge^k \Phi(n, \omega) \|$ is subadditive over θ, i.e.

$$f_{n+m}^{(k)}(\omega) \leq f_n^{(k)}(\theta^m \omega) + f_m^{(k)}(\omega),$$

and that $f_n^{(k)} \in L^1(\mathbb{P})$. □

4.2. Definition (Lyapunov spectrum). Let Φ be a linear cocycle satisfying the conditions of Theorem 4.1. Denote by $\lambda_1 > \ldots > \lambda_p$ the *different* numbers in the sequence $\Lambda_1 \geq \ldots \geq \Lambda_d$, and by d_i the frequency of the appearance of λ_i in this sequence. The set

$$\mathcal{S}(\theta, \Phi) = \{(\lambda_i, d_i) : i = 1, \ldots, p\}$$

is called the *Lyapunov spectrum* of Φ. ∎

4.3. THEOREM (FURSTENBERG-KESTEN THEOREM, TWO-SIDED TIME $\mathbb{T} = \mathbb{Z}$).
Let $\Phi(n, \omega)$ be a linear RDS over the ergodic DS $(\Omega, \mathcal{F}, \mathbb{P}, (\theta^n)_{n \in \mathbb{Z}})$ with time $\mathbb{T} = \mathbb{Z}$ generated by $A : \Omega \to Gl(d, \mathbb{R})$ which satisfies (4.1). Then the Lyapunov spectrum $\mathcal{S}(\theta^{-1}, \Phi(-\cdot))$ of the cocycle $\Psi(n, \omega) := \Phi(-n, \omega)$ over θ^{-1} is related to the Lyapunov spectrum $\mathcal{S}(\theta, \Phi)$ of Φ by

$$\mathcal{S}(\theta^{-1}, \Phi(-\cdot)) = -\mathcal{S}(\theta, \Phi) := \{(-\lambda_i, d_i) : i = 1, \ldots, p\}.$$

Proof. The crucial observations are that
(i) if $\delta_1 \geq \ldots \geq \delta_d > 0$ are the singular values of $A \in Gl(d, \mathbb{R})$ then $1/\delta_d \geq \ldots \geq 1/\delta_1 > 0$ are the singular values of A^{-1},
(ii) the cocycle property relates positive and negative times via $\Phi(-n, \omega) = \Phi(n, \theta^{-n}\omega)^{-1}$.
Hence

$$\delta_k(\Phi(-n, \omega)) = \delta_{d+1-k}^{-1}(\Phi(n, \theta^{-n}\omega)).$$

Now take $\mathbb{E}\log(\cdot)$ on both sides and use Theorem 4.1(iii). □

4.4. Remark. It follows from the Furstenberg-Kesten theorem that

$$\lambda_p \leq \bar{\lambda} := \frac{1}{d} \sum_{i=1}^{p} d_i \lambda_i = \frac{1}{d} \mathbb{E}\log |\det A| \leq \lambda_1.$$

$\bar{\lambda}$ is called the *average Lyapunov exponent*. ∎

4.3 The multiplicative ergodic theorem

The MET adds to the Furstenberg-Kesten theorem information about the Lyapunov exponents $\lambda(x)$ of *orbits* $\Phi(n, \omega)x$ of the cocycle, and on subspaces of $\mathbb{R}^d$ on which $\lambda(x)$ (is in fact a limit and) is less than or equal to a certain value. The construction of those subspaces is the 'hard work' of the proof of the MET.

We only present two-sided versions of the MET here.

4.5. THEOREM (MULTIPLICATIVE ERGODIC THEOREM, TIME $\mathbb{T} = \mathbb{Z}$).
Let $(\Omega, \mathcal{F}, \mathbb{P}, (\theta^n)_{n \in \mathbb{Z}})$ be an ergodic DS, and let the random variable $A : \Omega \to Gl(d, \mathbb{R})$ satisfy

$$\log^+ \|A(\cdot)\| \in L^1(\mathbb{P}) \quad and \quad \log^+ \|A(\cdot)^{-1}\| \in L^1(\mathbb{P}).$$

Consider the matrix cocycle

$$\Phi(n, \omega) = \begin{cases} A(\theta^{n-1}\omega) \cdots A(\omega), & n > 0, \\ I, & n = 0, \\ A(\theta^n \omega)^{-1} \cdots A(\theta^{-1}\omega)^{-1}, & n < 0. \end{cases}$$

Then there exists an invariant set $\tilde{\Omega} \in \mathcal{F}$ of full measure, constants $\lambda_1 > \ldots > \lambda_p$ and $d_i \in \mathbb{N}$ with $\sum_{i=1}^{p} d_i = d$ such that for each $\omega \in \tilde{\Omega}$ the following holds:

(i) There exists a splitting

$$\mathbb{R}^d = E_1(\omega) \oplus \cdots \oplus E_p(\omega)$$

of $\mathbb{R}^d$ into random measurable subspaces $E_i(\omega)$ (so-called Oseledets splitting) with $\dim E_i(\omega) = d_i$ which is invariant, i.e.

$$\Phi(n,\omega)E_i(\omega) = E_i(\theta^n \omega), \quad n \in \mathbb{Z}.$$

The splitting is dynamically characterized as follows:

$$\lim_{n \to \pm\infty} \frac{1}{n} \log \|\Phi(n,\omega)x\| = \lambda_i \iff x \in E_i(\omega) \setminus \{0\}.$$

Convergence is uniform with respect to $x \in E_i(\omega) \cap S^{d-1}$.

(ii) For each fixed $x \in \mathbb{R}^d \setminus \{0\}$, the Lyapunov exponents $\lambda^{\pm}(\omega,x)$ of the orbit $(\Phi(n,\omega)x)_{n \in \mathbb{Z}}$ exist as limits, and e.g. for $x = \oplus x_i(\omega,x)$, $x_i(\omega,x) \in E_i(\omega)$,

$$\lambda^+(\omega,x) = \lambda_{i_0(\omega,x)}, \quad i_0(\omega,x) := \min\{i : x_i(\omega,x) \neq 0\}.$$

(iii) We have

$$\mathcal{S}(\theta,\Phi) = \{(\lambda_i, d_i) : i = 1, \ldots, p\},$$

where $\mathcal{S}(\theta,\Phi)$ is the Lyapunov spectrum of Φ.

Proof. We can only briefly sketch the main steps of the proof. For details see Arnold [4], Chapter 3.

(i) We first look at the forward cocycle $(\Phi(n,\omega))_{n \in \mathbb{Z}^+}$. The Furstenberg-Kesten theorem obviously implies that the eigenvalues of $(\Phi(n,\omega)^*\Phi(n,\omega))^{1/2n}$ converge to the numbers $\exp \lambda_i$. Much more work is necessary to prove that even

$$\lim_{n \to \infty} (\Phi(n,\omega)^*\Phi(n,\omega))^{1/2n} =: \Psi(\omega) > 0$$

exists. The eigenvalues of $\Psi(\omega)$ are of course the nonrandom numbers $\exp \lambda_i$ with multiplicities d_i. If $U_i(\omega)$ denotes the corresponding eigenspace, put

$$V_i^+(\omega) := U_p(\omega) \oplus \cdots \oplus U_i(\omega), \quad i = 1, \ldots, p,$$

so that

$$V_p^+(\omega) \subset \ldots \subset V_1^+(\omega) = \mathbb{R}^d$$

defines a filtration (or flag). This filtration is invariant:

$$\Phi(n,\omega)V_i^+(\omega) = V_i^+(\theta^n \omega), \quad n \in \mathbb{Z}.$$

For each $x \neq 0$ the Lyapunov exponent $\lambda(\omega,x)$ of $(\Phi(n,\omega))_{n \in \mathbb{Z}^+}$ exists as a limit, and

$$\lambda(\omega,x) = \lambda_i \iff x \in V_i^+(\omega) \setminus V_{i+1}^+(\omega).$$

For a 'pedagogical' proof of those facts in dimension $d = 2$ see Arnold [4], section 3.5.

(ii) The same reasoning applied to the backward cocycle $(\Phi(-n,\omega))_{n\in\mathbb{Z}^+}$ over θ^{-1} yields a matrix $\Psi(\omega)^- > 0$ with eigenvalues $\lambda_i^- = -\lambda_{p+1-i}$, multiplicities $d_i^- = d_{p+1-i}$, $p^- = p$, and filtration

$$V_p^-(\omega) \subset \ldots \subset V_1^- = \mathbb{R}^d.$$

(iii) We now construct the spaces $E_i(\omega)$ by intersecting the forward and backward filtrations as follows:

$$E_i(\omega) := V_i^+(\omega) \cap V_{p+1-i}^-(\omega), \quad i = 1, \ldots, p$$

(in particular, $E_1(\omega) = V_p^-(\omega)$ and $E_p(\omega) = V_p^+(\omega)$). $\qquad\square$

For completeness we state the MET for continuous time.

4.6. THEOREM (MET FOR CONTINUOUS TIME $\mathbb{T} = \mathbb{R}$). *Let $(\Phi(t,\omega))_{t\in\mathbb{R}}$ be a continuous linear cocycle over the ergodic DS $(\Omega, \mathcal{F}, \mathbb{P}, (\theta(t))_{t\in\mathbb{R}})$. Assume $\alpha^\pm \in L^1(\mathbb{P})$, where*

$$\alpha^\pm := \sup_{0 \le t \le 1} \log^+ \|\Phi(t,\omega)^{\pm 1}\|.$$

Then all statements of the MET for time $\mathbb{T} = \mathbb{Z}$ hold with n and θ^n replaced with t and $\theta(t)$, and the invariant set $\tilde{\Omega} \in \mathcal{F}$ of full measure is now invariant with respect to $(\theta(t))_{t\in\mathbb{R}}$.

Proof. The integrability conditions $\alpha^\pm \in L^1$ make sure that
    (i) the discrete time MET holds for Φ,
    (ii) nothing 'bad' happens between discrete times. $\qquad\square$

4.4 Examples

As the Lyapunov spectrum and the invariant splitting depend on the whole history of the cocycle and are constructed by a limit procedure it is not surprising that explicit results can be obtained only for particular cases, some of which are collected in this subsection.

4.7. Example (Products of 2×2 triangular matrices).
Let $A : \Omega \to Gl(2, \mathbb{R})$, where

$$A(\omega) = \begin{pmatrix} a(\omega) & c(\omega) \\ 0 & b(\omega) \end{pmatrix}, \quad a(\omega) \ne 0, \ b(\omega) \ne 0.$$

Those matrices form a subgroup of $Gl(2, \mathbb{R})$. Put $a_k := a \circ \theta^k$, $b_k := b \circ \theta^k$, and $c_k := c \circ \theta^k$, and introduce the scalar cocycle α_n on $\mathbb{T} = \mathbb{Z}$ generated by a,

$$\alpha_n = \begin{cases} a_{n-1} \ldots a_0, & n > 0, \\ 1, & n = 0, \\ a_n^{-1} \ldots a_{-1}^{-1}, & n < 0, \end{cases}$$

and similarly the scalar cocycle β_n on $\mathbb{T} = \mathbb{Z}$ generated by b. Then the cocycle $\Phi(n, \omega)$ on $\mathbb{T} = \mathbb{Z}$ generated by A can be succinctly written as

$$\Phi(n, \cdot) = \begin{pmatrix} \alpha_n & \alpha_n \sum_{k=0}^{n-1} c_k \frac{\beta_k}{\alpha_{k+1}} \\ 0 & \beta_n \end{pmatrix}, \quad n > 0,$$

and

$$\Phi(n, \cdot) = \begin{pmatrix} \alpha_n & -\alpha_n \sum_{k=-1}^{n} c_k \frac{\beta_k}{\alpha_{k+1}} \\ 0 & \beta_n \end{pmatrix}, \quad n < 0.$$

Verify the following facts:

(i) $\log^+ \|A^{\pm 1}\| \in L^1 \iff \log|a|,\ \log|b|$ and $\log^+|c| \in L^1$. Assume this from now on.

(ii) By our assumptions,

$$\frac{1}{n} \sum_{0}^{n-1} \log|a_k| \to \mathbb{E}\log|a| =: \alpha, \quad \frac{1}{n} \sum_{0}^{n-1} \log|b_k| \to \mathbb{E}\log|b| =: \beta,$$

hence

$$\frac{1}{n} \log|\det \Phi(n, \cdot)| \to \gamma^{(2)} = \Lambda_1 + \Lambda_2 = 2\bar{\lambda} = \alpha + \beta.$$

(iii) The Lyapunov exponent of $\Phi(n, \omega)_{11}$ is α, that of $\Phi(n, \omega)_{22}$ is β, and that of $\Phi(n, \omega)_{12}$ is less than or equal to $\max(\alpha, \beta)$. By using Euclidean norm in $\mathbb{R}^{2 \times 2}$ we obtain

$$\frac{1}{n} \log \|\Phi(n, \omega)\| \to \gamma^{(1)} = \Lambda_1 = \max(\alpha, \beta),$$

hence for $\alpha \neq \beta$

$$\lambda_1 = \max(\alpha, \beta) > \bar{\lambda} = \frac{1}{2}(\alpha + \beta) > \lambda_2 = \min(\alpha, \beta).$$

For $\alpha = \beta$, $\lambda_1 = \bar{\lambda} = \alpha = \beta$ with multiplicity $d_1 = 2$.

(iv) Oseledets splitting:

Case 1: $\alpha = \beta$: In this case, the filtrations and the splitting are trivial, $E_1 = V_1^+ = V_1^- = \mathbb{R}^2$.

Case 2: $\lambda_1 = \alpha > \beta = \lambda_2$: The Oseledets space $E_1 = V_2^- = \mathbb{R}e_1$ is deterministic. To find $E_2 = \mathbb{R}\binom{u}{1}$ choose u in such a way that $\Phi(n, \omega)\binom{u(\omega)}{1}$ has Lyapunov exponent β. It turns out that

$$u(\omega) = -\sum_{k=0}^{\infty} c_k(\omega) \frac{\beta_k(\omega)}{\alpha_{k+1}(\omega)}$$

is the correct choice. Check that the expression makes sense, and

$$\Phi(n, \omega)\binom{u(\omega)}{1} = \begin{pmatrix} -\alpha_n \sum_{k=n}^{\infty} c_k \frac{\beta_k}{\alpha_{k+1}} \\ \beta_n \end{pmatrix} = \beta_n \binom{u(\theta^n \omega)}{1}$$

has Lyapunov exponent β, by which we have identified $E_2(\omega) = \mathbb{R}\binom{u(\omega)}{1}$.

Case 3: $\lambda_1 = \beta > \alpha = \lambda_2$: We obtain the deterministic space $E_2 = \mathbb{R}e_1$ and $E_1 = \mathbb{R}\binom{u}{1}$, where

$$u(\omega) = \sum_{k=-1}^{-\infty} c_k(\omega) \frac{\beta_k(\omega)}{\alpha_{k+1}(\omega)}.$$

∎

4.8. Example (Linear random differential equation in $\mathbb{R}^d$).
The equation

$$\dot{x} = A(\theta(t)\omega)x$$

generates a continuous linear cocycle $\Phi(t,\omega)$ on $\mathbb{T} = \mathbb{R}$ provided $A \in L^1$ (cf. section 2.2). Check that this condition also implies the integrability conditions $\alpha^\pm \in L^1$ of the MET for $\mathbb{T} = \mathbb{R}$.

Liouville's formula implies that the average Lyapunov exponent is

$$\bar{\lambda} = \frac{1}{d}\text{trace}\,\mathbb{E}(A).$$

A prototypical and well-investigated example (see the articles in Arnold and Wihstutz [13] or Arnold, Crauel and Eckmann [7]) is the damped linear oscillator with a random restoring force,

$$\ddot{y} + 2\beta\dot{y} + (1 + \sigma f(\theta(t)\omega))y = 0, \quad \beta, \sigma > 0, \ f \in L^1.$$

Putting $x_1 = y$, $x_2 = \dot{y}$,

$$\dot{x} = A(\theta(t)\omega)x = \begin{pmatrix} 0 & 1 \\ -1 - \sigma f(\theta(t)\omega) & -2\beta \end{pmatrix} x.$$

In this case $\bar{\lambda} = -\beta$.

∎

4.9. Example (RDE with 2×2 triangular right-hand side).
Let $\dot{x} = A(\theta(t)\omega)x$ in $\mathbb{R}^2$ with $A : \Omega \to \mathbb{R}^{2\times 2}$ given by

$$A(\omega) = \begin{pmatrix} a(\omega) & c(\omega) \\ 0 & b(\omega) \end{pmatrix}.$$

(i) We have $A \in L^1$ if and only if $a, b, c \in L^1$.

(ii) Put $\alpha := \mathbb{E}a$, $\beta := \mathbb{E}b$,

$$\alpha_t(\omega) := \int_0^t a(\theta(s)\omega\,ds, \quad \beta_t(\omega) := \int_0^t b(\theta(s)\omega)\,ds, \quad c_t(\omega) := c(\theta(t)\omega).$$

The cocycle has the form

$$\Phi(t,\omega) = \begin{pmatrix} e^{\alpha_t} & e^{\alpha_t} \int_0^t e^{\beta_s - \alpha_s} c_s\,ds \\ 0 & e^{\beta_t} \end{pmatrix}.$$

(iii) From $\det \Phi(t,\omega) = \exp(\alpha_t + \beta_t)$ we obtain

$$\frac{1}{t}\log|\det \Phi(t,\omega)| \to \gamma^{(2)} = \Lambda_1 + \Lambda_2 = 2\bar{\lambda} = \alpha + \beta.$$

Since the Lyapunov exponent of $\Phi(t,\omega)_{11}$ is α, that of $\Phi(t,\omega)_{22}$ is β and the one of $\Phi(t,\omega)_{12}$ is less than or equal to $\max(\alpha,\beta)$, we obtain (using the Euclidean norm in $\mathbb{R}^{2\times 2}$)

$$\frac{1}{t}\log\|\Phi(t,\omega)\| \to \gamma^{(1)} = \Lambda_1 = \max(\alpha,\beta),$$

hence for the case $\alpha \neq \beta$

$$\lambda_1 = \max(\alpha,\beta) > \bar{\lambda} = \frac{1}{2}(\alpha+\beta) > \lambda_2 = \min(\alpha,\beta).$$

For $\alpha = \beta$, $\lambda_1 = \bar{\lambda} = \alpha = \beta$ with multiplicity $d_1 = 2$.

(iv) Oseledets splitting:

Case 1: $\alpha = \beta$: In this case, the filtrations and the splitting are trivial, $E_1 = V_1^+ = V_1^- = \mathbb{R}^2$. But assume for example that $a = b$, $c = 0$, $\alpha = \beta = 0$, and the law of the iterated logarithm holds for α_t. Then the orbits $\Phi(t,\omega)x = \exp(\alpha_t(\omega))x$ typically undergoe fluctuations of the order $\exp\sqrt{2t\log\log t}$. Note, in contrast, that the deterministic "cocycle" $\exp(tA)$ has at most polynomial growth inside an eigenspace corresponding to an eigenvalue with vanishing real part.

Case 2: $\lambda_1 = \alpha > \beta = \lambda_2$: $E_1 = V_2^- = \mathbb{R}e_1$ is deterministic, and

$$E_2(\omega) = \mathbb{R}\begin{pmatrix} u(\omega) \\ 1 \end{pmatrix}, \quad u(\omega) = -\int_0^\infty e^{\beta_t(\omega)-\alpha_t(\omega)}c_t(\omega)\,dt.$$

The orbit of $\begin{pmatrix} u \\ 1 \end{pmatrix}$ is

$$\Phi(t,\omega)\begin{pmatrix} u(\omega) \\ 1 \end{pmatrix} = \begin{pmatrix} -e^{\alpha_t(\omega)}\int_t^\infty e^{\beta_s(\omega)-\alpha_s(\omega)}c_s(\omega)\,ds \\ e^{\beta_t(\omega)} \end{pmatrix} = e^{\beta_t(\omega)}\begin{pmatrix} u(\theta(t)\omega) \\ 1 \end{pmatrix}.$$

That u makes sense and the orbit of $\begin{pmatrix} u \\ 1 \end{pmatrix}$ has Lyapunov exponent β is a consequence of the following simple lemma.

4.10. LEMMA. *Let $u \in L^1$ and $v_t(\omega)$ be continuous with $v_t(\omega)/t \to \lambda < 0$ $\mathbb{P}$-a. s. Then*

$$\int_0^\infty e^{v_t(\omega)}u(\theta(t)\omega)\,dt \quad \text{exists } \mathbb{P}\text{-a. s. },$$

and

$$\limsup_{t\to\infty} \frac{1}{t}\log\left|\int_t^\infty e^{v_s(\omega)}u(\theta(s)\omega)\,ds\right| \le \lambda.$$

Case 3: $\lambda_1 = \beta > \alpha = \lambda_2$: Now $E_2 = V_2^+ = \mathbb{R}e_1$ is deterministic, and

$$E_1(\omega) = \mathbb{R}\begin{pmatrix} u(\omega) \\ 1 \end{pmatrix}, \quad u(\omega) = \int_{-\infty}^0 e^{\beta_t(\omega)-\alpha_t(\omega)}c_t(\omega)\,dt.$$

■

Lecture 5: Aspects of the theory of smooth random dynamical systems

We will now present some results from the theory of smooth nonlinear RDS φ which are based on the MET applied to the linearization $T\varphi$ of φ. Hence our first task is to formulate the MET for this situation.

5.1 The multiplicative ergodic theorem for RDS on manifolds

Let M be a smooth manifold, and φ be a $\mathcal{C}^1$ RDS on M over θ. We can differentiate (linearize) $\varphi(t, \omega)$ at $x \in M$ to obtain a linear mapping

$$T\varphi(t, \omega, x) : T_x M \to T_{\varphi(t, \omega)x} M.$$

Differentiating the cocycle property for φ at x and the chain rule yield

$$
\begin{aligned}
T\varphi(t + s, \omega, x) &= T\varphi(t, \theta(s)\omega, \varphi(s, \omega)x) \circ T\varphi(s, \omega, x) \\
&= T\varphi(t, \Theta(s)(\omega, x)) \circ T\varphi(s, \omega, x),
\end{aligned}
$$

i.e. $T\varphi$ is a continuous linear cocycle on the tangent bundle TM over the skew-product flow Θ (*not* θ!).

If M is a Riemannian manifold, we can define the *Lyapunov exponent* of the tangent vector $v \in T_x M \setminus \{0\}$ at (ω, x) by

$$\lambda(\omega, x, v) := \lim_{t \to \infty} \frac{1}{t} \log \|T\varphi(t, \omega, x)v\|.$$

5.1. THEOREM (MET ON MANIFOLDS, TIME $\mathbb{T} = \mathbb{Z}$ OR $\mathbb{R}$). *Let φ be a $\mathcal{C}^1$ RDS with time $\mathbb{T} = \mathbb{Z}$ or $\mathbb{R}$ on a Riemannian manifold M of dimension d. Let μ be an ergodic invariant measure of φ, such that $\alpha^{\pm} \in L^1(\mu)$, where*

$$
\alpha^{\pm}(\omega, x) := \begin{cases}
\log^+ \|T\varphi(1, \omega, x)^{\pm 1}\|, & \mathbb{T} = \mathbb{Z}, \\
\sup_{0 \le t \le 1} \log^+ \|T\varphi(t, \omega, x)^{\pm 1}\|, & \mathbb{T} = \mathbb{R}.
\end{cases}
$$

Then there exist

- *a Θ-invariant set $\Delta \in \Omega \times M$ of full μ measure,*

- *a finite list of fixed numbers*

$$\lambda_1 > \lambda_2 > \ldots > \lambda_p$$

(the Lyapunov exponents of φ under μ) with multiplicities $d_1 + d_2 + \cdots + d_p = d$ (the λ_i and d_i form the Lyapunov spectrum of φ under μ),

such that for each fixed $(\omega, x) \in \Delta$ the following holds:

(i) There is a measurable splitting

$$T_x M = E_1(\omega, x) \oplus E_2(\omega, x) \oplus \cdots \oplus E_p(\omega, x)$$

with $\dim E_i(\omega, x) = d_i$ which is invariant, i.e.

$$T\varphi(t, \omega, x) E_i(\omega, x) = E_i(\Theta(t)(\omega, x)).$$

The splitting is dynamically characterized by

$$\lim_{t \to \pm\infty} \frac{1}{t} \log \|T\varphi(t, \omega, x)v\| = \lambda_i \iff v \in E_i(\omega, x) \setminus \{0\}.$$

(ii) For each fixed $v \in T_x M \setminus \{0\}$, the Lyapunov exponent $\lambda(\omega, x, v)$ of v exists as a limit, and

$$\lambda(\omega, x, v) = \lambda_{i_0(\omega, x, v)}, \quad i_0(\omega, x, v) := \min\{i : proj_{E_i(\omega, x)} v \neq 0\}.$$

Proof. We can reduce this case to the one on a trivial bundle treated in Lecture 4 as follows:

(i) There is a global *measurable* trivialization of the tangent bundle TM, i.e. there is a bimeasurable bijection

$$\Psi : TM \to M \times \mathbb{R}^d$$

satisfying $\pi_M \circ \Psi = \text{id}_M \circ \pi$ (π the canonical projection of TM to M, π_M the projection from $M \times \mathbb{R}^d$ to M), such that it is a linear isomorphism on fibers, i.e.

$$\psi(x) := \Psi|_{T_x M} : T_x M \to \{x\} \times \mathbb{R}^d$$

is a linear isomorphism. Indeed, by definition of a smooth bundle, TM can be locally trivialized by a diffeomorphism. Now pick a countable covering of locally trivializing maps, "disjointify" them and piece them together.

(ii) Moreover, if M is a Riemannian manifold with Riemannian structure $\langle \cdot, \cdot \rangle_x$ then Ψ can be chosen to be isometric on fibers, i.e.

$$\psi(x) : (T_x M, \langle \cdot, \cdot \rangle_x) \to (\{x\} \times \mathbb{R}^d, \langle \cdot, \cdot \rangle_S)$$

is an isometry ($\langle \cdot, \cdot \rangle_S$ the standard scalar product in $\mathbb{R}^d$). See e.g. Klingenberg [53], Theorem 1.8.20.

(iii) Now consider the linear cocycle

$$\Phi(t, \omega, x) := \psi(\varphi(t, \omega)x) \circ T\varphi(t, \omega, x) \circ \psi(x)^{-1}$$

on $\mathbb{R}^d$ over the metric DS $(\Omega \times M, \mathcal{F} \otimes \mathcal{B}, (\Theta(t))_{t \in \mathbb{T}}, \mu)$. Φ and $T\varphi$ are linearly isomorphic ("cohomologous"), and the isomorphism preserves Lyapunov exponents since $\|\psi(x)\| = 1$.

(iv) Now apply the MET Theorem 4.5 or 4.6 to Φ (which satisfies the corresponding integrability conditions). We obtain the Lyapunov exponents λ_i and the splitting

$$E_i^{T\varphi}(\omega, x) = \psi(x)^{-1} E_i^{\Phi}(\omega, x).$$

The moral of the MET is that for a $\mathcal{C}^1$ RDS φ with invariant measure μ, things are as nice and as simple as for a fixed point in the deterministic autonomous case: The λ_i are the stochastic analogues of the (real parts of) eigenvalues, and the E_i are the analogues of the generalized eigenspaces.

On the basis of this theorem we can develop a local theory of smooth RDS as in the deterministic case. This will be exemplified in this and the next lecture.

We stress that the Lyapunov spectrum of φ depends crucially on the invariant measure μ.

Stochastic analogues of the imaginary parts of eigenvalues are the so-called *rotation numbers* for which there is also an MET, see Arnold and San Martin [11].

5.2 Invariant manifolds. Pesin's formula

Let φ be a $\mathcal{C}^1$ RDS on a Riemannian manifold M with (ergodic) invariant measure μ satisfying the integrability conditions of the MET Theorem 5.1.

Now choose a non-void subset $\Lambda \subset \{\lambda_1, \ldots, \lambda_p\}$ and consider on the Θ-invariant set Δ of "good" points (ω, x) of the MET the measurable subbundle

$$E_\Lambda(\omega, x) := \oplus_{\lambda \in \Lambda} E_i(\omega, x)$$

of TM. This subbundle is invariant with respect to the linearized cocyle,

$$T\varphi(t, \omega, x)E_\Lambda(\omega, x) = E_\Lambda(\Theta(t)(\omega, x)).$$

The problem is now to find (local) random submanifolds $M_\Lambda(\omega, x)$ of M with the following properties:

1. $T_x M_\Lambda(\omega, x) = E_\Lambda(\omega, x)$,

2. $\varphi(t, \omega)M_\Lambda(\omega, x) = M_\Lambda(\Theta(t)(\omega, x))$ locally.

If Λ consists of the negative/zero/positive Lyapunov exponents then M_Λ is the stable/center/unstable manifold, etc. For the deterministic case of smooth ergodic theory see Pesin [65], [66] and Pugh and Shub [68], for the random case Ruelle [70], Carverhill [30], Boxler [28] (for stochastic center manifolds) and Dahlke [36].

In particular, the stability of an RDS under an invariant measure is reduced to the problem whether $\lambda_1 < 0$. If yes, then for each $(\omega, x) \in \Delta$ there exists a random open neighborhood $U(\omega, x)$ of x such that a second trajectory of φ starting at $y \in U(\omega, x)$ satisfies

$$d(\varphi(t, \omega)x, \varphi(t, \omega)y) \to 0 \quad \text{exponentially fast}$$

for $t \to \infty$.

We have "hyperbolicity" of φ under μ if $\lambda_p < 0 < \lambda_1$.

Pesin's formula

One of the deepest and most exciting results in smooth ergodic theory is the fact that if M is a compact Riemannian manifold, $\varphi \in \text{Diff}^2(M)$ and ρ a φ-invariant measure which is absolutely continuous with respect to Riemannian volume then

$$h_\rho(\varphi) = \int_M \sum d_i(x) \lambda_i^+(x) \rho(dx)$$

(so-called *Pesin's formula*), where $x^+ := \max(x, 0)$ and $h_\rho(\varphi)$ denotes the metric entropy of φ.

Pesin's formula for $\mathcal{C}^2$ RDS φ on a compact Riemannian manifold M under an invariant measure μ should read

$$h_\mu(\varphi) = \int_{\Omega \times M} \sum d_i(\omega, x) \lambda_i^+(\omega, x) d\mu(\omega, x),$$

where $h_\mu(\varphi) = h_\mu(\Theta | \mathcal{F} \otimes M)$ denotes the *fiber* (or *relativized*) *entropy* (see Bogenschütz [25]). This formula was proved by Ledrappier and Young [58] for the particular case of iterations of independent and identically distributed random diffeomorphisms $\varphi(\omega) \in \text{Diff}^2(M)$ and for an invariant measure μ such that $\rho = \mathbb{E}\mu$ is a stationary measure for the corresponding Markov chain with transition probability

$$P(x, B) := \mathbb{P}\{\omega : \varphi(\omega)x \in B\}$$

(see Lecture 3, Example 3.3 and Exercise 3.4), and ρ is absolutely continuous with respect to Riemannian volume.

Recently Bahnmüller [19] and Liu Pei-Dong [60] proved Pesin's formula for general $\mathcal{C}^2$ RDS and invariant μ for which μ_ω is absolutely continuous with respect to Riemannian volume, or has the so-called *Sinai-Bowen-Ruelle property* (absolute continuity on the unstable manifolds), respectively.

5.3 Normal forms 1: Continuous coordinate transformations

Let now $X = \mathbb{R}^d$. The theorem of Hartman and Grobman from the early 60ies states that a $\mathcal{C}^1$ diffeomorphism with hyperbolic fixed point 0 (i.e. $A = D\varphi(0)$ is hyperbolic) is topologically equivalent to its linearization, i.e. there exists a homeomorphism h such that

$$A = h^{-1} \circ \varphi \circ h$$

(there are global and local versions of this theorem, see Irwin [47]).

Let φ be a $\mathcal{C}^1$ RDS with discrete time $\mathbb{T} = \mathbb{Z}$ (say), time-one map $\varphi(\omega) := \varphi(1, \omega)$ and $\varphi(\omega)0 = 0$. The point 0 is called *hyperbolic* if the cocycle generated by $A(\omega) := D\varphi(\omega, 0)$ is hyperbolic, i.e. has only non-vanishing Lyapunov exponents. We now seek a *random* homeomorphism $h(\omega)$ such that

$$A(\omega) = h(\theta\omega)^{-1} \circ \varphi(\omega) \circ h(\omega). \tag{5.1}$$

The crucial difference between the deterministic and stochastic case is that in equation (5.1) the DS θ appears ($\varphi(\omega)$ maps the fiber over ω to the fiber over $\theta\omega$, while $h(\omega)$ is a coordinate transformation of the same fiber) which makes the stochastic normal form problem infinite-dimensional and hence richer.

A random Hartman-Grobman theorem is now available for $\mathcal{C}^1$ RDS for $\mathbb{T} = \mathbb{Z}$ and for $\mathbb{T} = \mathbb{R}$ and random differential equations, see Wanner [80]. Wanner also recovers the invariant manifolds tangent to the Oseledets spaces.

5.4 Normal forms 2: Smooth coordinate transformations

Smooth normal form theory is one of the fundamental building blocks of the theory of nonlinear DS. The deterministic theory was founded by Poincaré, see e.g. V. I. Arnold [18].

Even for a $\mathcal{C}^\infty$ RDS with hyperbolic fixed point 0, the Hartman-Grobman theorem only gives a random *homeomorphism*. If we want to have an $h(\omega) \in$ Diff$^\infty(\mathbb{R}^d)$ (which also is *near identity*, i.e. satisfies $Dh(\omega, 0) = $ id), certain obstacles called *resonances* appear which prevent us from transforming away certain terms of the Taylor expansion of $\varphi(\omega)$ at 0. Those resonances take the form of algebraic relations between Lyapunov exponents. The task is now to choose $h(\omega)$ such that

$$\tilde{\varphi}(\omega) = h(\theta\omega)^{-1} \circ \varphi(\omega) \circ h(\omega)$$

is "as simple as possible", the best-possible case being $\tilde{\varphi}(\omega) = D\varphi(\omega, 0)$. Which terms of the Taylor expansion of φ survive depends on the interplay of the linear cocycle generated by $D\varphi(\omega, 0)$ and the metric DS θ. See Arnold and Xu [14], [17]. In [15] we calculate the random normal form for the noisy Duffing-Van der Pol oscillator using MAPLE.

5.5 Topological classification of linear cocycles. Structural stability

The random Hartman-Grobman theorem allows to linearize hyperbolic cocycles — but when are two *linear* hyperbolic cocycles topologically equivalent? More precisely, given two random matrices $A, B : \Omega \to Gl(d, \mathbb{R})$ satisfying the integrability conditions of the MET (Theorem 4.5) and generating hyperbolic cocycles. When does there exist a random homeomorphism $h : \Omega \to $ Homeo$(\mathbb{R}^d)$ with $h(\omega)0 = 0$ such that

$$B(\omega) = h(\theta\omega)^{-1} \circ A(\omega) \circ h(\omega) \text{ ?}$$

The question is answered by the following deep theorem of Nguyen Dinh Cong [63], [62].

5.2. THEOREM (TOPOLOGICAL CLASSIFICATION OF LINEAR COCYCLES).
Let $A, B : \Omega \to Gl(d, \mathbb{R})$ be random matrices satisfying the integrability conditions of the MET for $\mathbb{T} = \mathbb{Z}$ and generating hyperbolic cocycles Φ_A and Φ_B with

stable and unstable spaces $E_A^{s,u}$ and $E_B^{s,u}$. Define $C_{A,B}^s$ to be the set of all $\omega \in \Omega$ for which the mappings $A^s(\omega) : E_A^s(\omega) \to E_A^s(\theta\omega)$ and $B^s(\omega) : E_B^s(\omega) \to E_B^s(\theta\omega)$ are not both orientation preserving or both orientation reversing. The set $C_{A,B}^u$ is analogously defined.

Then the hyperbolic cocycles generated by A and B are topologically equivalent if and only if

1. $\dim E_A^s(\omega) = \dim E_B^s(\omega)$ (hence $\dim E_A^u(\omega) = \dim E_B^u(\omega)$),

2. the sets $C_{A,B}^s$ and $C_{A,B}^u$ are both coboundaries of θ.[7]

The proof of this theorem uses random norms (see Arnold [4], section 3.9) and is technically beyond these notes.

Nguyen Dinh Cong's theorem is a non-trivial generalization of the well-known deterministic topological classification theorem for matrices by Robbin (see Irwin [47]). Robbin's theorem can be obtained as a particular case of the above theorem by taking a one-point Ω and by noting that the empty set $\emptyset$ is always a coboundary. This gives $4d$ different classes of hyperbolic matrices. However, if θ is e.g. irrational rotation on (S^1, Leb) there are infinitely many classes of hyperbolic cocycles which are not pairwise topologically equivalent.

Nguyen Dinh Cong also proved in [62] that a linear cocycle is structurally stable (with respect to the random norm associated with the cocycle) if and only if it is hyperbolic.

Lecture 6: Towards a theory of stochastic bifurcation

6.1 What is stochastic bifurcation theory?

Deterministic bifurcation theory, one of the core areas of the theory of smooth DS, studies *qualitative changes* in parametrized families φ_α of DS, e.g. those generated by a family of ordinary differential equations

$$\dot{x} = f^\alpha(x).$$

The vague notion of *qualitative change* has been successfully formalized by the concepts *topological equivalence (conjugacy)* and *structural stability*. The point α_0 is a bifurcation point if the family φ_α is *not* structurally stable at α_0 (see Guckenheimer and Holmes [44], page 119).

Stochastic bifurcation theory (more precisely: bifurcation theory of RDS) studies *qualitative changes* in parametrized families φ_α of RDS, e.g. those generated by a family of stochastic differential equations

$$dx = f_0^\alpha(x) + \sum_{j=1}^{m} f_j^\alpha(x) \circ dW_t^j. \tag{6.1}$$

---

[7]A set $A \in \mathcal{F}$ is a coboundary of θ if there exists a set $C \in \mathcal{F}$ such that $A = C \,\Delta\, \theta C$.

How can we formalize *qualitative change* in this situation?
Two approaches have been pursued so far.

Phenomenological approach

Scientists have been observing *qualitative changes* of probability densities p_α which are stationary for (6.1), in other words are solutions of the Fokker-Planck equation

$$L_\alpha^* p_\alpha = 0, \quad L_\alpha = f_0^\alpha + \frac{1}{2} \sum_{j=1}^m (f_j^\alpha)^2 \tag{6.2}$$

(we have written L_α in Hörmander form). For example, transitions from one-peak to two-peak ot crater-like densities have been observed experimentally, numerically and analytically (see Horsthemke and Lefever [46] or Ebeling et al. [38]).

This concept can be formalized with the ideas of Zeeman [83], [82]: Call two probability densities p, q *equivalent*, if there are two diffeomorphisms β, γ such that $p = \beta \circ q \circ \gamma$. Then the transition point α_0 from a density with one peak to one with two peaks is a point of structural instability with respect to the above equivalence. We call those phenomena *phenomenological* or *P-bifurcations*.

The drawback of P-bifurcation is (see Arnold [2], Arnold and Boxler [5]) that it is a *static* concept in several respects:

(i) The stationary density $p_\alpha(x)dx$ measures the proportion of time spent by a typical solution of (6.1) in the volume element dx. Hence p_α has "forgotten" any dynamics.

(ii) p_α is generated by the one-point motion $t \mapsto \varphi_\alpha(t,\omega)x$ and can thus principally *not* be related to the stability of φ_α which is determined by the behavior of the two-point motion $t \mapsto (\varphi_\alpha(t,\omega)x, \varphi_\alpha(t,\omega)y)$ for $x \neq y$, in infinitesimal form by $D\varphi_\alpha(t,\omega,x)v$, hence by the Lyapunov exponents λ_i. See Baxendale [22] for an example of a family φ_α generated by (6.1) for particular vector fields which changes from 'exponentially stable' (all Lyapunov exponents negative) to 'exponentially unstable' (the top Lyapunov exponent positive), whereas the Fokker-Planck equation (6.2) is independent of the parameter α.

(iii) Considering only stationary measures (and not the invariant measures in the sense of Lecture 3, of which there are typically many more) is a too narrow point of view, leads to 'missing branches' in bifurcation diagrams and prevents us from understanding stochastic bifurcation.

Dynamical approach

As an invariant measure of an RDS is the stochastic analogue of a deterministic fixed point, it is reasonable to study as a first step the bifurcation of invariant measures.

6.1. Definition (Dynamical bifurcation). Given a family $(\varphi_\alpha)_{\alpha \in \mathbb{R}}$ of RDS in $\mathbb{R}^d$. Assume that φ_α has the invariant measure ('reference measure') μ_α. Then $(\alpha_0, \mu_{\alpha_0})$ is called a *dynamical* or *D-bifurcation point*, if for each α in a neighborhood of α_0 there is a φ_α-invariant measure $\nu_\alpha \neq \mu_\alpha$ for which $\nu_\alpha \to \mu_{\alpha_0}$ weakly (see Lecture 3) as $\alpha \to \alpha_0$. ∎

D-bifurcation is related to the stability of φ_α under μ_α via the Lyapunov exponents $\lambda_i(\mu_\alpha)$ of the linearization $D\varphi_\alpha(t,\omega,x)$.

In certain cases loss of stability of the reference measure leads to the bifurcation of a new invariant measure. For $d=1$ see Arnold and Boxler [5] and Xu [81], for general d see Baxendale [23].

Conversely, a very strong argument in favor of the concept of D-bifurcation is the following necessary condition.

6.2. THEOREM (NECESSARY CONDITION FOR STOCHASTIC BIFURCATION).
Let φ_α be a family of $\mathcal{C}^1$ RDS in $\mathbb{R}^d$ with $\varphi_\alpha(t,\omega)0 = 0$, i.e. with reference measure $\mu_\alpha = \delta_0$. Assume that the linear RDS $D\varphi(t,\omega,0)$ satisfies the integrability conditions of the MET.
Let α_0 be a D-bifurcation point, i.e. there is $\nu_\alpha \neq \delta_0$ with $\nu_\alpha \to \delta_0$ for $\alpha \to \alpha_0$. If ν_α is 'small enough', then necessarily $D\varphi_{\alpha_0}$ has a vanishing Lyapunov exponent.

The proof is given by Arnold and Xu [16] and is based on the local form of the random Hartman-Grobman theorem (see Lecture 5) by which the precise meaning of 'small enough' is also determined.

Interesting relations between P-bifurcation and D-bifurcation were discovered by Baxendale [23], [24].

6.2 A one-dimensional example

The stochastic versions of transcritical, pitchfork and saddle-node bifurcations are known to exist in dimension $d=1$ (for the white-noise case see Arnold and Boxler [5], for the real-noise case see Xu [81] — he found scenarios which do *not* exist in the deterministic case!).

Let us briefly review the transcritical case.

The stochastic differential equation

$$dx = (\alpha x - x^2)dt + \sigma x \circ dW \qquad (6.3)$$

($\alpha \in \mathbb{R}$ is the bifurcation parameter, $\sigma > 0$ is a strength parameter of the noise) can be explicitly solved to give the family of $\mathcal{C}^\infty$ RDS

$$x \mapsto \varphi_\alpha(t,\omega)x = \begin{cases} 0, & x = 0, \\ \dfrac{\exp(\alpha t + \sigma W_t)}{\frac{1}{x} + \int_0^t \exp(\alpha s + \sigma W_s)ds}, & x \neq 0. \end{cases}$$

From this we can read-off all invariant measures. We find
(i) $\mu_\alpha = \delta_0$ for all α,
(ii) $\nu_{\alpha,\omega} = \delta_{\kappa_\alpha(\omega)}$ for all $\alpha \neq 0$, where

$$\kappa_\alpha(\omega) = \begin{cases} -(\int_0^\infty \exp(\alpha t + \sigma W_t)dt)^{-1}, & \alpha < 0, \\ (\int_{-\infty}^0 \exp(\alpha t + \sigma W_t)dt)^{-1}, & \alpha > 0. \end{cases}$$

We have $\mathbb{E}\kappa_\alpha = \alpha$.

Solving the Fokker-Planck equation

$$L^*p = -((\alpha x + \frac{\sigma^2}{2}x - x^2)p)' + \frac{\sigma^2}{2}(x^2 p)'' = 0$$

yields

(i) $p_\alpha = \delta_0$ for all α,

(ii) for $\alpha > 0$ we find

$$q_\alpha(x) = \begin{cases} N_\alpha x^{2\alpha/\sigma^2 - 1} \exp(-\frac{2x}{\sigma^2}), & x > 0, \\ 0, & x \leq 0. \end{cases}$$

We do not find a solution for $\alpha < 0$ corresponding to $\mathbb{E}\nu_{\alpha,\omega}(dx) = q_\alpha(x)dx$, the deeper reason being that for $\alpha < 0$, κ_α is not $\mathcal{F}^0_{-\infty}$-measurable, see Lecture 3. Clearly q_α undergoes a P-bifurcation at $\alpha = \sigma^2/2$.

The stability of $\mu_\alpha = \delta_0$ can be determined by linearizing equation (6.3) at $x = 0$,

$$dv = \alpha v dt + \sigma v \circ dW,$$

which has the solution

$$\Phi_\alpha(t,\omega)v = v \exp(\alpha t + \sigma W_t),$$

hence the Lyapunov exponent

$$\lambda(\mu_\alpha) = \lim_{t \to \infty} \frac{1}{t} \log \|\Phi(t,\omega)v\| = \alpha.$$

For ν_α we obtain

$$\lambda(\nu_\alpha) = -\alpha,$$

so we have a D-bifurcation at $\alpha = 0$, where the reference measure μ_α looses its stability.

Why do we have a P-bifurcation of the new branch ν_α at $\alpha = \sigma^2/2$? This has been shown by Baxendale [23], [24] to be a large deviations phenomenon. Define the p'th moment Lyapunov exponent of a linear RDS Φ by

$$g(p) := \lim_{t \to \infty} \frac{1}{t} \log \mathbb{E}\|\Phi(t,\omega)v\|^p, \quad p \in \mathbb{R}.$$

In our example

$$g(p) = \alpha p + \frac{\sigma^2}{2}p^2.$$

Baxendale proved that the new branch ν_α of invariant measures undergoes a P-bifurcation at a parameter value at which the non-trivial zero of the function $g(p)$ corresponding to the reference measure $\mu_\alpha = \delta_0$ is equal to $-d = -\dim \mathbb{R}^d$, in our example $\alpha = \sigma^2/2$.

The real challenge for stochastic bifurcation theory is to explain and analyze what 'stochastic Hopf bifurcation' could mean. For extensive simulations of the noisy Duffing-Van der Pol oscillator see Schenk [74]. Theoretical work is in progress.

Further problems. Outlook

Topological random dynamical systems

In these lectures we have mainly discussed problems from the theory of *smooth* RDS which is based on the MET.

However, with the notions of metric and topological fiber entropy and fiber pressure we can develop a *thermodynamical formalism* for topological RDS with all its typical ingredients (variational principle, equilibrium and Gibbs states, Ruelle-Perron-Frobenius theory), see e.g. Ledrappier and Walters [57], Kifer [51], [52], Ferrero and Schmitt [40], Bogenschütz [25] and Bogenschütz and Gundlach [26].

This formalism has applications to the dynamics of populations in a random environment (Arnold, Demetrius and Gundlach [8]).

For expanding random mappings, Markov partitions (and hence in particular their cardinality) depend on chance. For a symbolic representation we thus need randomly many symbols (Bogenschütz and Gundlach [26]).

This leads to the largely open problems of which RDS deserve the name *canonical*, of describing and classifying RDS through their isomorphism invariants (e.g. its fiber entropy).

Measurability and dynamics

The interplay of measurability and dynamics genuinely belongs to the theory of *random* DS and has no deterministic analogue. Here is a sample result: If an invariant measure μ_ω of a $\mathcal{C}^1$ RDS φ is *not* measurable with respect to the 'past' or the 'future' of the system (described by sub-σ-algebras $\mathcal{F}^0_{-\infty}$, $\mathcal{F}^\infty_0 \subset \mathcal{F}$, see Lecture 3), then φ has *saddle behavior* with respect to μ_ω, i.e. it has positive *and* negative Lyapunov exponents (Crauel [34]).

For RDS generated by stochastic differential equations many problems are blocked by the following 'collision' of stochastic analysis and multiplicative ergodic theory: One of the fundamental objects provided by the MET are the random Oseledets spaces $E_i(\omega, x)$ (see Theorem 5.1). As clearly seen from their construction, they are typically not measurable with respect to the past, hence with respect to the information available in stochastic analysis at time $t = 0$. This calls for utilizing anticipative calculus (Arnold [3], Arnold and Imkeller [9], [10]).

Bibliography

[1] H. Amann. *Gewöhnliche Differentialgleichungen*. Walter de Gruyter, Berlin, 1983.

[2] L. Arnold. Lyapunov exponents of nonlinear stochastic systems. In F. Ziegler and G. I. Schuëller, editors, *Nonlinear stochastic dynamic engineering systems*, IUTAM Symposium Innsbruck 1987, pages 181–201. Springer, Berlin Heidelberg New York, 1988.

[3] L. Arnold. Anticipative problems in the theory of random dynamical systems. Report 294, Institut für Dynamische Systeme, Universität Bremen, 1993.

[4] L. Arnold. *Random dynamical systems*. Preliminary version, 1994.

[5] L. Arnold and P. Boxler. Stochastic bifurcation: instructive examples in dimension one. In M. Pinsky and V. Wihstutz, editors, *Diffusion processes and related problems in analysis, volume II: Stochastic flows*, volume 27 of *Progress in Probability*, pages 241–255. Birkhäuser, Boston Basel Stuttgart, 1992.

[6] L. Arnold and H. Crauel. Iterated function systems and multiplicative ergodic theory. In M. Pinsky and V. Wihstutz, editors, *Diffusion processes and related problems in analysis, volume II: Stochastic flows*, volume 27 of *Progress in Probability*, pages 283–305. Birkhäuser, Boston Basel Stuttgart, 1992.

[7] L. Arnold, H. Crauel, and J.-P. Eckmann, editors. *Lyapunov Exponents. Proceedings, Oberwolfach 1990*, volume 1486 of *Lecture Notes in Mathematics*. Springer, Berlin Heidelberg New York, 1991.

[8] L. Arnold, L. Demetrius, and M. V. Gundlach. Evolutionary formalism for products of positive random matrices. *Annals of Applied Probability*, 1994. To appear.

[9] L. Arnold and P. Imkeller. Stratonovich calculus with spatial parameters and anticipative problems in multiplicative ergodic theory. Report 288, Institut für Dynamische Systeme, Universität Bremen, 1993.

[10] L. Arnold and P. Imkeller. Furstenberg-Khasminskii formulas for Lyapunov exponents via anticipative calculus. Report, Institut für Dynamische Systeme, Universität Bremen, 1994.

[11] L. Arnold and L. San Martin. A multiplicative ergodic theorem for rotation numbers. *Journal of Dynamics and Differential Equations*, 1:95–119, 1989.

[12] L. Arnold and M. Scheutzow. Perfect cocycles through stochastic differential equations. *Probab. Th. Rel. Fields*, 1994 (to appear).

[13] L. Arnold and V. Wihstutz, editors. *Lyapunov Exponents. Proceedings, Bremen 1984*, volume 1186 of *Lecture Notes in Mathematics*. Springer, Berlin Heidelberg New York, 1986.

[14] L. Arnold and Xu Kedai. Normal forms for random diffeomorphisms. *J. Dynamics and Differential Equations*, 4:445–483, 1992.

[15] L. Arnold and Xu Kedai. Simultaneous normal form and center manifold reduction for random differential equations. In C. Perelló, C. Simó, and J. Solá-Morales, editors, *EQUADIFF-91*, volume 1, pages 68–80. World Scientific, Singapore, 1993.

[16] L. Arnold and Xu Kedai. Invariant measures for random dynamical systems, and a necessary condition for stochastic bifurcation from a fixed point. *Random & Computational Dynamics*, 2:165–182, 1994.

[17] L. Arnold and Xu Kedai. Normal forms for random differential equations. *Journal of Differential Equations*, 1994 (to appear).

[18] V. I. Arnold. *Geometrical methods in the theory of ordinary differential equations*. Springer, Berlin Heidelberg New York, 1983.

[19] J. Bahnmüller. A Pesin formula for random dynamical systems. Preprint, 1994.

[20] M. F. Barnsley. *Fractals everywhere*. Academic Press, New York, 1988.

[21] P. Baxendale. Brownian motion in the diffeomorphism group I. *Compositio Mathematica*, 53:19–50, 1984.

[22] P. Baxendale. Asymptotic behaviour of stochastic flows of diffeomorphisms. In K. Itō and T. Hida, editors, *Stochastic processes and their applications*, volume 1203 of *Lecture Notes in Mathematics*, pages 1–19. Springer, Berlin Heidelberg New York, 1986.

[23] P. Baxendale. Invariant measures for nonlinear stochastic differential equations. In Arnold et al. [7], pages 123–140.

[24] P. Baxendale. A stochastic Hopf bifurcation. *Probab. Th. Rel. Fields*, 1994 (to appear).

[25] T. Bogenschütz. Entropy, pressure, and a variational principle for random dynamical systems. *Random & Computational Dynamics*, 1:99–116, 1992.

[26] T. Bogenschütz and V. M. Gundlach. Symbolic dynamics for expanding random dynamical systems. *Random & Computational Dynamics*, 1:219–227, 1992.

[27] P. Bougerol and J. Lacroix. *Products of random matrices with applications to Schrödinger operators*. Birkhäuser, Boston Basel Stuttgart, 1985.

[28] P. Boxler. A stochastic version of center manifold theory. *Probab. Th. Rel. Fields*, 83:509–545, 1989.

[29] J. R. Brown. *Ergodic theory and topological dynamics*. Academic Press, New York, 1976.

[30] A. Carverhill. Flows of stochastic dynamical systems: ergodic theory. *Stochastics*, 14:273–317, 1985.

[31] I. P. Cornfeld, S. V. Fomin, and Y. G. Sinai. *Ergodic theory*. Springer, Berlin Heidelberg New York, 1982.

[32] H. Crauel. Extremal exponents of random dynamical systems do not vanish. *Journal of Dynamics and Differential Equations*, 2:245–291, 1990.

[33] H. Crauel. Markov measures for random dynamical systems. *Stochastics and Stochastics Reports*, 37:153–173, 1991.

[34] H. Crauel. Non-markovian invariant measure are hyperbolic. *Stochastic Processes and their Applications*, 45:13–28, 1993.

[35] H. Crauel. Attractors for random dynamical systems. Habilitationsschrift (in preparation), 1994.

[36] S. Dahlke. *Invariante Mannigfaltigkeiten für Produkte zufälliger Diffeomorphismen*. PhD thesis, Institut für Dynamische Systeme, Universität Bremen, 1989.

[37] J. Dieudonné. *Grundzüge der modernen Analysis, Volume 9*. VEB Deutscher Verlag der Wissenschaften, Berlin, 1987.

[38] W. Ebeling, H. Herzel, W. Richert, and L. Schimansky-Geier. Influence of noise on Duffing-Van der Pol oscillators. *Zeitschrift f. Angew. Math. u. Mechanik*, 66:141–146, 1986.

[39] K. D. Elworthy. *Stochastic differential equations on manifolds*. Cambridge University Press, Cambridge, 1982.

[40] P. Ferrero and B. Schmitt. Produits aléatoires d'opérateurs matrices de transfert. *Probab. Th. Rel. Fields*, 79:227–248, 1988.

[41] H. Furstenberg. Noncommuting random products. *Trans. Am. Math. Soc.*, 108:377–428, 1963.

[42] H. Furstenberg and H. Kesten. Products of random matrices. *Ann. Math.*, 31:457–469, 1960.

[43] I. Y. Goldsheid and G. A. Margulis. Lyapunov indices of a product of random matrices. *Russian Mathematical Surveys*, 44:11–71, 1989.

[44] J. Guckenheimer and P. Holmes. *Nonlinear oscillations, dynamical systems, and bifurcation of vector fields.* Springer, Berlin Heidelberg New York, 1983.

[45] Y. Guivarc'h and A. Raugi. Frontière de Furstenberg, propriétés de contraction et théorème de convergence. *Z. Wahrscheinlichkeitstheorie Verw. Gebiete*, 69:187–242, 1985.

[46] W. Horsthemke and R. Lefever. *Noise-induced transitions.* Springer, Berlin Heidelberg New York, 1984.

[47] M. C. Irwin. *Smooth dynamical systems.* Academic Press, New York, 1981.

[48] R. Johnson. On a Floquet theory for almost periodic, two-dimensional linear systems. *Journal of Differential Equations*, 37:184–205, 1980.

[49] R. Johnson, K. Palmer, and G. R. Sell. Ergodic properties of linear dynamical systems. *SIAM J. Math. Anal.*, 18:1–33, 1987.

[50] R. Z. Khasminskii. *Stochastic stability of differential equations.* Sijthoff and Noordhoff, Alphen, 1980. (Translation of the Russian edition, Nauka, Moscow 1969).

[51] Y. Kifer. *Ergodic theory of random transformations.* Birkhäuser, Boston Basel Stuttgart, 1986.

[52] Y. Kifer. Equilibrium states for random expanding transformations. *Random & Computational Dynamics*, 1:1–31, 1992.

[53] W. Klingenberg. *Riemannian geometry.* Walter de Gruyter, Berlin, 1982.

[54] H. Kunita. *Stochastic flows and stochastic differential equations.* Cambridge University Press, Cambridge, 1990.

[55] Y. Le Jan. Équilibre statistique pour les produits de difféomorphismes aléatoires indépendants. *Ann. Inst. H. Poincaré Probab. Statist.*, 23:111–120, 1987.

[56] F. Ledrappier. Quelques propriétés des exposants caractéristiques. *École d'Été de Probabilités de Saint-Flour*, pages 305–396, 1984. Springer Lecture Notes in Mathematics, Volume 1097.

[57] F. Ledrappier and P. Walters. A relativised variational principle for continuous transformations. *J. London Math. Soc.*, 16:568–576, 1977.

[58] F. Ledrappier and L.-S. Young. Entropy formula for random transformations. *Probab. Th. Rel. Fields*, 80:217–240, 1988.

[59] M. A. Liapounoff. *Problème général de la stabilité du mouvement*. Annales Fac. Sciences Toulouse 9 (1907) (Translation of the Russian edition, Kharkov 1892). Reprinted by Princeton University Press, Princeton, N. J., 1949 and 1952.

[60] Liu Pei-Dong. *Smooth ergodic theory of random dynamical systems*. Manuscript, Peking University, 1994.

[61] R. Mañé. *Ergodic theory and differentiable dynamics*. Springer, Berlin Heidelberg New York, 1987.

[62] Nguyen Dinh Cong. Structural stability of linear random dynamical systems. Report 312, Institut für Dynamische Systeme, Universität Bremen, June 1994.

[63] Nguyen Dinh Cong. Topological classification of linear hyperbolic cocycles. Report 305, Institut für Dynamische Systeme, Universität Bremen, May 1994.

[64] V. I. Oseledets. A multiplicative ergodic theorem. Lyapunov characteristic numbers for dynamical systems. *Trans. Moscow Math. Soc.*, 19:197–231, 1968.

[65] Y. B. Pesin. Families of invariant manifolds corresponding to nonzero characteristic exponents. *Math. USSR Izvestija*, 10:1261–1305, 1976.

[66] Y. B. Pesin. Characteristic Lyapunov exponents and smooth ergodic theory. *Russian Mathematical Surveys*, 32:55–114, 1977.

[67] M. A. Pinsky and V. Wihstutz, editors. *Diffusion processes and related problems in analysis, volume II: Stochastic flows*. Birkhäuser, Boston Basel Stuttgart, 1992.

[68] C. Pugh and M. Shub. Ergodic attractors. *Trans. Am. Math. Soc.*, 312:1–54, 1989.

[69] D. Ruelle. *Thermodynamic formalism*. Addison-Wesley, Reading, Mass., 1978.

[70] D. Ruelle. Ergodic theory of differentiable dynamical systems. *Publ. Math., Inst. Hautes Etud. Sci.*, 50:275–306, 1979.

[71] D. Ruelle. *Elements of differentiable dynamics and bifurcation theory*. Academic Press, New York, 1989.

[72] R. Sacker and G. Sell. *Lifting properties in skew product flows with applications to differential equations*. Memoirs of the American Mathematical Society, Volume 190, 1977.

[73] R. Sacker and G. Sell. A spectral theory for linear differential systems. *Journal of Differential Equations*, 27:320–358, 1978.

[74] K. R. Schenk. Bifurcation scenarios of the noisy Duffing-Van der Pol oscillator. Report 295, Institut für Dynamische Systeme, Universität Bremen, 1993.

[75] G. Sell. The structure of a flow in the vicinity of an almost periodic motion. *Journal of Differential Equations*, 27:359–393, 1978.

[76] M. Shub. *Global stability of dynamical systems.* Springer, Berlin Heidelberg New York, 1987.

[77] Y. G. Sinai, editor. *Dynamical systems II.* Encyclopaedia of Mathematical Sciences, Volume 2. Springer, Berlin Heidelberg New York, 1989.

[78] W. Szlenk. *An introduction to the theory of smooth dynamical systems.* Wiley, New York, 1984.

[79] P. Walters. *An introduction to ergodic theory.* Springer, Berlin Heidelberg New York, 1982.

[80] T. Wanner. Linearization of random dynamical systems. In C. Jones, U. Kirchgraber, and H. O. Walther, editors, *Dynamics Reported.* Springer, Berlin Heidelberg New York, 1994.

[81] Xu Kedai. Bifurcations of random differential equations in dimension one. *Random & Computational Dynamics*, 1:277–305, 1993.

[82] E. C. Zeeman. On the classification of dynamical systems. *Bull. London Math. Soc.*, 20:545–557, 1988.

[83] E. C. Zeeman. Stability of dynamical systems. *Nonlinearity*, 1:115–155, 1988.

Geometric Singular Perturbation Theory

Christopher K.R.T. Jones
Division of Applied Mathematics
Brown University
Providence, RI 02912

Contents

The research described here was supported by the National Science Foundation under grant DMS-9403774 and the Office of Naval Research under grant N00014-92-J-1401. This manuscript was written, and some of the research described herein was completed, partly while the author was supported by a Humboldt Award at the Universität Stuttgart and partly while a Visiting Professor of the Consiglio Nazionale Delle Richerche at Università di Roma II. The author wishes to thank Jonathan Rubin for his careful reading of the original manuscript.

Chapter 1

Introduction

The goal of these lectures is an exposition of the geometric approach to singular perturbation problems. Singularly perturbed equations gain their special structure from the presence of differing time scales. The fundamental tool in their analysis, from the perspective taken here, is the set of theorems due to Fenichel. The first step is then to explain these theorems and their significance. At the same time, new proofs of Fenichel's three main results will be outlined.

1.1 Background and motivation

The basic equations we consider are of the form

$$\begin{aligned} x' &= f(x, y, \epsilon) \\ y' &= \epsilon g(x, y, \epsilon), \end{aligned} \tag{1.1}$$

where $' = \frac{d}{dt}$, $x \in \mathbf{R}^n$, $y \in \mathbf{R}^l$ and ϵ is a real parameter. We shall compile various hypotheses about the system (1.1), which are denoted with the letter H.

(H1) The functions f and g are both assumed to be C^∞ on a set $U \times I$ where $U \subset \mathbf{R}^N$ is open, with $N = n + l$, and I is an open interval, containing 0.

Note that we are assuming full smoothness on the nonlinear terms which is unnecessary but greatly simplifies the discussion. If less smoothness is present in a given problem the precise smoothness required can be easily retraced through the proofs.

System (1.1) can be reformulated with a change of time-scale as

$$\begin{aligned} \epsilon \dot{x} &= f(x, y, \epsilon) \\ \dot{y} &= g(x, y, \epsilon), \end{aligned} \tag{1.2}$$

where $\dot{} = \frac{d}{d\tau}$ and $\tau = \epsilon t$. The time scale given by τ is said to be slow whereas that for t is fast, as long as $\epsilon \neq 0$ the two systems are equivalent. Thus we call

(1.1) *the fast system* and (1.2) *the slow system.* We have two distinguished limits for these equations, one naturally associated with each scaling as $\epsilon \to 0$. In (1.1) letting $\epsilon \to 0$ we obtain the system

$$
\begin{aligned}
x' &= f(x, y, 0) \\
y' &= 0.
\end{aligned}
\tag{1.3}
$$

According to (1.3) the variable x will vary while y will remain constant. Thus x is called the fast variable. If we let $\epsilon \to 0$ in (1.2), the limit only makes sense if $f(x, y, 0) = 0$ and is thus given by

$$
\begin{aligned}
f(x, y, 0) &= 0 \\
\dot{y} &= g(x, y, 0).
\end{aligned}
\tag{1.4}
$$

One thinks of the condition $f(x, y, 0) = 0$ as determining a set on which the flow is given by $\dot{y} = g(x, y, 0)$. It is natural to attempt to solve x in terms of y from the equation $f(x, y, 0) = 0$ and plug it into the second equation of (1.4) (the reader should check that the dimensions are right to expect such a solution if non-degeneracy conditions hold). Notice that this set is exactly the set of critical points for (1.3). We thus have the "formal" picture that (1.3) has large sets of critical points and that (1.4) blows the flow on this set up to produce non-trivial behavior.

In either limiting formulation, one pays a price. On the (large) set $f(x, y, 0) = 0$ *the flow is trivial* for (1.3). Whereas under (1.4) the flow is non-trivial on this set, but *the flow is not defined off this set.* The primary mathematical goal of *geometric singular perturbation theory*, henceforth denoted by the acronym GSP, is to realize both these aspects (i.e., fast and slow) simultaneously. This apparently contradictory aim will be accomplished within the phase space of (1.1) (or, equivalently, (1.2)) for ϵ *non-zero but small.*

There are two basic reasons why GSP is a powerful tool for analyzing high-dimensional systems:

1. In many applications, quantities will vary on *widely differing time scales*, and thus are naturally formulated in the form (1.1).

2. It affords a reduction of a possibly high-dimensional system, such as (1.1), into the lower-dimensional systems (1.3) and (1.4).

The first reason given above justifies the theory from an applied point of view, but also offers us the opportunity of invoking many different applications as examples to guide the theory. The second rationale means that we can hold the hope of analyzing singularly perturbed systems of twice the size of those analyzed without this theory. For example if $n = l = 2$, we would study 2-dimensional systems, the analysis of which is well understood, and, through GSP, make conclusions about 4-dimensional systems, which are ostensibly far less tractable. Moreover, the resulting behavior is not restricted to being merely a shadow of that present in (1.3) and (1.4), for new structures can result from the patching together of solutions of (1.3) and (1.4). The early examples will not reflect such effects, but we will progressively develop richer dynamical behavior. The later applications will exhibit *intrinsically higher-dimensional phenomena*, despite their being rendered susceptible to analysis by reduction to low-dimensional systems. It should be noted here that the theory of singular perturbations commands a large literature that does not fit into the category of GSP as discussed here. Indeed, this classical theory predates GSP and the interested reader is referred to [11] and [44] for good expositions of this theory.

The phenomena that we will isolate for (1.1) will generally involve the construction of specific orbits, such as *homoclinic, heteroclinic, or periodic orbits*. These are constructed by *following certain invariant manifolds* (for instance, stable and unstable manifolds of critical points) through their ambient phase space and using the reductions i.e., (1.3) and (1.4), to keep track of their position and configuration at different points of their travel. These "special" orbits may be of significance due to their rôle in the overall dynamics of the equation, for instance homoclinic orbits are often a signature of chaotic motion, or as special solutions, such as travelling waves, of a related partial differential equation.

Summary of goals

- Determination of flow near sets $f(x, y, 0) = 0$ for (1.1):
  - Fenichel's theorems,
  - Fenichel coordinates and normal form,
  - slow manifold flow.

- Effective tracking of invariant manifolds through the phase space of (1.1):
  - use of differential forms,
  - transversality,
  - exchange lemmas.

- Applications to the existence and properties of special orbits of (1.1):
  - perturbed slow structures,
  - travelling waves and stability,
  - homoclinic abundance.

1.2 Fenichel's first theorem

The set of critical points $f(x, y, 0) = 0$ for (1.3) is formed by solving n equations in $\mathbf{R}^N$, where $N = n + l$, and thus is expected to be, at least locally, an l-dimensional manifold. Indeed, it is natural to expect it to have a parametrization by the variable y. We shall thus assume that we are given an l-dimensional manifold, possibly with boundary, M_0 which is contained in the set $\{f(x, y, 0) = 0\}$. The fundamental hypothesis on M_0 will be that, as a set of critical points, the directions normal to the manifold will correspond to eigenvalues that are not neutral. In the following, the notation $\mathcal{R}(\lambda)$ denotes the real part of λ.

Definition 1 *The manifold M_0 is said to be normally hyperbolic if the linearization of (1.1) at each point in M_0 has exactly l eigenvalues on the imaginary axis $\mathcal{R}(\lambda) = 0$.*

Fenichel's first theorem asserts the existence of a manifold that is a perturbation of M_0. It will be connected with the flow of (1.1) when $\epsilon \neq 0$. We need a definition to clarify this connection. The notation $x \cdot t$ is used to denote the application of a flow after time t to the intial condition x. The existence of a flow for (1.1) follows from the basic theorems of ODE.

Definition 2 *A set M is locally invariant under the flow from (1.1) if it has neighborhood V so that no trajectory can leave M without also leaving V. In other words, it is locally invariant if for all $x \in M$, $x \cdot [0,t] \subset V$ implies that $x \cdot [0,t] \subset M$, similarly with $[0,t]$ replaced by $[t,0]$ when $t < 0$.*

Fenichel's theorems will actually address the perturbation of a subset of $\hat{M}_0$, because of technical difficulties near the boundary.

(H2) The set M_0 is a compact manifold, possibly with boundary, and is normally hyperbolic relative to (1.3).

The set M_0 will be called the critical manifold. We are now in a position to state the first theorem that Fenichel proved, under the hypotheses (H1) and (H2).

Theorem 1 (Fenichel's Invariant Manifold Theorem 1) *If $\epsilon > 0$, but sufficiently small, there exists a manifold M_ϵ that lies within $O(\epsilon)$ of M_0 and is diffeomorphic to M_0. Moreover it is locally invariant under the flow of (1.1), and C^r, including in ϵ, for any $r < +\infty$.*

This theorem follows from Fenichel's early work, [15], as singular perturbations are a special case of the more general decomposition by exponential rates that he considered in that context. However, his later paper, see [18], specifically addresses singular perturbations. There are a number of alternative formulations and proofs of this basic theorem, see, for instance, the work of Sakamoto [51].

The manifold M_ϵ will be called the slow manifold. It should be noted that the only connection to the flow is through the statement that the perturbed manifold M_ϵ is locally invariant. This seems weak but, in fact, is not as it entails that we can restrict the flow to this manifold, which is lower-dimensional, in order to find interesting structures. The fact that the manifold is *locally invariant*, and not *invariant*, is due to the (possible) presence of the boundary and the resulting possibility that trajectories may fall out of M_ϵ by escaping through the boundary. This cannot be avoided as most applications do indeed supply us with manifolds that have boundaries. A comment is also in order about normal hyperbolicity; there are many applications in which interesting phenomena occur because normal hyperbolicity of manifolds breaks down, such as in relaxation-oscillations, but we shall not consider such cases in these lectures. As is the case for center manifolds, it should be noted here that the exponent r cannot be set to $+\infty$.

In order to significantly simplify the notation, as well as the structure of the proofs, we shall restrict attention throughout these lectures to the case that M_0 is given as the graph of a function of x in terms of y. That is we assume there is a function $h^0(y)$, defined for $y \in K$, with K being a compact domain in $\mathbf{R}^l$, and so that

$$M_0 = \left\{ (x,y) : x = h^0(y) \right\}.$$

This is a natural assumption as it can always be satisfied for M_0 *locally*. Indeed, on account of normal hyperbolicity (H2) the matrix

$$D_x f(\hat{x}, \hat{y}, 0)$$

is invertible for any $(\hat{x}, \hat{y}) \in M_0$ and hence x can locally be solved for y by the Implicit Function Theorem. We are thus just assuming that such a solution can be made *globally* over M_0.

Thus, consider $x = h^0(y)$ wherein $y \in K$ and make the following assumption.

(H3) The set M_0 is given as the graph of the C^∞ function $h^0(y)$ for $y \in K$. The set K is a compact, simply connected domain whose boundary is an $(l-1)$-dimensional C^∞ submanifold.

Under the hypotheses (H1)-(H3), we can restate Fenichel's first theorem in terms of the graph of a function.

Theorem 2 *If $\epsilon > 0$ is sufficiently small, there is a function $x = h^\epsilon(y)$, defined on K, so that the graph*

$$M_\epsilon = \left\{ (x,y) : x = h^\epsilon(y) \right\},$$

is locally invariant under (1.1). Moreover h^ϵ is C^r, for any $r < +\infty$, jointly in y and ϵ.

Remark The diffeomorphism between M_ϵ and M_0 follows easily in this formulation through the diffeomorphism of the graph to K.

An equation on M_ϵ can easily be calculated using Theorem 1. We substitute the function $h^\epsilon(y)$ into (1.1) and see that the y equation will decouple from that of the x equation. We thus obtain an equation for the variation of the variable y. Since y parametrizes the manifold M_ϵ, this equation will suffice to describe the flow on M_ϵ. It is given by

$$y' = \epsilon g(h^\epsilon(y), y, \epsilon). \tag{1.5}$$

In the alternative slow scaling we can recast (1.6) as

$$\dot{y} = g(h^\epsilon(y), y, \epsilon), \tag{1.6}$$

which has the distinct advantage that a limit exists as $\epsilon \to 0$, given by

$$\dot{y} = g(h^0(y), y, 0), \tag{1.7}$$

which naturally describes a flow on the critical manifold M_0, and is exactly the second equation in (1.2). Using this theorem and this resulting equation (1.6), the problem of studying (1.1), at least on M_ϵ is reduced to a *regular perturbation problem*. In the next three sections we shall give examples in which this is applied.

1.3   An equation from phase-field theory

An equation with spatial derivatives of even powers in formulated by Caginalp and Fife [7] to describe the behavior of phase transitions. In a model case in which a scalar equation can be used as a reasonable model, Gardner and Jones [19] studied the stability of travelling waves. As an example of the above theory, it will be shown here how to construct the basic travelling wave. Consider the equation

$$\frac{\partial \phi}{\partial t} = \epsilon^4 \frac{\partial^6 \phi}{\partial x^6} + \epsilon^2 A \frac{\partial^4 \phi}{\partial x^4} + \frac{\partial^2 \phi}{\partial x^2} + f(\phi), \tag{1.8}$$

where $f(\phi) = \phi(\phi - a)(1 - \phi)$ is the bistable nonlinearity with $a < \frac{1}{2}$, and $A > 0$. The parameter ϵ is intended to be small, see [19], and thus (1.8) is easily seen to be a perturbation of the well-known, scalar bistable reaction-diffusion equation. We shall seek travelling wave solutions of (1.8), namely $\phi(\xi) = \phi(x - ct)$, satisfying

$$\phi(\xi) \to \begin{cases} 0 & \text{as} \quad \xi \to -\infty \\ 1 & \text{as} \quad \xi \to +\infty. \end{cases} \tag{1.9}$$

The wave is then seen to satisfy the ODE

$$-c\phi^{(1)} = \epsilon^4 \phi^{(6)} + \epsilon^2 A \phi^{(4)} + \phi^{(2)} + f(\phi), \tag{1.10}$$

where the derivatives are taken with respect to the travelling wave variable ξ. The equation (1.10) can be rewritten as a system of six equations

$$\begin{aligned}
\dot{u}_1 &= u_2 \\
\dot{u}_2 &= u_3 \\
\epsilon \dot{u}_3 &= u_4 \\
\epsilon \dot{u}_4 &= u_5 \\
\epsilon \dot{u}_5 &= u_6 \\
\epsilon \dot{u}_6 &= -Au_5 - u_3 - cu_2 - f(u_1),
\end{aligned} \tag{1.11}$$

where $\dot{} = \frac{d}{d\tau}$. This system is already formulated in slow variables and we have replaced ξ by τ as the independent variable to conform to our established notation. The correspondence with our notation is:

$$x = \begin{pmatrix} u_3 \\ u_4 \\ u_5 \\ u_6 \end{pmatrix}$$

is the fast variable, and

$$y = \begin{pmatrix} u_1 \\ u_2 \end{pmatrix}$$

is the slow variable. The critical manifold M_0 can be taken as any compact subset of

$$\{u_4 = u_5 = u_6 = 0, \ u_3 = -cu_2 - f(u_1)\},$$

which shall be chosen to be large enough to contain any of the dynamics of interest. The eigenvalues of the linearization at any point of M_0, other than the double eigenvalue at 0 are seen to be solutions of the quartic

$$\mu^4 + A\mu^2 + 1 = 0,$$

which are not pure imaginary if $0 < A < 2$.

The equations for the slow flow on the critical manifold M_0 are given by

$$\begin{aligned} \dot{u}_1 &= u_2 \\ \dot{u}_2 &= -cu_2 - f(u_1). \end{aligned} \tag{1.12}$$

The slow manifold M_ϵ, which exists by virtue of Theorem 1, is given by the equations

$$(u_3, u_4, u_5, u_6) = h^\epsilon(u_1, u_2) = (-cu_2 - f(u_1), 0, 0, 0) + O(\epsilon).$$

and the equations on M_ϵ are

$$\begin{aligned} \dot{u}_1 &= u_2 \\ \dot{u}_2 &= -cu_2 - f(u_1) + O(\epsilon). \end{aligned} \tag{1.13}$$

It is a well-known fact that (1.12) has a heteroclinic orbit connecting the critical point $(0,0)$ at $-\infty$ with $(1,0)$ at $+\infty$, for a particular value of c, say $c = c^*$. One checks easily that $(0,0)$ and $(1,0)$ are still critical points of (1.13), for ϵ sufficiently small (why?). The strategy is then to show that there is a $c = c(\epsilon)$ defined for ϵ small, with $c(0) = c^*$, at which there is such a heteroclinic orbit for (1.13). The idea is to show that the heteroclinic orbit for (1.12) exists by virtue of a transverse intersection of stable and unstable manifolds and thus perturbs.

Appending an equation for c to (1.12), the heteroclinic, on M_0, can be viewed as the intersection of the unstable manifold of the curve of critical points $\{(0,0,c) : |c - c^*| \text{ small }\}$, say W^- with the stable manifold of the curve

$\{(1,0,c) : |c - c^*| \text{ small }\}$, say W^+, see Figure 1. this intersection will be viewed in the plane $u = a$. In $u = a$, W^- is given by the graph of a function, say $u_2 = h^-(c)$, and W^+ is given by the graph of another function, say $u_2 = h^+(c)$. These curves are each monotone, which is the usual proof of the uniqueness of the wave and its speed, see for instance Aronson and Weinberger [2]. The transverse intersection is related to this fact and it will be shown specifically in Chapter 4 that the following quantity

$$\left(\frac{\partial h^-}{\partial c} - \frac{\partial h^+}{\partial c} \right) |_{c=c^*} \neq 0. \tag{1.14}$$

This is a Melnikov type calculation.

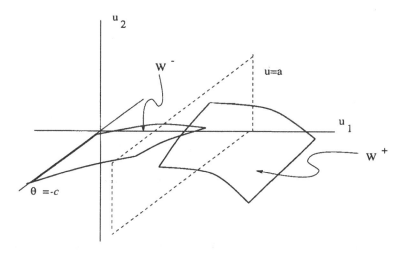

Figure 1
The intersection of the unstable and stable manifolds.

The next step is to check whether this intersection perturbs to M_ϵ. Indeed, on M_ϵ the relevant unstable and stable manifolds will again be given as graphs. The manifold $W^- \cap \{u_1 = a\}$ will be given by $u_2 = h^-(c, \epsilon)$ and $W^+ \cap \{u_1 = a\}$ will be given by $u_2 = h^+(c, \epsilon)$. An intersection point is found by solving these equations simultaneously for $u_2 = u_2^*(\epsilon)$ and $c = c^*(\epsilon)$. This will follow from the Implicit Function Theorem if the determinant of the matrix

$$\begin{pmatrix} 1 & \frac{\partial h^-}{\partial c} \\ 1 & \frac{\partial h^+}{\partial c} \end{pmatrix},$$

at $c = c^*$, and $\epsilon = 0$, is non-zero. But this is exactly the statement (1.14).

1.4 A travelling wave in semiconductor theory

An example due to Szmolyan, see [53], concerning the problem of finding a travelling wave in a system of equations governing the behavior of a semiconductor material will be presented. In the following, E is the electric field, n is the total concentration of *all* the electrons and u is the concentration of one individual species of electrons. There are two species of electrons present, and so the concentration of the other species is $n - u$. The concentration variable u satisfies a second order equation. We bypass the PDE's and formulate immediately the travelling wave equations, which are thus a system of four equations, given by

$$
\begin{aligned}
u' &= w \\
w' &= \epsilon(\nu_1(E) - c)w + \epsilon^2 \nu_1'(E)u\tfrac{n-1}{\lambda^2} + (1 + \alpha(E))u - n \\
E' &= \epsilon\tfrac{n-1}{\lambda^2} \\
n' &= \epsilon\left(\nu_1(E)u + \nu_2(E)(n - u) - cn + \gamma\right),
\end{aligned}
\tag{1.15}
$$

where $\nu_i(E)$ and $\alpha(E)$ are phenomenologically determined, positive, smooth functions, the exact structure of which is not directly important. The nature of a certain combination of these functions that appears in the slow equations will be of most relevance. Of interest will be orbits of (1.15) that are homoclinic to critical points. These correspond to travelling waves of the original PDE that decay to a fixed constant state at $\pm\infty$. Clearly the variables E and n are slow, while u and w are fast. We thus have the correspondence $x = (u, w)$ and $y = (E, n)$ with the notation above. The critical manifold M_0 will be given by the equations

$$
w = 0, \quad u = \frac{n}{1 + \alpha(E)}
\tag{1.16}
$$

and is thus easily seen to be defined, for our purposes, on any compact domain of $\mathbf{R}^2$. The eigenvalues of the linearization at any critical point of M_0, apart from the double eigenvalue at 0, are $\pm(1 + \alpha(E))^{\frac{1}{2}}$. Since α is positive, these are then non-zero and the uniform normal hyperbolicity assumption is satisfied.

The limiting slow equations i.e., equations (1.7) for this example, are given by

$$
\begin{aligned}
\dot{E} &= \tfrac{n-1}{\lambda^2} \\
\dot{n} &= (G(E) - c)\,n + \gamma,
\end{aligned}
\tag{1.17}
$$

where

$$
G(E) = \frac{\nu_1(E) + \alpha(E)\nu_2(E)}{1 + \alpha(E)}.
\tag{1.18}
$$

The graph of $G(E)$ will be important and is given in Figure 2.

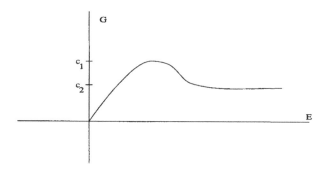

Figure 2
The graph of the function G.

The equations on the perturbed manifold M_ϵ, which exists by Theorem 1 for ϵ positive, but sufficiently small, are given by

$$
\begin{aligned}
\dot{E} &= \tfrac{n-1}{\lambda^2} \\
\dot{n} &= (G(E) - c)\, n + \gamma + O(\epsilon).
\end{aligned}
\tag{1.19}
$$

The goal is to find an orbit for (1.19) that is homoclinic to a rest state. This orbit will be the desired travelling wave, as it is a homoclinic orbit for the system that happens to live on M_ϵ. The strategy is to find a homoclinic orbit for (1.17) and prove that it perturbs to such for (1.19) when ϵ is sufficiently small.

We first analyze (1.17) when γ is also set equal to 0. In this case a simple transformation converts it into a Hamiltonian system, namely we let $m = \log n$, and from (1.17) we obtain

$$
\begin{aligned}
\dot{E} &= \tfrac{e^m - 1}{\lambda^2} \\
\dot{m} &= G(E) - c.
\end{aligned}
\tag{1.20}
$$

Since the variable n is a concentration, we are only interested in solutions with $n > 0$, and thus the transformation is valid for the solutions that will ultimately be of interest to us. It is easily checked that the function

$$
H(E, m) = \frac{e^m - m}{\lambda^2} - \Gamma(E),
$$

where $\Gamma'(E) = G(E) - c$, is a Hamiltonian for (1.20). If c is in the interval (c_1, c_2), see Figure 2, then (1.20) has 2 critical points, the left one being a saddle, which we denote $\hat{y}$, and the right one a center. It is an exercise (left to the reader-see [53]) to check that, from the Hamiltonian above, one can conclude the existence of an orbit homoclinic to $\hat{y}$, see Figure 3. In principle, this could be concluded from sketching the level curves of H, but this can be avoided by invoking some qualitative arguments about the nature of the level curve containing the saddle.

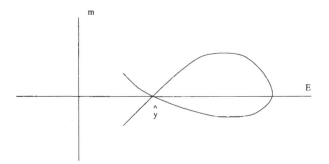

Figure 3
The homoclinic orbit in M_0.

One would not expect this homoclinic orbit to survive a perturbation, such as to equation (1.19), as it is not caused by a transversal intersection. However, we have not used the parameter γ. Undoing the transformation we see that (1.17) will also have a homoclinic orbit to a saddle, which, with an abuse of notation, we continue to denote by $\hat{y}$, but now $\hat{y}$ has its second coordinate given by $n = 1$. Since this point is a saddle an application of the Implicit Function Theorem shows that there is a nearby saddle critical point for γ sufficiently small (note that n stays equal to 1). We denote this curve of critical points by $\mathcal{C}$. The next step is to show that the unstable manifold of the curve $\mathcal{C}$ intersects its stable manifold transversely in (E, n, γ)-space at $\gamma = 0$ (we append the equation $\gamma' = 0$). There is then a hope of its perturbing to M_ϵ.

As is common in transversality arguments, we consider the intersection of these manifolds with the set $n = 1$. Again by the Implicit Function Theorem, it is easily checked that, for γ sufficiently small, these intersections are, indeed, curves. We denote them by

$$E = h^-(\gamma) \text{ and } E = h^+(\gamma),$$

for $W^u(\mathcal{C}) \cap \{n = 1\}$ and $W^s(\mathcal{C}) \cap \{n = 1\}$ respectively, see Figure 4. The intersection is transversal if

$$M = \left(\frac{\partial h^+}{\partial \gamma} - \frac{\partial h^-}{\partial \gamma} \right) |_{\gamma=0} \neq 0. \tag{1.21}$$

This is a type of Melnikov calculation and can be checked to hold. The reader is asked to have faith that such a result holds, or can consult [53].

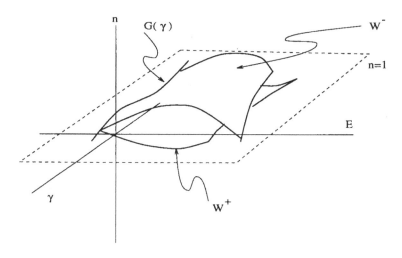

Figure 4
The The stable and unstable manifolds in $n = 1$.

It remains to show that, for some γ, there is a homoclinic orbit for (1.19). By the same argument as above the saddle perturbs for all γ and ϵ sufficiently small. We now think of the system (1.19) with equations for γ and ϵ appended. The unstable manifold of this surface of critical points, call it again W^u, will intersect its stable manifold W^s when $\gamma = 0$. We wish to find a curve of intersections given by γ as a function of ϵ. If W^u is given by $E = h^-(\gamma, \epsilon)$, and W^s by $E = h^+(\gamma, \epsilon)$, inside the set $n = 1$. We need to simultaneously solve the equations

$$
\begin{aligned}
E - h^-(\gamma, \epsilon) &= 0 \\
E - h^+(\gamma, \epsilon) &= 0,
\end{aligned}
\tag{1.22}
$$

by E and γ as functions of ϵ. This can be achieved, by the Implicit Function Theorem, exactly when the determinant of the matrix

$$
\begin{pmatrix}
1 & \frac{\partial h_-}{\partial \gamma} \\
1 & \frac{\partial h_+}{\partial \gamma}
\end{pmatrix}
$$

is non-zero. But this is, again, exactly the condition $M \neq 0$. The transversality condition thus indeed supplies us with a homoclinic orbit.

A note on notation used in the examples is in order here. Many of the examples have the goal of finding travelling waves of a certain PDE. We usually use the variable ξ for the travelling wave variable i.e., $\xi = x - ct$, as the variable t has a meaning in the original PDE. However, we shall abuse the notation here and always revert to independent variables that conform to our general framework, once the ODE's in a given example are derived. As a result the variable t may have nothing to do with "time" in the original PDE.

The above examples are pleasing applications of GSP but do not use the perturbation that is supplied by the original equations to the equation on the slow manifold. Indeed, all the information is present in the limiting slow equation and it is merely checked that the object of interest, namely the homoclinic or heteroclinic orbit, is not destroyed by the perturbation. In the next subsection, we consider an application that actually uses the perturbing terms i.e., the order ϵ terms for the equations on M_ϵ to create the homoclinic orbit.

1.5   Solitary waves of the KdV-KS equation

We shall base this section on a paper by Ogawa, see [47]. The results, and approach, are very similar in spirit to the work of Ercolani et al., see [13], except that in the latter work periodic orbits are considered. The basic equations are a perturbed form of the Korteweg-deVries equations. The higher order terms perturbing the KdV part are characteristic of the Kuramoto-Sivashinsky equations, and the full model has arisen in a number of places, including in models of shallow water on tilted planes, see [56]. The partial differential equations are then

$$U_t + UU_x + U_{xxx} + \epsilon(U_{xx} + U_{xxxx}) = 0, \tag{1.23}$$

where $x \in \mathbf{R}$ and $t \geq 0$. We seek travelling wave solutions of (1.23). These will be solutions of (1.23) that are functions of the single variable $\xi = x - ct$. We are specifically interested in those that are asymptotic to the rest state $u = 0$ as $\xi \to \pm\infty$, these will then be solitary waves. The wave $U = U(\xi)$ must satisfy the ODE

$$- cU^{(1)} + UU^{(1)} + U^{(3)} + \epsilon\left(U^{(2)} + U^{(4)}\right) = 0, \tag{1.24}$$

where $^{(1)} = \frac{d}{d\xi}$. Using the boundary condition at $-\infty$, (1.24) can be integrated once to yield the equation

$$- cU + \frac{U^2}{2} + U^{(2)} + \epsilon\left(U^{(1)} + U^{(3)}\right) = 0. \tag{1.25}$$

This, in turn, we rewrite as a system of ODE's, wherein $u = U/c$

$$
\begin{aligned}
\dot{u} &= v \\
\dot{v} &= w \\
\epsilon\dot{w} &= \frac{1}{\sqrt{c}}\left(u - \frac{u^2}{2} - w - \frac{\epsilon}{\sqrt{c}}v\right),
\end{aligned}
\tag{1.26}
$$

where $\dot{} = \frac{d}{d\tau}$ and $\tau = \sqrt{c}\xi$. Note the location of the small parameter ϵ means that (1.26) is already formulated on a slow time scale. The corresponding fast equations are

$$
\begin{aligned}
u' &= \epsilon v \\
v' &= \epsilon w \\
w' &= \frac{1}{\sqrt{c}}\left(u - \frac{u^2}{2} - w - \frac{\epsilon}{\sqrt{c}}v\right).
\end{aligned}
\tag{1.27}
$$

The critical manifold M_0 is given by the conditions $w = u - \frac{u^2}{2}$ suitably restricted to any compact domain K of (u, v) space. Since (1.27) has only three equations, there is only one normal direction to this 2-dimensional manifold. The derivative of the fast part in the w-direction is $-1/\sqrt{c}$ and hence M_0 is normally hyperbolic, in fact attracting. Fenichel's Invariant Manifold Theorem then guarantees the existence of M_ϵ. The flow on M_ϵ is found by writing out (1.6) for this case

$$\begin{aligned}\dot{u} &= v \\ \dot{v} &= u - \frac{u^2}{2} + O(\epsilon).\end{aligned} \tag{1.28}$$

This equation has the limiting form, on M_0, of

$$\begin{aligned}\dot{u} &= v \\ \dot{v} &= u - \frac{u^2}{2},\end{aligned} \tag{1.29}$$

which can be easily analyzed as it is a simple one-degree of freedom Hamiltonian system. Indeed, (1.29) has a homoclinic orbit to the critical point $(0,0)$, see Figure 5.

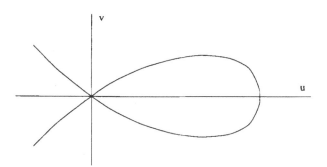

Figure 5
The homoclinic orbit in M_0.

However, it is inevitably not transversal. Moreover, the constant c offers no relief to this dilemma as it does not enter into (1.29). This is the point at which this example departs from its similarity to those of the preceding sections. We must then consider the $O(\epsilon)$ terms in (1.28). We know that M_ϵ is given by a function $w = h(u, v, \epsilon)$ and, by smoothness, can be expanded in ϵ, so that

$$w = u - \frac{u^2}{2} + \epsilon h_1(u, v) + O(\epsilon^2). \tag{1.30}$$

We need to calculate the term $h_1(u, v)$, which is also likely to depend on the parameter c. The only remaining information about M_ϵ is the local invariance relative to the equation and this must then be used to evaluate h_1.

To this end, we differentiate (1.30)

$$w' = u' - uu' + \epsilon \left(\frac{\partial h_1}{\partial u} u' + \frac{\partial h_1}{\partial v} v' \right) + O(\epsilon^2). \tag{1.31}$$

We substitute the expressions for u', v' and w', from (1.28), and also the expression for w, given by (1.30) into (1.31), and, after cancelling the $O(1)$ terms, we have

$$\frac{1}{\sqrt{c}}\left(-(\epsilon h_1 + O(\epsilon^2)) - \frac{\epsilon}{\sqrt{c}}v\right) = \epsilon v - \epsilon u v + O(\epsilon^2). \tag{1.32}$$

Equating the terms of $O(\epsilon)$ in (1.32) we obtain

$$h_1 = \sqrt{c}\left(uv - v\left(1 + \frac{1}{\sqrt{c}}\right)\right). \tag{1.33}$$

We cannot expect (1.28) to have a homoclinic orbit forced merely by adding the $O(\epsilon)$ term. The parameter c will also need to be used, in the jargon: it is a codimension two problem. We shall augment the system (1.28) with both equations for ϵ and c

$$\begin{aligned}
\dot{u} &= v \\
\dot{v} &= u - \frac{u^2}{2} + \epsilon\sqrt{c}\left(u - \left(1 + \frac{1}{\sqrt{c}}\right)\right)v + O(\epsilon^2) \\
\dot{\epsilon} &= 0 \\
\dot{c} &= 0.
\end{aligned} \tag{1.34}$$

We seek homoclinic orbits for (1.34) with small ϵ. These will be found at values of c that depend on ϵ. From the original equations one can see that 0 remains a critical point and must lie on M_ϵ (why?). We thus look for orbits homoclinic to 0. The critical point 0 can, in reference to (1.34), be construed as a surface of critical points, say $\mathcal{S}$, parametrized by c, ϵ. This in turn spawns an unstable manifold $W^u(\mathcal{S})$ and stable manifold $W^s(\mathcal{S})$ which meet in a curve at $\epsilon = 0$, namely the homoclinic orbits found already, see Figure 5. In the set $v = 0$ we parametrize W^u and W^s respectively, near the intersection away from the critical point, as $u = h^-(c, \epsilon)$ and $u = h^+(c, \epsilon)$.

We next define

$$d(c, \epsilon) = h^-(c, \epsilon) - h^+(c, \epsilon),$$

and observe that zeroes of d render homoclinic orbits. Since there are homoclinic orbits independently of c when $\epsilon = 0$, we have that $d(c, 0) = 0$, and thus that $d(c, \epsilon) = \epsilon \tilde{d}(c, \epsilon)$. The Melnikov function is here given by

$$\tilde{d}(c, 0) = M(c) = \left(\frac{\partial h^+}{\partial \epsilon} - \frac{\partial h^-}{\partial \epsilon}\right)|_{\epsilon=0}. \tag{1.35}$$

It is a simple application of the Implicit Function Theorem to see that there is a curve of homoclinic orbits given by $c = c(\epsilon)$ for ϵ small, if there exists a c $(= c(0))$, at which

$$M(c) = 0 \text{ and } M'(c) \neq 0. \tag{1.36}$$

The function $M(c)$ can be calculated explicitly, see Lecture 3 below, as

$$M(c) = \frac{1}{\alpha}\left\{\frac{1}{\sqrt{c}}\left\{c\int_{-\infty}^{+\infty}\ddot{u}^2 - \int_{-\infty}^{+\infty}\dot{u}^2\right\}\right\}, \tag{1.37}$$

where u comes from the underlying, already known, homoclinic orbit and $\alpha \neq 0$. It is clear then that (1.36) at a unique value of c.

It is interesting to note in this application that the perturbing terms supply a speed selection that is not evident without them. The reduced equations i.e., those on M_0, are the travelling wave equations for the KdV equation. No particular speed for this equation is determined by the travelling waves as they exist at every speed. However, when the perturbing terms, supplied by the Kuramoto-Sivashinsky formulation, are added a specific wave speed is selected.

Chapter 2

Invariant Manifold Theorems

Fenichel's Theorem as stated in the first lecture gives us only part of the picture in a neighborhood of a slow manifold. The existence of the slow manifold is guaranteed by the Theorem and the equation on the manifold can be computed as shown in the examples. However, at this point, we know nothing of the flow off the slow manifold and this must now be addressed. Our goal here is the derivation of a normal form, that we shall call "Fenichel Normal Form" for the equations near a slow manifold. This goal will be reached in the third lecture. In this lecture, Fenichel's Second Theorem that describes the stable and unstable manifolds of a slow manifold will be presented. These are perturbations of the stable and unstable manifolds, respectively, of the critical manifold. They are related to those invariant manifolds in the same way that the slow manifold is related to the critical manifold.

2.1 Stable and unstable manifolds

The slow manifold discussed in the first lecture possesses attendant stable and unstable manifolds that are perturbations of the corresponding manifolds when $\epsilon = 0$. The following theorem holds under (H1)-(H3) and its conclusion is depicted in Figure 6.

Theorem 3 (Fenichel Invariant Manifold Theorem 2) *If $\epsilon > 0$ but sufficiently small, there exist manifolds $W^s(M_\epsilon)$ and $W^u(M_\epsilon)$ that lie within $O(\epsilon)$ of, and are diffeomorphic to, $W^s(M_0)$ and $W^u(M_0)$ respectively. Moreover, they are each locally invariant under (1.1), and C^r, including in ϵ, for any $r < +\infty$.*

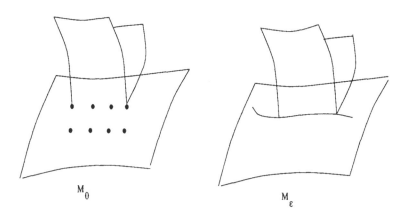

Figure 6
The slow manifold and its stable and unstable manifolds.

The Fenichel Normal Form will be produced through a series of coordinate changes. Initially, these will be made with the purpose of facilitating the proofs of the theorems. The final set of coordinate changes will follow from the theorems themselves. We shall restrict, as stated earlier, to the case of critical manifolds that are given as graphs.

Without loss of generality, we can assume that $h^0(y) = 0$ for all $y \in K$. Indeed, we can replace x by $\tilde{x} = x - h^0(y)$ and recompute the equations. For each point $y \in K$, there are subspaces $S(y)$ and $U(y)$, corresponding, respectively, to stable and unstable eigenvalues. Since the eigenvalues are bounded uniformly away from the imaginary axis over K, the dimensions of $S(y)$ and $U(y)$ are independent of y. Let $\dim S(y) = m$ and $\dim U(y) = k$. Since K is simply-connected by (H3), we can smoothly choose bases for $S(y)$ and $U(y)$. Changing the coordinates to be in terms of these new bases, we can set $x = (a, b)$, where $a \in \mathbf{R}^k$ and $b \in \mathbf{R}^m$, so that our equations have the form

$$
\begin{aligned}
a' &= A(y)a + F_1(x, y, \epsilon) \\
b' &= B(y)b + F_2(x, y, \epsilon) \\
y' &= \epsilon g(x, y, \epsilon),
\end{aligned}
\tag{2.1}
$$

where the spectrum of the matrix $A(y)$ lies in the set $\{\lambda : \mathcal{R}(\lambda) > 0\}$ and the spectrum of the matrix $B(y)$ lies in $\{\lambda : \mathcal{R}(\lambda) < 0\}$. Both F_1 and F_2 are higher order in x and ϵ; to be precise, we have the estimates

$$
|F_i| \le \gamma(|x| + \epsilon),
\tag{2.2}
$$

$i = 1, 2$ and γ can be taken to be as small as desired by restricting to a set with $|a|$ and $|b|$ small.

With this notation established, we can determine $W^s(M_\epsilon)$ and $W^u(M_\epsilon)$ as graphs and give the following restatement of Theorem 3

Theorem 4 *If $\epsilon > 0$ is sufficiently small, then, for some $\Delta > 0$,*

(a) there is a function $a = h_s(b,y,\epsilon)$ defined for $y \in K$ and $|b| \leq \Delta$, so that the graph

$$W^s(M_\epsilon) = \{(a,b,y) : a = h_s(b,y,\epsilon)\}$$

is locally invariant under (2.1). Moreover, $h_s(b,y,\epsilon)$ is C^r in (b,y,ϵ) for any $r < +\infty$.

(b) there is a function $b = h_u(a,y,\epsilon)$ defined for $y \in K$ and $|a| \leq \Delta$, so that the graph

$$W^u(M_\epsilon) = \{(a,b,y) : b = h_u(a,y,\epsilon)\}$$

is locally invariant under (2.1). Moreover, $h_u(a,y,\epsilon)$ is C^r in (a,y,ϵ) for any $r < +\infty$.

These theorems also apply when $\epsilon = 0$ and render the stable and unstable manifolds of the known critical manifold, the existence of which is also guaranteed by the usual stable and unstable manifold theorems at critical points (their smooth variation in y requires a little work to show but follows from Theorem 4). These latter two theorems then assert that these manifolds perturb. At this point, there is little justification for naming these manifolds as stable and unstable, other than their status gained as perturbations of the $\epsilon = 0$ case. It will be seen below, see Theorem 5, that they enjoy certain decay and growth estimates respectively.

Theorem 1 can be concluded from Theorem 3 by taking the intersection of $W^s(M_\epsilon)$ with $W^u(M_\epsilon)$. Locally, the Implicit Function Theorem gives the intersection as a graph, and these functions can be patched together since K is a compact set. Moreovwe, we need only give the construction of the stable manifold, as that of the unstable manifold follows immediately by a reversal of time. The proof to be given is very geometric in flavor and is based on the use of cones. The immediate result will be of a Lipschitz manifold and the smoothness proof for these manifolds will only be sketched here.

It is appropriate at this point to say some words about the history of these invariant manifold theorems, although one can only address such a task in an incomplete manner. There are two approaches taken to proving invariant manifold theorems and both have an extensive history. The first is that due to Hadamard, see [21], and relies on the geometry present in the splitting due to the decay rates. The second approach is due to Perron, see [48], and is based on proving the existence of the invariant manifold as a fixed point of a certain integral equation. Fenichel adopted the Hadamard approach in his seminal papers, see [15, 16, 17, 18]. These lectures are, to a great extent, based on his adaptation of the method to the case of singularly perturbed ODE's [18]. Simultaneous to Fenichel's work, Hirsch, Pugh and Shub [25] used the more analytic approach to achieve related results. Sakamoto [51] used the Lyapounov-Perron approach to derive Fenichel's results. An extensive exposition of Fenichel's Theorems, as well as their proofs, are given by Wiggins [58]. Other results in this direction have been obtained by many different authors including, but not limited to, Knobloch, Lin and Szmolyan.

The proofs given here are an extreme geometric version of Fenichel's. Many of the ideas lying behind the proofs were learnt by the author from Conley in his lectures on dynamical systems at Wisconsin. They have also been used by McGehee [42] and Bates and Jones [3] in the case of single fixed points.

2.2 Preparation of equations

The set K will need to be somewhat enlarged. Since M_0 is given by h^0, which is assumed to be C^∞ on K, which is compact, a set $\hat{K}$ can be found so that $K \subset \text{int} \hat{K}$ and h^0 is defined, and C^∞ for all $y \in \hat{K}$. Moreover, $\hat{M}_0 = \{(x,y) : x = h^0(y), ; y \in \hat{K}\}$ is a set of critical points and we can choose $\hat{K}$ so that $\hat{M}_0$ is normally hyperbolic.

The equations will be further prepared before the proofs can be given. The coordinates given above in terms of a and b separate the stable and unstable parts, but do not necessarily give good estimates for decay and growth. We set the quantities $\lambda_+ > 0$ and $\lambda_- < 0$ so that

$$\lambda_+ < \mathcal{R}(\lambda) \text{ for any } \lambda \in \sigma(A(y)) \text{ and } y \in \hat{K}, \tag{2.3}$$

$$\lambda_- > \mathcal{R}(\lambda) \text{ for any } \lambda \in \sigma(B(y)) \text{ and } y \in \hat{K}, \tag{2.4}$$

We shall refine the coordinates for a and b so that appropriate decay estimates on the linear parts are exposed.

Lemma 1 *Coordinates can be chosen so that, in the new inner product, the following estimates hold*

$$< a, A(y)a > \geq \lambda_+ < a, a >, \tag{2.5}$$

$$< b, B(y)b > \leq \lambda_- < b, b > . \tag{2.6}$$

Proof The coordinates can be found locally using ϵ-Jordan form. These can be patched together over all of K using a partition of unity.

In all proofs of the Center Manifold Theorem a modification has to be made to the equation in the center directions. This serves the purpose of mitigating its neutral character. We must perform the same modification here to deal with the slow directions, which are, effectively, center directions. The set $\hat{K}$ can further be chosen so that its boundary is given by the condition $\hat{\nu}(y) = 0$ for some C^∞ function $\hat{\nu}(y)$ and $\hat{\nu}(y)$ satisfies $\nabla \hat{\nu}(y) \neq 0$ for all $y \in \partial \hat{K}$. The function $\hat{\nu}(y)$ is assumed to have been normalized so that $\nabla \hat{\nu}(y) = n_y$ is a unit outward normal for $\partial \hat{K}$. We let $\rho(y)$ be a C^∞ function that has the following values

$$\rho(y) = \begin{cases} 1 & \text{if} \quad y \in \hat{K}^c, \\ 0 & \text{if} \quad y \in K, \end{cases} \tag{2.7}$$

The existence of such a function can be achieved locally with a C^∞ bump function and then the full function is created with a partition of unity. We now modify the third equation of (2.1) by adding the term $\delta\rho(y)n_y$, where δ is some number that remains to be chosen.

We shall need to append an equation for the small parameter ϵ. However, for the purpose of making estimates later, we shall actually use a multiple of ϵ as the new auxiliary variable. Thus, we set $\epsilon = \eta\sigma$ and append the equation $\eta' = 0$ to the system (2.1). We then arrive at the system

$$
\begin{aligned}
a' &= A(y)a + F_1(x, y, \epsilon) \\
b' &= B(y)b + F_2(x, y, \epsilon) \\
y' &= \eta\sigma g(x, y, \epsilon) + \delta\rho(y)n_y \\
\eta' &= 0,
\end{aligned}
\tag{2.8}
$$

in which it is understood that x is a function of a and b, and ϵ is a function of η. Clearly if Theorem 4 is restated with ϵ replaced by η and proved in that formulation, the original version of Theorem 4 can easily be recaptured by substituting ϵ back in.

2.3 Proof of Theorem 4

We are now ready for the proof of Theorem 4. The strategy will be to find a function $a = h_s(b, y, \eta)$ defined for $y \in \hat{K}$ and then restrict it to K, where the new equation agrees with the old as $\rho = 0$ in K.

Proof The first step is to set the neighborhood of $\hat{M}_0$ in which we shall work. The neighborhood will be called $\hat{D}$ and is determined by the conditions:

$$
y \in \hat{K}, \; |a| \le \Delta, \; |b| \le \Delta, \; \eta \in [0, \eta_0].
$$

Define the set

$$
\Gamma_s = \left\{ (a, b, y, \eta) : (a, b, y, \eta) \cdot t \in \hat{D}, \text{ for all } t \ge 0 \right\}.
\tag{2.9}
$$

We shall prove that Γ_s is the graph of a function given by a in terms of the remaining variables and this will be the function $h_s(b, y, \eta)$. The next step is to show that Γ_s contains the graph of a function. Set $\zeta = (b, y, \eta)$ and let $\hat{D}_{\hat{\zeta}}$ denote the cross-section of $\hat{D}$ at fixed $\zeta = \hat{\zeta}$.

$$
\hat{D}_{\hat{\zeta}} = \left\{ (a, \hat{\zeta}) : |a| \le \Delta \right\},
$$

as depicted in Figure 7.

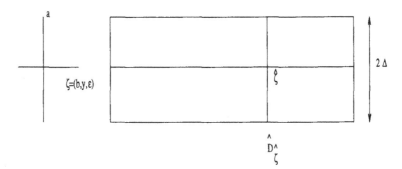

Figure 7
The neighborhood $\hat{D}$ and the cross-section.

We need to show that there is at least one point $(a, \hat{\zeta}) \in \hat{D}$ for which $(a, \hat{\zeta}) \cdot t \in \hat{D}$ for all $t \geq 0$. To achieve this, the Wazewski Principle is used. Let $\hat{D}^+$ be the immediate exit set of $\hat{D}$ and $\hat{D}^0$ be the eventual exit set. If $\hat{D}^+$ is closed relative to $\hat{D}^0$ then $\hat{D}$ is called Wazewski set and the map $W : \hat{D}^0 \to \hat{D}^+$, that takes each point to the first from which it exits $\hat{D}$, is continuous. We need to check the boundary of $\hat{D}$ and find the immediate exit set.

$|a| = \Delta$:
$$\begin{aligned}
< a, a >' \; &= 2 \{< a, A(y)a > + < a, F_1 >\}, \\
&\geq 2 \{\lambda_+ \Delta^2 - \Delta \gamma (\Delta + \epsilon_0)\},
\end{aligned} \tag{2.10}$$

using Lemma 1 and (2.2). If ϵ_0 is chosen less than Δ, we see that

$$< a, a >' \geq 2(\lambda_+ - 2\gamma)\Delta^2 > 0, \tag{2.11}$$

if γ is chosen small enough, which can be achieved by choosing Δ and ϵ_0 sufficiently small. The set $|a| = \Delta$ is then a part of the immediate exit set, as $|a|$ increases there.

$|b| = \Delta$:
It can been similarly that, with small enough Δ and ϵ_0,

$$< b, b >' < 0.$$

$y \in \partial \hat{K}$:
$$< y', n_y > = \epsilon < g(x, y, \epsilon), n_y > + \delta < n_y, n_y >,$$

since on $\partial \hat{K}$ $\rho = 1$. Setting $M = \sup_{\hat{D}} \{|g|, |Dg|\}$, the above can be estimated as

$$< y', n_y > \geq \delta - \epsilon_0 M > 0,$$

if $\delta > \epsilon_0 M$, which can be assumed as δ was arbitrary.

$\eta = 0$ or $\eta = \eta_0$:
both of these sets are invariant and thus render neither entrance nor exit sets.

The immediate exit set is thus seen to be the set $|a| = \Delta$, and $\hat{D}$ is easily checked to be a Wazewski set. The set $\hat{D}_{\hat{\zeta}}$ is a ball of dimension k. Suppose that $\Gamma_s \cap \hat{D}_{\hat{\zeta}} = \emptyset$, then $\hat{D}_{\hat{\zeta}} \subset \hat{D}^0$ and $\hat{D}_{\hat{\zeta}}$ lies in the domain of the Wazewski map W, so, restricting W, we have

$$W : \hat{D}_{\hat{\zeta}} \to \hat{D}^+ = \{|a| = \Delta\}.$$

If we follow this by a projection $\pi(a, \zeta) = a$, we see that $\pi \circ W$ maps a k- ball onto its boundary, while keeping that boundary fixed. This contradicts the No-Retract Theorem, which is equivalent to the Brouwer Fixed-Point Theorem, see, for instance, [41]. Thus there is a point in $\hat{D}_{\hat{\zeta}} \cap \Gamma_s$. Since $\hat{\zeta}$ was arbitrary, this gives, at least, one value for a as a function of (b, y, η), and we name it $h_s(b, y, \eta)$.

The next step will be to show that the graph of the above derived function is all of Γ_s. At the same time, it will be shown that the function is, in fact, Lipschitz with Lipschitz constant equal to 1. A comparison between the growth rates in different directions will be derived in the next lemma. Let $(a_i(t), \zeta_i(t))$, $i = 1, 2$ be two solutions of (2.8), set $\Delta a = a_2(t) - a_1(t)$ and $\Delta \zeta = \zeta_2(t) - \zeta_1(t)$. Further, we define

$$M(t) = |\Delta a|^2 - |\Delta \zeta|^2.$$

Lemma 2 *If $M(t) = 0$ then $M'(t) > 0$, as long as the two solutions stay in $\hat{D}$, unless $\Delta a = 0$.*

Proof The lemma follows from estimates that we will make on each of the quantities $< \Delta a, \Delta a >$ etc. The equation for Δa is

$$\Delta a' = A(y_2)a_2 - A(y_1)a_1 + F_1(x_2, y_2, \eta_2\sigma) - F_1(x_1, y_1, \eta_1\sigma), \qquad (2.12)$$

which we rewrite as

$$\Delta a' = A(y_2)\Delta a + [A(y_2) - A(y_1)] a_1 + \Delta_x F_1 + \Delta_y F_1 + \Delta_\epsilon F_1, \qquad (2.13)$$

where

$$\Delta_x F_1 = F(x_2, y_2, \epsilon_2) - F_1(x_1, y_2, \epsilon_2),$$
$$\Delta_y F_1 = F_1(x_1, y_2, \epsilon_2) - F_1(x_1, y_1, \epsilon_2)$$

and

$$\Delta_\epsilon F_1 = F_1(x_1, y_1, \epsilon_2) - F_1(x_1, y_1, \epsilon_1).$$

Using the fact that F_1 involves only higher order terms, one can derive the estimates

$$|\Delta_x F_1| \leq \gamma |\Delta x|, \qquad (2.14)$$

and

$$|\Delta_\epsilon F_1| \le \sigma\gamma\, |\Delta\eta|\,, \tag{2.15}$$

wherein γ can be made as small as desired by reducing the defining parameters of $\hat{D}$. Since $F_1(0, y, 0) = 0$, we can write

$$F_1(x, y, \epsilon) = x\tilde{F}_1(x, y, \epsilon) + \epsilon\hat{F}_1(x, y, \epsilon),$$

from which we obtain the estimate

$$|\Delta_y F_1| \le C\left\{\Delta\, |\Delta y| + \epsilon_0\, |\Delta y|\right\}. \tag{2.16}$$

We can estimate $< \Delta a, \Delta a >' = 2 < \Delta a', \Delta a >$ by taking the inner product of (2.13) with Δa. Each term can then be estimated using Lemma 1, (2.14), (2.16), (2.15) and the continuity of A in y. We then obtain

$$< \Delta a', \Delta a > \;\;\ge \lambda_+ |\Delta a|^2 - \{c_1\Delta\, |\Delta y|\,|\Delta a| + \gamma\,|\Delta x|\,|\Delta a| + \atop c_2(\Delta + \epsilon_0)\,|\Delta y|\,|\Delta a| + \sigma\gamma\,|\Delta\eta|\,|\Delta a|\}, \tag{2.17}$$

for some constants c_1 and c_2. We can bound $|\Delta x|$ by $c_3\,(|\Delta a| + |\Delta b|)$ and each term with $|\Delta b|$, $|\Delta y|$ or $|\Delta\eta|$ by $|\Delta\zeta|$. The estimate (2.17) can then be written as

$$< \Delta a, \Delta a >' \ge 2\,(\lambda_+ - \beta_1)\,|\Delta a|^2 - \beta_2\,|\Delta a|\,|\Delta\zeta|\,, \tag{2.18}$$

where β_1 and β_2 can be made small. If $M(t) = 0$, we can then replace $|\Delta\zeta|$ throughout by $|\Delta a|$. The net result is that

$$< \Delta a, \Delta a >' \ge 2\,\{\lambda_+ - (\beta_1 + \beta_2)\}\,|\Delta a|^2\,. \tag{2.19}$$

Note that if Δ and ϵ_0 are chosen sufficiently small, then the coefficient of $|\Delta a|^2$ in (2.19) can be made positive, say greater than $\lambda_+ - \beta$.

The estimate on $|\Delta\zeta|$ must be broken down into pieces. In a similar fashion to the above we can estimate

$$< \Delta b, \Delta b >' \le 2\,\{\lambda_- + \alpha\}\,|\Delta a|^2\,, \tag{2.20}$$

where α can be made as small as desired by choosing the parameters of the neighborhood small. We can write, as above,

$$\Delta y' = \;\; \sigma\Delta\eta g(x_2, y_2, \eta_2\sigma) + \eta\sigma\,\{\Delta_x g + \Delta_y g + \Delta_\epsilon g\} \atop + \delta\,(\rho(y_2)n_{y_2} - \rho(y_1)n_{y_1})\,. \tag{2.21}$$

Moreover, the terms in parentheses can be estimated as follows:

$$|\Delta_x g| \le \hat{\gamma}\,|\Delta x|\,, \tag{2.22}$$

$$|\Delta_y g| \le M\,|\Delta y|\,, \tag{2.23}$$

$$|\Delta_\epsilon g| \le \sigma\hat{\gamma}\,|\Delta\eta|\,, \tag{2.24}$$

where $\hat{\gamma}$ can be made as small as desired by reducing the size of the neighborhood. The following estimate can then be deduced

$$
\begin{aligned}
< \Delta y', \Delta y > \quad &\leq \sigma M \left|\Delta \eta\right| \left|\Delta y\right| + \eta_1 \sigma \left\{\hat{\gamma} c_3 (\left|\Delta a\right| + \left|\Delta b\right|) + \right. \\
&\left. M \left|\Delta y\right| + \sigma \hat{\gamma} \left|\Delta \eta\right|\right\} \left|\Delta y\right| + c_4 \left|\Delta y\right|^2 ,
\end{aligned}
\tag{2.25}
$$

and, again in a similar fashion to the above estimates, we can conclude that

$$
< \Delta y, \Delta y >' \leq 2\sigma \hat{C} \left|\Delta a\right|^2 .
\tag{2.26}
$$

Combining the estimates (2.19), (2.20), (2.26) and the fact that $\Delta \eta' = 0$, we conclude that

$$
M'(t) \geq \left\{\lambda_+ - \beta - \left(\lambda_- + \alpha + \sigma \hat{C}\right)\right\} \left|\Delta a\right|^2 ,
\tag{2.27}
$$

when $M(t) = 0$. The coefficient of the right hand side is

$$
\lambda_+ - \lambda_- - \left(\beta + \alpha + \sigma \hat{C}\right) .
$$

Since α and β can be made as small as desired by adjusting the parameters of the neighborhood and σ can be made small, this quantity is positive and the Lemma follows.

The decomposition of ϵ into η and σ is used at this last step. It seems somewhat artificial but, in fact, is not, as it corresponds to imposing an "ϵ-Jordan form" on the neutral directions coming from y and ϵ.

The Lemma can be interpreted in terms of "moving cones". This notion will be important in the next lecture and so it is worth exposing further at this point. Define the cone

$$
\mathcal{C} = \{(a, \zeta) : \left|a\right| \geq \left|\zeta\right|\} ,
\tag{2.28}
$$

and then Lemma 2 can be restated in terms of $\mathcal{C}$ as follows: If $z_2 \in z_1 + \mathcal{C}$ then $z_2 \cdot t \in z_1 \cdot t + \mathcal{C}$ so long as $z_2 \cdot t$ and $z_1 \cdot t$ stay in $\hat{D}$, see Figure 8.

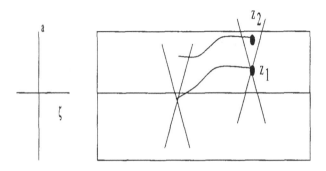

Figure 8
The moving cones.

The proof of Theorem 4 can now be completed, at least for the Lipschitz case. The set Γ_s will be our stable manifold. We have shown that it contains the graph of a function, which we denote by $a = h_s(b, y, \eta)$. Suppose that Γ_s contains more than one point with the same values of b, y and η. There would then be a_1 and a_2 so that both (a_1, b, y, η) and (a_2, b, y, η) lie in Γ_s. At $t = 0$, we would then have $|\Delta a| \geq |\Delta \zeta|$. By Lemma 2,

$$|\Delta a(t)| \geq |\Delta \zeta(t)|,$$

for all $t \geq 0$. In the estimate (2.18) we can then replace $|\Delta \zeta|$ by $|\Delta a|$ to obtain

$$\left\{ |\Delta a|^2 \right\}' \geq (\lambda_+ - \beta) |\Delta a|^2.$$

From which it can be easily concluded that Δa grows exponentially, which contradicts the hypothesis that both points stay in $\hat{D}$ for all $t \geq 0$.

The same argument can be used to show that h_s is also Lipschitz. If (a_1, ζ_1) and (a_2, ζ_2) are both in Γ_s and $|a_2 - a_1| \geq |\zeta_2 - \zeta_1|$, then $|a_2 - a_1|$ can be seen to grow exponentially, contradicting the hypothesis again that both points lie in Γ_s. We have now shown that the set Γ_s is the graph of a Lipschitz function. This manifold is $W^s(M_\epsilon)$, when y is restricted to the set K, in which the modified equation agrees with the original.

2.4 Decay estimates

Some justification should be given to the terminology "stable manifold". Since the base manifold M_ϵ no longer consists of equilibria, we cannot characterize $W^s(M_\epsilon)$ as the stable manifold of a set of critical points, however we can say that the solutions in $W^s(M_\epsilon)$ will decay to M_ϵ at an exponential rate, with the caveat that the decay will only last as long as the solution under consideration stays in the neighborhood D. In the following $d(.,.)$ is the Euclidean distance.

Theorem 5 *There are* $\kappa_s > 0$ *and* $\alpha_s < 0$ *so that if* $v \in W^s(M_\epsilon)$ *and* $v \cdot [0, t] \subset D$, *with* $t > 0$, *then*

$$d(v \cdot t, M_\epsilon) \leq \kappa_s \exp\{\alpha_s t\}. \tag{2.29}$$

Furthermore, there are $\kappa_u > 0$ *and* $\alpha_u > 0$ *so that if* $v \in W^u(M_\epsilon)$ *and* $v \cdot [t, 0] \subset D$, *with* $t < 0$, *then*

$$d(v \cdot t, M_\epsilon) \leq \kappa_u \exp\{\alpha_u t\}. \tag{2.30}$$

The proof follows from the results of the next lecture.

Chapter 3

Fenichel Normal Form

In this lecture, the third Theorem of Fenichel will be presented. This result gives a more detailed picture of the structure of the flow on the stable and unstable manifolds. An application will be given to settling in a cellular flow field in which this extra structure is used to show that the full flow in question is slaved in a precise manner to a two-dimensional flow. This two-dimensional flow can be analysed to reveal the salient features of the flow.

Of central interest will be, however, the derivation of a normal form for singularly perturbed equations in the neighborhood of a slow manifold. The derivation of this normal form, which we call Fenichel Normal Form, will rest on Fenichel's Third Theorem and will be the main goal of this chapter.

We shall thus focus on the proof of the existence of the stable manifold in Theorem 3, the proof for the unstable manifold follows immediately by a reversal of time.

3.1 Smoothness of invariant manifolds

The change of variables will need to be smooth and thus we need that the manifolds constructed in the Fenichel Theorems are smooth. In order to prove the smoothness of the invariant manifolds, the variational equation is used. This procedure will be sketched without details. The equation of variations of (2.1) (with an equation for ϵ and the modification added), which was the last version of the equation before Fenichel's Theorems were used, is given by

$$
\begin{aligned}
\delta a' &= A(y)\delta a + D\left(A(y)a\right)\delta y + DF_1\delta z \\
\delta b' &= B(y)\delta b + D\left(B(y)b\right)\delta y + DF_2\delta z \\
\delta y' &= \epsilon Dg\delta z + g\delta\epsilon + \delta\rho'(y)\delta y \\
\delta\epsilon' &= 0,
\end{aligned}
\tag{3.1}
$$

where $\delta z = (\delta a, \delta b, \delta y, \delta \epsilon)$. We imagine coupling this with the underlying equation to achieve a system in $\mathbf{R}^{2N+2}$. The linearization of this big system at a point in M_0 is easily seen to be in block form, each block having dimension $N + 1$. This matrix has the form

$$
\begin{pmatrix}
A(y) & & & & & & \\
 & B(y) & & & & & \\
 & & 0 & \sigma g(\hat{x}, \hat{y}, 0) & & & \\
 & & & 0 & & & \\
 & & & & A(y) & & \\
 & & & & & B(y) & \\
 & & & & & & 0 \quad \sigma g(\hat{x}, \hat{y}, 0) \\
 & & & & & & \quad 0
\end{pmatrix},
\tag{3.2}
$$

with 0's in all the vacant places.

A perusal of (3.2) gives that the block associated with $(\delta a, \delta b, \delta y, \delta \epsilon)$ is, in fact, exactly the same as the linearization of (2.1) at a point in M_0. The strategy for smoothness is then to reconstruct the invariant manifolds for this larger system, taking care to balance the unboundedness due to the variational equation by its linearity. In particular, for the stable manifold, this renders a function

$$
\delta a = H_s(a, b, y, \epsilon, \delta b, \delta y, \delta \epsilon),
$$

and the proof follows by showing that H_s is, in fact, the derivative of h_s i.e.,

$$
H_s(a, b, \epsilon, \delta b, \delta y, \delta \epsilon) = Dh_s(a, b, y, \epsilon)(\delta a, \delta b, \delta y, \delta \epsilon).
$$

This is achieved by using the characterization of the manifolds in terms of the set Γ_s and its uniqueness properties.

3.2 Straightening the invariant manifolds

The first part of Fenichel Normal Form can be implemented from the theorems already proved. Indeed, we shall straighten out the stable and unstable manifolds of the slow manifold M_ϵ. Using the functions that give these manifolds, we shall transform them to coordinate planes. First, set

$$
\begin{aligned}
a_1 &= a - h_s(b, y, \epsilon), & b_1 &= b, \\
y_1 &= y, & \epsilon_1 &= \epsilon,
\end{aligned}
\tag{3.3}
$$

which has the effect of transforming $W^s(M_\epsilon)$ to the subspace $a_1 = 0$. This transformation is invertible by inspection and is as smooth as h_s. Next, set

$$
\begin{aligned}
a_2 &= a_1, & b_2 &= b_1 - h_u(a_1 + h_s(b_1, y_1, \epsilon_1), b_1, \epsilon_1), \\
y_2 &= y_1, & \epsilon_2 &= \epsilon_1,
\end{aligned}
\tag{3.4}
$$

which has the effect of moving $W^u(M_\epsilon)$ to the subspace $b_2 = 0$. This latter transformation can also be checked to be invertible, using the fact that $W^s(M_\epsilon)$ is tangent to $b = 0$ along M_0, and obviously as smooth as h_u and h_s. We shall drop the subscripts and revert to the notation (a, b, y, ϵ) for a point in the new coordinate system. We thus have that the sets $a = 0$ and $b = 0$ are invariant in D, it thus must follow that $a = 0$ implies $a' = 0$ and $b = 0$ implies $b' = 0$. This imposes a certain character on the equations. Indeed, the variable a can be factored out of the equation for a', and analogously for b'. We thus arrive at

$$
\begin{aligned}
a' &= \Lambda(a, b, y, \epsilon)a \\
b' &= \Gamma(a, b, y, \epsilon)b \\
y' &= \epsilon g(a, b, y, \epsilon),
\end{aligned}
\tag{3.5}
$$

where Λ and Γ are matrices with $\Lambda(0, 0, y, 0) = A(y)$ and $\Gamma(0, 0, y, 0) = B(y)$. Note that the function g has been transformed appropriately and, with an abuse of notation, is also denoted by g. The matrix Λ therefore inherits the spectral properties of A and Γ those of B, if a, b and ϵ are all sufficiently small. As a side benefit of this coordinate change, we have obtained that M_ϵ is given by $a = b = 0$.

3.3 Fenichel fibering

The above change of coordinates has refined the stable and unstable directions to the point that estimates can be easily invoked from the linearized system. We still need to refine the equations for the slow directions. This involves using what has become known as "Fenichel Fibering", see, for instance, Wiggins [57]. It can be motivated by asking a question: we have seen that the stable and unstable manifolds of M_0 perturb to analogous objects when ϵ is sufficiently small, do the individual stable and unstable manifolds of points in M_0 also perturb? The answer would appear to be negative as the base points themselves do not perturb as critical points. However, this judgement is premature.

A minor technical difficulty arises here on account of the modification that we have performed to the equations. The equations (2.8) agree in D with the original equation (1.1) and we can restrict our attention to that set. However, points may leave D but re-enter it at a later time. Once trajectories of (2.8) have left D, their evolution is no longer governed by the original equation and they are no longer of interest. We must, therefore, restrict attention to the solutions while they are only in D. To facilitate this discussion, we need a definition.

Definition 3 *The forward evolution of a set $A \subset D$ restricted to D is given by the set*

$$
A \cdot_D t = \{x \cdot t : x \in A \text{ and } x \cdot [0, t] \subset D\}.
$$

With this definition in hand, we can state Fenichel's Third Invariant Manifold Theorem. In the following $v_\epsilon \in M_\epsilon$ is smooth in ϵ, including $\epsilon = 0$, and we are assuming (H1)-(H3).

Theorem 6 (Fenichel Invariant Manifold Theorem 3) *For every $v_\epsilon \in M_\epsilon$, there is an m-dimensional manifold*

$$
W^s(v_\epsilon) \subset W^s(M_\epsilon),
$$

and an l-dimensional manifold

$$W^u(v_\epsilon) \subset W^u(M_\epsilon),$$

lying within $O(\epsilon)$ of, and diffeomorphic to, $W^s(v_0)$ and $W^u(v_0)$ respectively. Moreover, they are C^r for any r, including in v and ϵ. The family $\{W^s(v_\epsilon) : v_\epsilon \in M_\epsilon\}$ is invariant in the sense that

$$W^s(v_\epsilon) \cdot_D t \subset W^s(v_\epsilon \cdot t), \tag{3.6}$$

if $v_\epsilon \cdot s \in D$ for all $s \in [0,t]$, and the family $\{W^u(v_\epsilon) : v_\epsilon \in M_\epsilon\}$ is invariant in the sense that

$$W^u(v_\epsilon) \cdot_D t \subset W^u(v_\epsilon \cdot t), \tag{3.7}$$

if $v_\epsilon \cdot s \in D$ for all $s \in [t,0]$.

Naturally the Theorem will again be proved in the case that M_0 is given by a function over K, and we shall produce a function to describe the fiber $W^s(v_\epsilon)$. It will also be assumed that the above coordinate changes have been made so that M_ϵ is given by $a = b = 0$, and $W^s(M_\epsilon)$ is given by $a = 0$. Some further technical difficulties are caused by the fact that the flow has been modified near the boundary of $\hat{D}$. To alleviate this difficulty we shall assume that there is a compact set $\tilde{K}$ so that $K \subset \text{int } \tilde{K} \subset \tilde{K} \subset \text{int } \hat{K}$ and $\rho = 0$ on $\tilde{K}$. The original equations are then seen to hold on the (larger) set $\tilde{K}$.

Just as for Theorem 1 and Theorem 3, we shall actually prove, and later use, this theorem in the case that the invariant manifolds can be given by graphs. We are assuming the form of the equations given by (3.5).

Theorem 7 *If $\epsilon > 0$ but sufficiently small, then*

(a) *in $a = 0$ (which is $W^s(M_\epsilon)$) there is, for each $v = v_\epsilon = (\hat{y}, \epsilon) \in M_\epsilon$, a function $y = h_s^v(b)$ defined for $|b| \leq \Delta$, so that the graphs*

$$W^s(v) = \{(0, b, y, \epsilon) : y = h_s^v(b)\}$$

form a locally invariant family as in (3.6). Moreover, $h_s^v(b)$ is C^r in v and ϵ jointly for any $r < +\infty$.

(b) *in $b = 0$ (which is $W^u(M_\epsilon)$) there is, for each $v = v_\epsilon = (\hat{y}, \epsilon) \in M_\epsilon$, a function $y = h_u^v(b)$ defined for $|a| \leq \Delta$, so that the graphs*

$$W^u(v) = \{(a, 0, y, \epsilon) : y = h_u^v(a)\}$$

form a locally invariant family as in (3.7). Moreover, $h_u^v(b)$ is C^r in v and ϵ jointly for any $r < +\infty$.

Proof of Theorem 7 We work entirely inside $W^s(M_\epsilon)$, which has become $a = 0$. The arguments for $W^u(M_\epsilon)$ are analogous. In $a = 0$, the variables (b, y, η) will suffice. The splitting of decay rates between b and $\zeta = (y, \eta)$ will be crucial. Define the cone

$$\mathcal{C} = \left\{ (b, \zeta) : |b| \geq \left| \zeta - \hat{\zeta} \right| \right\},$$

where $\hat{\zeta} = (\hat{y}, \hat{\eta})$ is fixed. Set $v = (0, \hat{\zeta}) = (0, \hat{y}, \hat{\eta})$ and $\hat{\eta} = \epsilon\sigma$. In the, by now familiar, strategy we shall characterize $W^s(v)$ as

$$\Gamma_v = \{ u = (b, y, \eta) : u \cdot t \in v \cdot t + \mathcal{C}, \text{ for all } t \geq 0 \}. \tag{3.8}$$

We use the modified equations exactly as in the proof of Theorem 4 so that $W^s(M_\epsilon)$ is positively invariant and (3.8) is well-defined. We need to show that the set Γ_v is the graph of a Lipschitz function. As in the proof of Theorem 3, we take cross-sections, but in this case of the cone $v + \mathcal{C}$. Fix $\hat{b}$ and set

$$S_{\hat{b}} = \left\{ (b, y, \eta) \in v + \mathcal{C} : b = \hat{b} \right\}.$$

The first task is to show that there is a point $u \in S_{\hat{b}}$ for which $u \cdot t \in v \cdot t + \mathcal{C}$ for all $t \geq 0$. A lemma analogous to Lemma 2 can be proved here and will be stated without proof. If $(b_i(t), \zeta_i(t))$ are solutions of (2.8) with $i = 1, 2$, set $M(t) = |b_2(t) - b_1(t)|^2 - |\zeta_2(t) - \zeta_1(t)|^2$.

Lemma 3 *If $M(t) = 0$ then $M'(t) > 0$, unless $b_2 = b_1$.*

One can now apply the flow to the set $\Sigma = S_{\hat{b}}$. By a similar topological argument to that used in the proof of Theorem 4, it can be seen that, for each $t \geq 0$ there is at least one point in the set $\{v \cdot t + \mathcal{C}\} \cap \Sigma \cdot t$. Call this point u_t and consider the set $\{u_t \cdot (-t) : 0 < t < +\infty\} \subset \Sigma$. Since Σ is compact, we can find a sequence $t_n \to +\infty$ so that $u_{t_n} \cdot (-t_n)$ converges to, say, $\hat{u}$. One can then see that $\hat{u}$ has the desired property, namely that $\hat{u} \cdot t \in v \cdot t + \mathcal{C}$ for all $t \geq 0$. This argument constructs a point $\zeta = (y, \eta)$ for each $b = \hat{b}$ so that $|b| \leq \Delta$. The y component of ζ is $h_s^v(b)$. The uniqueness argument folows in the same way as in the proof of Theorem 4. Converting back to the variable ϵ gives the functions of Theorem 7

The invariance of the family follows from the cone characterization of the fiber, and the proof of the existence of Lipschitz fibers is complete.

The fibers give a very useful matching between the points in $W^s(M_\epsilon)$ and partners they have in M_ϵ. One can then see that the decay of points in $W^s(M_\epsilon)$ to M_ϵ is actually to the base point of the fiber, this gives a decay result with "asymptotic phase"; similarly for points in $W^u(M_\epsilon)$. The proof of Theorem 5 actually follows from Corollary 1

Corollary 1 *$\kappa_s > 0$ and $\alpha_s < 0$ so that if $u \in W^s(v)$ then*

$$|u \cdot t - v \cdot t| \leq \kappa_s \exp\{\alpha_s t\},$$

for all $t \geq 0$ for which $v \cdot [0, t] \subset D$ and $u \cdot [0, t] \subset D$. Furthermore, there are $\kappa_u > 0$ and $\alpha_u < 0$ so that if $u \in W^u(v)$ then

$$|u \cdot t - v \cdot t| \leq \kappa_u \exp\{\alpha_u t\},$$

for all $t \geq 0$ for which $v \cdot [t, 0] \subset D$ and $u \cdot [t, 0] \subset D$.

Proof A differential inequality on $|b_2(t) - b_1(t)|$, where $v \cdot t = (b_1(t), \varsigma_1(t))$ and $u \cdot t = (b_2(t), \varsigma_2(t))$, can be derived using the fact that $u \cdot t \in v \cdot t + K$ for all $t \geq 0$. This then leads to the decay estimate.

The cone characterization of the fibers is not the usual approach taken. Fenichel constructs a graph transform map and most other authors follow this lead. However, the cone approach has a very appealing intuition and gives a real characterization of the fibers in terms of the flow. I believe that Bates (private communication) was the first to observe the relevance of cones in this context. **Remark** We shall also use the fibers to construct stable and unstable manifolds of subsets of M_ϵ, so that if $A \subset M_\epsilon$,

$$W^u(A) = \cup_{v \in A} W^u(v), \tag{3.9}$$

$$W^s(A) = \cup_{v \in A} W^s(v). \tag{3.10}$$

3.4 Settling in a cellular flow field

The problem of the settling under the influence of gravity of particles, with small inertia, through a fluid flow field can be formulated as a singular perturbation problem. The underlying flow will be assumed to be two-dimensional and of a particular form, namely a cellular fluid flow. Stommel [52] studied this situation in the case of zero inertia and concluded that both suspension in the cells as well as settling could take place. In the following model, $y(t)$ is the position of the center of the particle in space $(y = (y_1, y_2))$.

$$\begin{aligned}
\dot{y}_1 &= v_1 \\
\dot{y}_2 &= v_2 \\
\epsilon \dot{v}_1 &= -v_1 + \sin y_1 \cos y_2 \\
\epsilon \dot{v}_2 &= -v_2 - W - \cos y_1 \sin y_2,
\end{aligned} \tag{3.11}$$

where $v(t)$ is therefore the velocity of the particle and W is the settling velocity scaled so that $0 < W < 1$. The small parameter $\epsilon > 0$ is the Stokes' number and measures the inertial response time of the medium to the particle.

The question of interest here is whether the suspension of particles can still occur if inertial effects are included. In the limit that Stommel studied inertial effects were neglected and the equations were those governing the motion of a fluid particle. In other words, the particle in the fluid was considered to be behaving as a fluid particle would under the combined influence of the fluid flow and the gravitational field. The introduction of inertial effects, when small, supplies a singular perturbation of the case considered by Stommel.

The critical manifold is any appropriate subset M_0 as follows

$$M_0 \subset \{v_1 = \sin y_1 \cos y_2, \ v_2 = -W - \cos y_1 \sin y_2\}. \tag{3.12}$$

The normal eigenvalues are both -1 and hence M_0 is not only normally hyperbolic, but, in fact, attracting. The $W^s(M_0)$ thus fills an entire neighborhood of M_0. Note that the equations (3.11) are periodic in both y_1 and y_2 and thus M_0 can be restricted to any domain that contains a fundamental domain and conclusions that are global in y_1, y_2 can be drawn. Setting $\epsilon = 0$ in (3.11), it is then not hard to see that any initial condition has its ω-limit set in the set of critical points. Thus the stable manifolds of these critical points fills the entire space. The slow manifold M_ϵ is then given by the equations

$$\begin{aligned} v_1 &= \sin y_1 \cos y_2 + O(\epsilon) \\ v_2 &= -W - \cos y_1 \sin y_2 + O(\epsilon), \end{aligned} \tag{3.13}$$

over an appropriate domain. It is also attracting and, if the flow is considered on the space with both y_1 and y_2 identified modulo 2π, is globally attracting. To see this last point, one follows a trajectory with given initial condition by using the approximation of the $\epsilon = 0$ flow to get close to M_ϵ i.e., until the trajectory lies in $W^s(M_\epsilon)$.

The equations on M_ϵ are given by

$$\begin{aligned} \dot{y}_1 &= \sin y_1 \cos y_2 + O(\epsilon) \\ \dot{y}_2 &= -W - \cos y_1 \sin y_2 + O(\epsilon), \end{aligned} \tag{3.14}$$

which reduce to the following on M_0

$$\begin{aligned} \dot{y}_1 &= \sin y_1 \cos y_2 \\ \dot{y}_2 &= -W - \cos y_1 \sin y_2. \end{aligned} \tag{3.15}$$

This latter system is exactly the Stommel model, confirming the expectation that the zero inertia case should appear as the singular limit of the small inertia case. An interesting point to note here is that the variables of physical space, namely y_1, y_2 parametrize the slow manifold and thus the analysis of the trajectories on this manifold has a pleasing interpretation in terms of the flow of particles in physical space.

The phase portrait of (3.15) was given by Stommel and is shown in Figure 9. The space is divided into two distinct types of behavior. Inside each cell, there is a region in which the particles are trapped, this is bounded by the heteroclinic orbits, see Figure 9. The other particle trajectories will settle through the cells and never be trapped. The size of the trapped region increases as W decreases to 0, giving in the limit the usual cellular flow, as expected as gravity is then no longer present, in which all trajectories are trapped. The issue is, to what extent, these proportions of trapped versus settling particles change when ϵ is introduced.

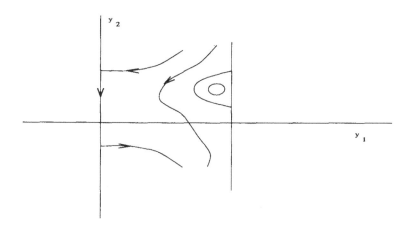

Figure 9
The Stommel flow.

The fate of the heteroclinic orbits surrounding the trapped region when ϵ is turned on must be determined. The following analysis is taken from the paper by Jones, Maxey and Rubin [31]. Indeed, there are two possible scenarios: if it opens so that the unstable manifold is "inside" the stable manifold then more particles can be trapped in a cell by staying inside the heteroclinic orbit. Moreover, particles that were settling can become trapped in the cell by being "caught" by a heteroclinic in some lower cell. However, if the heteroclinic breaks as shown in Figure 10 then particles that are settling will not be trapped in a cell but will continue to pass through. There is a possibility that some particles are still trapped. Indeed the critical point interior to the cell, which necessarily survives the perturbation (why?), will obviously be trapped, but if there is a surviving periodic orbit that surrounds this critical point then it will bound a region of trapped particles. This appears not to occur, as shown by numerical investigations.

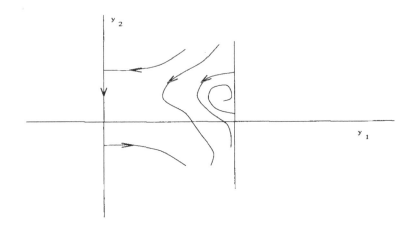

Figure 10
The case in which settling predominates.

As for the Ogawa waves, see Chapter 1, one must calculate the $O(\epsilon)$ terms in order to see whether the heteroclinic orbit persists. the equations up to $O(\epsilon)$ are given by

$$
\begin{aligned}
\dot{y}_1 &= \sin y_1 \cos y_2 + \epsilon \sin y_1 (W \sin y_2 + \cos y_1) + O(\epsilon^2) \\
\dot{y}_2 &= -W - \cos y_1 \sin y_2 + \epsilon \cos y_2 (W \cos y_1 + \sin y_2) + O(\epsilon^2).
\end{aligned}
\tag{3.16}
$$

We fix attention on the cell $0 \le y_1 \le \pi$, $0 \le y_2 \le \pi$. Let $y^-(\epsilon)$ be the critical point so that $y^-(0) = (0, \hat{y}_2)$ and $\hat{y}_2 \in [\frac{\pi}{2}, \pi]$ and $y^+(\epsilon)$ be the analogous point with $\hat{y}_2 \in [0, \frac{\pi}{2}]$. Further, let W^- be the unstable manifold of $y^-(\epsilon)$ and W^+ the stable manifold of $y^+(\epsilon)$. Consider the intersection of these manifolds with the set $y_1 = \frac{\pi}{2}$. Each will be given as the graphs of functions, so that $W^- \cap \{y_1 = \frac{\pi}{2}\}$ is given by $y_1 = h^-(\epsilon)$ and $W^+ \cap \{y_1 = \frac{\pi}{2}\}$ is given by $y_1 = h^+(\epsilon)$. To see which way the heteroclinic opens it suffices to calculate the quantity

$$
K = \left(\frac{\partial h^-}{\partial \epsilon} - \frac{\partial h^+}{\partial \epsilon} \right) |_{\epsilon=0}.
$$

Indeed, If $K > 0$ then the heteroclinic opens in such a way as to facilitate settling and inhibit trapping. It is shown in [31] that

$$
\frac{\partial h^-}{\partial \epsilon}|_{\epsilon=0} = W \int_{\tilde{y}_1}^{\pi} \bigg[\quad - \sin 2y_1 + y_1 (\cos^2 y_1 + 1) \\
+ W^2 y_1 \frac{\sin^2 y_1 - y_1^2}{\sin^2 y_1 \sqrt{(\sin^2 y_1 - W^2 y_1^2)}} \bigg] dy_1.
\tag{3.17}
$$

It is shown in [31] that $\frac{\partial h^-}{\partial \epsilon} > 0$, and, by symmetry, $\frac{\partial h^+}{\partial \epsilon} < 0$. It follows that $M > 0$ and the flow on M_ϵ is as shown in Figure 10.

The above analysis shows that on the slow manifold M_ϵ the settling of particles is facilitated. Indeed, it is more likely that particles with small inertia will settle than those with no inertia i.e., fluid particles. Thus a striking change to Stommel's conclusions occurs when inertia is taken into account.

Somewhat more about the flow on the slow manifold is shown in [31]. The slow manifold M_ϵ can be viewed as a torus by identifying both y_1 and y_2 modulo 2π. It is shown injonrubmax that on this torus there are only finitely many periodic orbits that wind around the torus. These are settling trajectories and are expected to be the asymptotic motion of the particles. If W is sufficiently small, it appears, from numerical computations, that there is a unique such periodic trajectory. In fact, there is one periodic orbit that has a special status as the fixed point of a certain map and all the others are obtained as period doubling bifurcations from this base orbit, see [31].

Since the slow manifold is attracting as discussed above, this predominance of settling will hold for the entire system. It is important to consider here exactly what more one obtains from knowing that the stable manifold, now filling a neighborhood of M_ϵ, is fibered by the individual stable manifolds. To each point $z = (y_1, y_2, v_1, v_2) \in \mathbf{R}^4$ the Fenichel fiber map assigns a point $\pi^-(z) \in M_\epsilon$ so that

$$\left| z \cdot t - \pi^-(z) \cdot t \right|$$

decays exponentially. This means that the point z will inherit all the asymptotic characteristics of the point $\pi^-(z)$. For instance, asymptotic settling rates can be concluded from the period of the attracting periodic orbits on M_ϵ. These settling rates will then also be valid for all initial conditions, including those off M_ϵ. If any non-trivial region of initial conditions is trapped in the cells, other than the critical point, on M_ϵ, then the corresponding region in $\mathbf{R}^4$ is determined to be the stable fibers to the points in this set. If, as expected, almost everything, except for the trapped critical points, settles in M_ϵ, then so does everything off M_ϵ except for the stable fiber of that critical point. When there is a unique periodic orbit on the torus (M_ϵ), almost all trajectories on M_ϵ will tend to this orbit, and thus so will almost all off M_ϵ. The only remaining possibility is that a periodic orbit surrounding the critical point persists then it, and its interior, will consist of trapped points. Thus the union of the stable fibers to this set would render the full set of initial conditions that lead to trapping.

We have used, in this example, the Fenichel Theorems to great effect in showing that even the smallest inertial effects of particles settling under a gravitational cellular flow field will encourage their settling and inhibit any potential trapping. Moreover, the Fenichel fibers have allowed us to see that the motion on M_ϵ determines all the asymptotic features of the system.

3.5 Normal form

Using the fibers constructed in Theorem 7, we can complete the transformation to Fenichel Normal Form. Indeed, Theorem 7 gives a map from (b, v) to a point $(b, h_s^v(b))$. The inverse of this map will send each point in $W^s(M_\epsilon)$ to the base point of its fiber. The fact that the inverse exists can be seen by observing that the construction of the fiber did not use the fact that $v_\epsilon \in M_\epsilon$, and could easily have been based also at the point $u = (b, \zeta)$, where $\zeta = h_s^v(b)$. It then follows that v would lie on the fiber for the point u, and would give the inverse of the map. We denote this inverse map $(b, y, \epsilon) \mapsto \hat{y}$ by π^-.

The final step in deriving the Fenichel Normal Form is to straighten out the fibers inside each of $W^s(M_\epsilon)$ and $W^u(M_\epsilon)$. To this end, set

$$
\begin{aligned}
a_3 &= a, & b_3 &= b, \\
y_3 &= \pi^-(b, y, \epsilon) & \epsilon_3 &= \epsilon,
\end{aligned}
\tag{3.18}
$$

so that the y-coordinate of each point is changed into that of its fiber base-point. The fibers on the unstable manifold are straightened out analogously by

$$
\begin{aligned}
a_4 &= a_3, & b_4 &= b_3, \\
y_4 &= \pi^+(a_3, y_3, \epsilon_3) & \epsilon_4 &= \epsilon_3,
\end{aligned}
\tag{3.19}
$$

which takes each point in $W^u(M_\epsilon) = \{b = 0\}$ to the base point of its unstable fiber.

The transformations to arrive at the coordinates $(a_4, b_4, y_4, \epsilon_4)$ have now modified the equations on both $W^s(M_\epsilon)$ ($a_4 = 0$) and $W^u(M_\epsilon)$ ($b_4 = 0$). We shall drop the subscripts and, with an abuse of notation, revert to the use of the original letters, with the understanding that the new coordinates are being used. On these sets, the slow flow has become independent of both a and b. It follows that if, either $a = 0$ or $b = 0$ we have $g(a, b, y, \epsilon)$ is a function only of y and ϵ, Thus we can write

$$
g(a, b, y, \epsilon) = h(y, \epsilon) + H(a, b, y, \epsilon)(a, b),
\tag{3.20}
$$

where $H(a, b, y, \epsilon)$ is a bilinear function of a and b.

Putting the pieces together, we can give the final Fenichel Normal Form for singularly perturbed equations in the neighborhood of a slow manifold

$$
\begin{aligned}
a' &= \Lambda(a, b, y, \epsilon)a \\
b' &= \Gamma(a, b, y, \epsilon)b \\
y' &= \epsilon \{h(y, \epsilon) + H(a, b, y, \epsilon)(a, b)\},
\end{aligned}
\tag{3.21}
$$

which holds in the set $D = \{(a, b, y, \epsilon) : |a| \leq \Delta, |b| \leq \Delta, y \in K, \epsilon \in [0, \epsilon_0]\}$. A simplified version of this normal form was derived by Jones and Kopell, see [28]. The current version was derived by Jones, Kaper and Kopell, see [27], and simultaneously discovered by Sandstede (private communication). Tin [54] has used it extensively and derived it in the more general context of perturbed invariant manifolds without a singular structure in which Fenichel first derived his Theorems.

Chapter 4

Tracking with Differential Forms

The Fenichel Normal Form will be a key in proving the Exchange Lemma. The Exchange Lemma concerns the passage of certain invariant sets (which will, in fact, be manifolds in their own right) near a slow manifold. In order to track the appropriate information about these invariant sets during the passage, we shall use differential forms to quantify this information.

4.1 Motivation

With the equations in Fenichel Normal Form (3.21) in a neighborhood D of M_ϵ, when ϵ is sufficiently small, a picture can be drawn in which the stable and unstable manifolds are coordinate planes and the spine, along which they intersect, is the manifold M_ϵ. Indeed, as shown in Figure 11, the set $a = 0$ is $W^s(M_\epsilon)$, $b = 0$ is $W^u(M_\epsilon)$ and $a = b = 0$ is M_ϵ. Of interest will be the situation in which a locally invariant manifold will be followed from some remote part of the phase space and studied as it passes near M_ϵ. We shall call this manifold the "shooting manifold" and usually denote it by Σ_ϵ. This manifold should not be confused with any of the manifolds that have been constructed above under the guise of the Fenichel Invariant Manifold Theorems. The purpose of following such a manifold will be to construct a homoclinic, or heteroclinic, orbit. The shooting manifold will then be the unstable manifold of a certain invariant set, such as a curve of critical points or periodic orbit. This invariant set will, in general, be unrelated to the slow manifold it is passing.

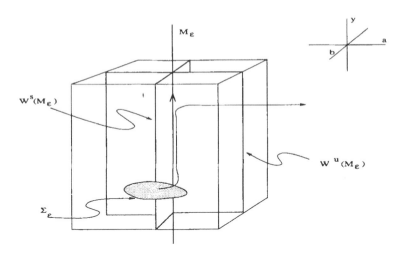

Figure 11
The passage near a slow manifold.

The shooting manifold will enter the neighborhood D and pass through D being modified by virtue of its flight near M_ϵ. It is viewed as entering with certain information and exiting with certain other information. This "exchange" of information will be the subject of the Exchange Lemma. It is instructive to consider what information is significant. A typical method for constructing homoclinic, or heteroclinic, orbits is by locating the (transverse) intersections of relevant stable and unstable manifolds. To determine a transverse intersection requires precise knowledge of the tangent spaces to the manifolds. It is the information encoded in the tangent spaces to the shooting manifold that will thus be of interest. The shooting manifold arrives at the boundary of D carrying certain tangent vectors and, during its passage through D, they will be exchanged for other tangent vectors. The goal of the ensuing analysis will thus be to see how to figure out the tangent vectors upon exit from D in terms of those at the entrance to D.

The general problem of following tangent spaces through phase space can be attacked in numerous ways. The approach adopted here will be to use differential forms that give "coordinates" to subspaces of a given space. These are particularly well suited to studies involving singular perturbations. Indeed, individual tangent vectors are hard to follow as vectors can switch from being predominantly in fast directions to the slow directions. The use of differential forms gives a satisfactory resolution to this problem as they afford a way of tracking the entire tangent space, without reference to individual vectors.

The consideration of an example will put this challenge of tracking invariant manifolds during their passage near a slow manifold in perspective. The canonical example that is the easiest to visualize is that of finding the travelling pulse solution in the FitzHugh-Nagumo equations and this will be described in the next section.

4.2 FitzHugh-Nagumo equations

The paradigmatic example for the construction of homoclinic orbits in singularly perturbed systems is the travelling pulse problem of the FitzHugh-Nagumo equations. These equations arose originally as a simplification to the Hodgkin-Huxley equations, formulated independently by FitzHugh and Nagumo. This is the first example in which both fast and slow structure will appear in the orbits of interest. The mathematical proof of the existence of these travelling pulses was originally given independently by Carpenter [10] and Hastings [24]. A geometric proof was later given by Langer [38], which paper spawned the Exchange Lemma through attempts to simplify difficult parts of Langer's proof and generalize the construction. It should be noted here that the geometric approach is needed in order to assess the stability of the travelling wave. The topological construction of the wave is inadequate for proving stability, see [26]. The full partial differential equations are

$$
\begin{aligned}
u_t &= u_{xx} + f(u) - w \\
w_t &= \epsilon(u - \gamma w),
\end{aligned}
\tag{4.1}
$$

where $f(u) = u(u - a)(1 - u)$ is the usual bistable nonlinearity and ϵ is a small parameter. The travelling pulse is a solution of (4.1) which is a function only of $\xi = x - ct$, and thus satisfies the system

$$
\begin{aligned}
u' &= v \\
v' &= -cv - f(u) + w \\
w' &= \tfrac{\epsilon}{\theta}(u - \gamma w) \\
c' &= 0,
\end{aligned}
\tag{4.2}
$$

so that (u, v) are fast variables and (w, c) are slow variables. The problem here is to construct an orbit homoclinic to the rest state $u = v = w = 0$. The first step is to construct a singular orbit consisting of fast transitions (heteroclinic orbits) between critical manifolds with intervening trajectories of the slow system. The critical manifolds must lie in the set $\{v = 0,\ w = f(u)\}$, which is the graph of a cubic inside the plane $v = 0$. Pieces of this critical set will form critical manifolds, namely in regions where $f'(u) \neq 0$. Of particular interest will be critical manifolds in the left branch of the cubic, say M_0^L, which is defined as the graph of the cubic restricted to an interval $[u_1, u_2]$ where $u_1 < 0$ and $f'(u) < 0$ on this interval, and also in the right branch, say M_0^R, defined similarly on $[u_3, u_4]$, see Figure 12. Both M_0^L and M_0^R are easily seen to be normally hyperbolic with 1-dimensional stable manifolds and 1-dimensional unstable manifolds.

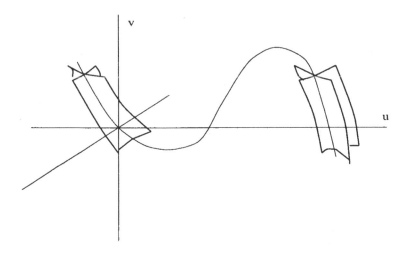

Figure 12
The slow manifolds and their stable
and unstable manifolds
for the FitzHugh-Nagumo system.

The singular homoclinic orbit, which exists at $\epsilon = 0$, will form a template on which the full orbit is built when $\epsilon > 0$. It consists of two fast pieces, $\mathcal{F}_1$, $\mathcal{F}_2$, and two slow pieces $\mathcal{S}_1$ and $\mathcal{S}_2$. These are determined as follows, see Figure 13.

$\mathcal{F}_1$ heteroclinic orbit when $\epsilon = 0$, with $w = 0$ and $c = c^*$, exactly as constructed for the scalar bistable reaction-diffusion equation,

$\mathcal{F}_2$ heteroclinic orbit when $\epsilon = 0$ from M_0^R to M_0^L, exists at given $c = c^*$ and fixed $w = w^*$ which is determined by the construction of this heteroclinic orbit,

$\mathcal{S}_1$ solution of limiting slow system $\dot{w} = -\frac{1}{c}(u - \gamma w)$ on M_0^R, connecting end of $\mathcal{F}_1$ to beginning of $\mathcal{F}_2$,

$\mathcal{S}_2$ solution of limiting slow system $\dot{w} = -\frac{1}{c}(u - \gamma w)$ on M_0^R, connecting end of $\mathcal{F}_1$ to rest state.

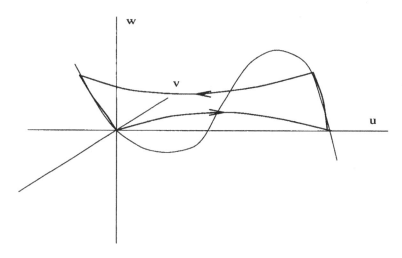

Figure 13
The singular homoclinic orbit.

The idea for constructing the true travelling pulse (homoclinic orbit) at a value of c near to c^* is to carry W_-^u, the unstable manifold of the curve of critical points $u = v = w = 0$ and $|c - c^*| \leq \delta$ for some fixed $\delta > 0$, around the phase space when $0 < \epsilon \ll 1$, see Figure 14.

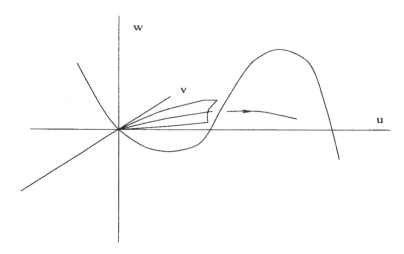

Figure 14
The shooting manifold.

It will be controlled by the information available from the $\epsilon = 0$ case. Indeed, the singular solution gives a template along which the manifold W_-^u is carried.

The ultimate goal will be to force it to intersect the stable manifold of the same curve of critical points.

The transversality shown above for the bistable equation shows that, when $\epsilon = 0$, W^u_- will transversely intersect $W^s(M^R_0)$. It will then also follow, by, for instance, the Implicit Function Theorem, that, when $\epsilon > 0$ but small, W^u_- transversely intersects $W^s(M^R_\epsilon)$. We need to see how W^u_- is affected during its time near the slow manifold. This is necessarily an $\epsilon \neq 0$ consideration, as when $\epsilon = 0$, W^u_- will stay in $w = 0$. The Exchange Lemma, see [29], [28], will precisely answer this question. It is expected that W^u_- will emerge from a neighborhood of M^R_0 near to the point where the singular orbit exits the neighborhood, at least if ϵ is small enough, which is given as the point at which $\mathcal{F}_2$ emerges. What is needed is the tangent space to W^u_- at this exit point as then the next stage of carrying the shooting manifold over the next fast jump can be realized and a transversality set up to see how the manifold intersects the stable manifold of the slow manifold M^L_0.

To see that the problem of determining the tangent space to W^u_- upon exit from D is non-trivial, one should consider the dimensions of the manifolds involved. The shooting manifold W^u_- is 2-dimensional, while the manifold M^R_0 is also 2- dimensional. But then the unstable manifold to the slow manifold, namely $W^u(M^R_0)$ is 3-dimensional. It is natural to expect that W^u_- will be crushed against the unstable manifold of M^R_0. This intuition comes from the λ-Lemma, see [49], which applies when M^R_0 is a single critical point. The λ-Lemma has been generalized to the case of critical manifolds, for the purpose of applying it to singular perturbation problems, by Deng, see [12], and this has the consequence that, indeed, W^u_- does become close to $W^u(M^R_0)$. However, this information is insufficient as $W^u(M^R_0)$ is 3-dimensional and W^u_- is only 2-dimensional. Clearly the fast unstable direction must be present in the tangent space to W^u_-. The question is: which slow direction is picked out by the tangent space to W^u_- at the exit point? The Exchange Lemma will give a precise answer to this question and we will then be able to finish the construction of the FitzHugh-Nagumo homoclinic orbit.

4.3 Variational equation and differential forms

The statement of the Exchange Lemma and the sketch of its proof will be postponed until after the technique of using differential forms has been introduced. Consider a general ODE

$$z' = F(z), \tag{4.3}$$

where F is a smooth (C^∞) function on an open subset U of $\mathbf{R}^N$. The variational equation of (4.3) can be written as

$$p' = DF(z)p, \tag{4.4}$$

where z satisfies (4.3) and $p \in \mathbf{R}^N$. The coordinates of p can be conveniently expressed using differential forms, $p_i = dz_i(p)$, recalling that dz_i is a linear form on tangent vectors. Since (4.4) can be rewritten, using the summation convention, as

$$p_i' = \partial_j F_i(z) p_j, \tag{4.5}$$

we can also write

$$dz_i(p)' = \partial_j F_i(z) dz_j(p). \tag{4.6}$$

Moreover, we will usually suppress the tangent vector p itself and write

$$dz_i' = \partial_j F_i(z) dz_j, \tag{4.7}$$

which ostensibly is nonsensical as the forms dz_i are constant, but whenever a derivative of dz_i appears, it should be understood as applied to whatever particular tangent vector is currently under consideration. We can also write (4.7) in shorthand as

$$dz' = DF(z) dz. \tag{4.8}$$

4.4 Tracking tangent spaces

We have seen that the variational equation can conveniently be expressed using differential 1-forms. We shall now put this to good use by using it as a vehicle to calculate equations on higher order forms. It is natural to compute such equations in order to ascertain how information about tangent spaces to invariant manifolds is carried under the flow. If invariant sets, or more specifically manifolds, are to be tracked under the influence of the flow, their configuration at a certain time is locally encoded in the tangent space to the invariant manifold at the relevant underlying point of the flow. The "coordinates" of a tangent space are given by projecting a (unit) cube in that space onto each of the coordinate subspaces of the same dimension as the tangent spaces themselves. The "volume" of the resulting object is the value of that coordinate. These quantities can be algebraically calculated using differential forms as they are indeed the values of the various k-forms for manifolds of dimension k. In other words, if Π is a k-dimensional subspace of $\mathbf{R}^N$, using coordinates $z = (z_1, \ldots, z_N)$, the coordinates of Π are given by

$$dz_{i_1} \wedge dz_{i_2} \wedge \ldots \wedge dz_{i_k}(\Pi),$$

for all choices of $(i_1, \ldots, i_k)$ (without repetition or permutation). If Π is spanned by $\{v_1, \ldots, v_k\}$ then

$$dz_{i_1} \wedge \ldots \wedge dz_{i_k}(\Pi) = \sum_{\pi}(-1)^{\mathrm{sgn}\pi} dz_{i_1}(v_{\pi(1)}) dz_{i_2}(v_{\pi(2)}) \ldots dz_{i_k}(v_{\pi(k)}), \tag{4.9}$$

where π is a permutation of $(1, \ldots, k)$. This is exactly the volume of the cube in Π, that is determined by the spanning vectors $v_1, \ldots, v_k$, projected onto the $(z_{i_1}, \ldots z_{i_k})$ subspace.

If $p \cdot t$ is a solution of (4.3) and $p \cdot t$ belongs to some invariant manifold, to which $\Pi(t)$ is the tangent space at $p \cdot t$, we should be able to calculate an equation for the coordinates of $\Pi(t)$. This can indeed be done and is related to the variational equation as a k-form version of it. The calculation of this equation in full generality is tedious and thus we shall carry out such a computation only on examples. Hopefully, these examples will make the general structure clear, in which there are, associated with any given equation, flows on the entire exterior algebra; in other words, flows on the space of k-forms, for any $1 \le k \le N$.

4.5 Example

We shall refresh this abstract discussion by considering a specific example and calculating the flow on tangent spaces using differential forms for it. We shall apply it to the derivation of an important transversality condition. The example will be that of the travelling wave for the bistable reaction-diffusion equation. This has appeared twice already in these lectures. In the FitzHugh-Nagumo travelling pulse problem, these equations arose as the $\epsilon = 0$, $w = 0$ subsystem that gave the fast jump $\mathcal{F}_1$. This fast jump is a heteroclinic orbit and a key point about the FitzHugh- Nagumo pulse is that it is constructed as the transverse intersection of stable and unstable manifolds. It has also appeared as the limiting slow equation for the phase field model considered in the first lecture, namely (1.12). Indeed, a transversality condition was given in that lecture for the heteroclinic orbit to persist to such an orbit on the slow manifold M_ϵ. This transversality condition will be derived using differential forms. This example has thus arisen as both the slow equation, for the phase field problem, and the fast equation, for the FitzHugh-Nagumo pulse!

Consider then the travelling wave problem for the bistable reaction-diffusion equation with an equation for the speed parameter appended

$$
\begin{aligned}
u' &= v \\
v' &= -cv - f(u) \\
c' &= 0,
\end{aligned}
\tag{4.10}
$$

where $f(u) = u(u - a)(1 - u)$ and $a < \frac{1}{2}$. We shall show that W^-, the unstable manifold of the curve $\{(0, 0, c) : c \text{ near } c*\}$, and W^+, the stable manifold of $\{(1, 0, c) : c \text{ near } c*\}$, intersect transversely at $q^* \in \{u = a\}$ at the value of c, say $c*$, at which the heteroclinic orbit exists.

The variational equation for (4.10), in the differential form notation of the above section, can be calculated as

$$
\begin{aligned}
du' &= dv \\
dv' &= -cdv - Df(u)du - vdc \\
dc' &= 0.
\end{aligned}
\tag{4.11}
$$

Since W^- and W^+ are both 2-dimensional manifolds, we will need to track 2-dimensional subspaces and thus should calculate the equations for the various different 2-forms. There are three different 2-forms, namely $du \wedge dv$, $dv \wedge dc$ and

$du \wedge dc$. The equations for the variation of these 2-forms can be calculated as shown in the preceding section. The product rule is used to see that

$$
\begin{aligned}
(du \wedge dv)' &= du' \wedge dv + du \wedge dv', \\
&= dv \wedge dv + du \wedge (-cdv - Df(u)du - vdc) \\
&= -cdu \wedge dv - vdu \wedge dc,
\end{aligned}
\tag{4.12}
$$

where it has been used that $du \wedge du = dv \wedge dv = 0$. The calculation of the equations for the remaining two forms is left as an exercise.

Recall the reduction of the transversality to calculating the sign of the quantity (rewritten in the current notation)

$$
\left(\frac{\partial h^-}{\partial c} - \frac{\partial h^+}{\partial c} \right) |_{c=c*},
\tag{4.13}
$$

where the intersections of $W^\pm$ with $\{u = a\}$ are given, respectively by $v = h^\pm(c)$. As a procedure for verifying transversality, the following has general applications, and a very definite structure to the argument, thus we divide it into steps that can be repeated in other contexts, such as for Melnikov calculations, see below.
Step 1 Observe first that the vectors

$$
\eta_1^\pm = \left(0, \frac{\partial h^\pm}{\partial c}, 1 \right)
$$

are tangent respectively to $W^\pm$. We seek a 2-form that renders a multiple of $\frac{\partial h^\pm}{\partial c}$ when applied to the tangent space of $W^\pm$. We know another vector that is tangent to both $W^\pm$, namely the vector field itself

$$
\eta_2 = (v, -cv - f(u), 0).
$$

We then see that

$$
= du \wedge dv(\eta_1^\pm, \eta_2) = v \frac{\partial h^\pm}{\partial c},
\tag{4.14}
$$

at $q = q^*$.
Step 2 Equation (4.12) would be very useful for evaluating the left hand side of (4.14) were the last term known. The difficulty is that the value of $\eta_1^\pm \cdot t$, where the flow referred to is that of the variational equation over the underlying heteroclinic orbit, is not known for $t \neq 0$. However we do know that

$$
dc(\eta_1^\pm \cdot t) = 1
$$

as the c-component is invariant from (4.11). Moreover $\eta_2 \cdot t$ is known for all t as it is exactly the vector field, that is $\eta_2 \cdot t = (v, -cv - f(u), 0)$. This is sufficient to compute $du \wedge dc$, indeed

$$
du \wedge dc(\eta_1^\pm \cdot t, \eta_2 \cdot t) = -v,
$$

and so, setting $\omega = du \wedge dv(\Pi^\pm(t))$, we obtain

$$
\omega' = -c\omega - v^2.
\tag{4.15}
$$

Step 3 It is an exercise to check that $\exp{(ct)}\,\omega \to 0$ as $t \to -\infty$. Equation (4.15) can then be explicitly solved to render

$$\omega = e^{-ct} \int_{-\infty}^{t} \left\{ -e^{ct} v^2 \right\} dt, \qquad (4.16)$$

which shows that $\omega(0) = -v(0)\frac{\partial h^-}{\partial c} < 0$. Since $v(t) > 0$ for all t, we conclude that

$$\frac{\partial h^-}{\partial c}\big|_{c=c^*,q=q^*} > 0.$$

A similar argument shows that

$$\frac{\partial h^+}{\partial c}\big|_{c=c^*,q=q^*} < 0.$$

Putting these two inequalities together, we have the desired result that

$$\left(\frac{\partial h^-}{\partial c} - \frac{\partial h^+}{\partial c} \right)\big|_{c=c^*,q=q^*} > 0, \qquad (4.17)$$

as desired.

4.6 Transversality for the KdV-KS waves

As a second application, the transversality condition involved in the existence of solitary waves in the KdV-KS equations as discussed in the first lecture, will be derived. This will give an application of the above prescription that is very close to the Melnikov method and the reader is invited to check that the usual Melnikov conditions can be derived using this differential form approach, see [20].

Consider the equations (1.34), for which an intersection of $W^u(\mathcal{S})$ and $W^s(\mathcal{S})$ is sought. We need to calculate $M(c)$ as defined in (1.36).

Step 1 A 3-form must be found that renders the quantities $\frac{\partial h^\pm}{\partial \epsilon}$ when applied to the tangent spaces $\Pi^\pm(0)$ to the invariant manifolds $W^u(\mathcal{S})$ and $W^s(\mathcal{S})$. There are three tangent vectors to $W^u(\mathcal{S})$ and $W^s(\mathcal{S})$ at $t = 0$ that are easily found. They are given by

$$\begin{aligned}
\eta_1 &= \left(\frac{\partial h^\pm}{\partial \epsilon}, 0, 1, 0 \right) \\
\eta_2 &= \left(v, u - \frac{u^2}{2}, 0, 0 \right) = (0, \alpha, 0, 0) \\
\eta_3 &= (0, 0, 0, 1),
\end{aligned} \qquad (4.18)$$

where $\alpha < 0$. It can then be checked that

$$du \wedge dv \wedge dc\,(\eta_1, \eta_2, \eta_3) = \alpha \frac{\partial h^\pm}{\partial \epsilon},$$

Step 2 The equation for the form $du \wedge dv \wedge dc$ can be calculated as

$$(du \wedge dv \wedge dc)^{\cdot} = \sqrt{c} \left(u - (1 + \frac{1}{c}) \right) v du \wedge d\epsilon \wedge dc.$$

Step 3 As in the previous section, the form $du \wedge dv \wedge dc$, when applied to the subspace $\Pi^{\pm}(t)$, can actually be calculated. Since

$$\eta_1 \cdot t = (*, *, 1, 0),$$

$$\eta_2 \cdot t = (v, u - \frac{u^2}{2}, 0, 0),$$

and

$$\eta_3 \cdot t = (*, *, 0, 1),$$

it can be seen that $du \wedge d\epsilon \wedge dc(\eta_1 \cdot t, \eta_2 \cdot t, \eta_3 \cdot t) = v$. It follows that

$$(du \wedge dv \wedge dc)^{\cdot} = -\sqrt{c} \left(u - (1 + \frac{1}{c}) \right) v^2.$$

From which, one obtains that

$$\alpha M(c) = -\sqrt{c} \left[\int_{-\infty}^{+\infty} (u - 1)v^2 - \frac{1}{c} \int_{-\infty}^{+\infty} v^2 \right].$$

The expression for $M(c)$ given in (1.37) is now easily derived using the equation.

Chapter 5

Exchange Lemma

With the above motivation, we are now in a position to give the simplest version of the Exchange Lemma. This will be the one originally formulated by Jones and Kopell, see [28]. In the next lecture, more sophisticated versions will be presented, but today's version is already powerful enough to prove a general theorem on the existence of heteroclinic orbits.

5.1 $k+1$ Exchange Lemma

In the following, we are assuming the standard singular perturbation set-up and that the slow manifold under consideration has a k-dimensional unstable manifold, as usual. We shall track another (locally) invariant manifold, say Σ_ϵ during its passage near the slow manifold M_ϵ; this should not be confused with the manifold M_ϵ itself, or its attendant stable, unstable manifolds or fibers. Recall that the manifold Σ_ϵ will be generated, most probably, at some other part of the phase space; the example of the FitzHugh-Nagumo equations should be kept in mind here, wherein the invariant manifold is the unstable manifold of a curve of critical points lying on a *different* slow manifold. We make the following hypothesis, in which the notation $\cap_T$ means that the intersection is transversal. The set D is the standard neighborhood of M_ϵ in when the coordinates of Fenichel Normal Form are used i.e.,

$$D = \{(a, b, y) : |a| \leq \Delta, |b| \leq \Delta, y \in K\}.$$

(H4) There is a $(k+1)$-dimensional, locally invariant manifold Σ_ϵ, defined for $0 < \epsilon \ll 1$, and smooth in ϵ, so that

$$\Sigma_0 \cap_T W^s(M_0) \neq \emptyset,$$

at a point $q \in \partial D$.

As a technical point, we shall always assume that Σ_0 is cut off so as to intersect $W^s(M_0)$ only in a neighborhood of q.

Since Σ_0 is $(k+1)$-dimensional and $W^s(M_0)$ is $(l+m)$-dimensional, in the intersection of these two manifolds we have available $k+l+m+1$ directions. Since the dimension of the phase space is $N = k+l+m$ the intersection will be 1-dimensional under the transversality hypothesis (H3). But this is optimal as both manifolds are locally invariant and hence their intersection must contain trajectories, which means that it must be, at least, 1-dimensional. From (H4) it then follows that $\Sigma_0 \cap W^s(M_0)$ must live in $W^s(J_0)$ for some point $J_0 \in M_0$. Alternatively, we can write

$$J_0 = \omega\left(\Sigma_0 \cap W^s(M_0)\right). \tag{5.1}$$

The geometry is shown in Figure 15. It is also useful to note here that $J_0 = \pi^-\left(\Sigma_0 \cap W^s(M_0)\right)$, where π^- is the map sending each point in $W^s(M_0)$ to the base point of its fiber.

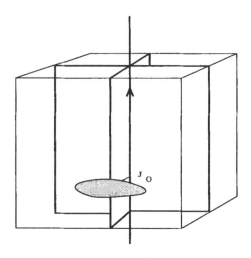

Figure 15
The point J_0.

Of central importance will be the trajectory of the limiting slow flow through the point J_0. Let $\hat{\tau}$ be some fixed (slow) time, choose $0 < \eta_- < \hat{\tau}$ and $\eta_+ > 0$. Set $I = [\hat{\tau} - \eta_-, \hat{\tau} + \eta_+]$. Assume that these quantities are chosen so that $J_0 \circ \tau$ is defined for all $\tau \in I$, where "$\circ$" refers to the action of the limiting slow flow i.e., that associated with (1.7) on M_0. In the following, we shall consider the set

$$J_0 \circ I \subset M_0. \tag{5.2}$$

When $\epsilon > 0$ but sufficiently small, the manifold Σ_ϵ will intersect $W^s(M_\epsilon)$ transversely. However the point $p_\epsilon \in \Sigma_\epsilon \cap W^s(M_\epsilon)$ will not ultimately be the point of interest as the trajectory emanating from this point will not leave D near

$W^u(M_\epsilon)$. Indeed, it will either never leave D or else leave it near the boundary of M_0. With reference to the FitzHugh-Nagumo example, if we wish to construct an orbit close to the singular orbit, the true orbit will have to leave D near the fast jump away from the critical manifold, and thus be close to $W^u(M_0)$. The notation q_ϵ will refer to a point in $\Sigma_\epsilon \cap \partial D$. This will be the point along the trajectory of which we shall track the manifold $\Sigma_\epsilon \cdot t$. We shall assume that the trajectory through q_ϵ does indeed leave D in forward time after (fast) time $T_\epsilon > 0$, so that $q_\epsilon \cdot T_\epsilon \in \partial D$. We need to able to control where the trajectory leaves D. Recall that the unstable manifold for a subset of M_0 is just the union of the unstable manifolds of its elements, see (3.10), which is here only used in the $\epsilon = 0$ case but also makes sense when $\epsilon \neq 0$.

Proposition 1 *Given* $r_0 \in W^u(J_0 \circ I) \cap \partial D$, *there is a* $q_\epsilon \in \Sigma_\epsilon \cap \partial D$ *and a* $T_\epsilon > 0$ *so that* $q_\epsilon \cdot T_\epsilon \in \partial D$ *and*

$$|q_\epsilon \cdot T_\epsilon - r_0| = O(\epsilon). \tag{5.3}$$

The proof of this Proposition uses the Wazewski Principle, in other words it is a topological shooting argument. The proof will be omitted. From this proposition we can see that trajectories can be found that exit D near prescribed points, for instance the point where a fast jump (when $\epsilon = 0$) will leave the neighborhood of the critical manifold. It does not, however, tell us how the manifold Σ_ϵ is configured in a neighborhood of this point. Thus, we can think of the above Proposition as a C^0-Exchange Lemma. The C^1-Exchange Lemma determines the configuration of the tangent space to Σ_ϵ upon exit from D. We shall call it the $(k + 1)$-Exchange Lemma as it refers to the case in which Σ_ϵ is $(k + 1)$-dimensional. The hypotheses (H1)-(H4) are assumed to hold.

Lemma 4 $((k + 1)$-**Exchange Lemma**$)$ *The manifold* $\Sigma_\epsilon \cdot T_\epsilon$ *is* C^1 $O(\epsilon)$ *close to* $W^u(J_0 \circ I)$ *in a neighborhood of the point* $\hat{q}_\epsilon = q_\epsilon \cdot T_\epsilon$.

Lemma 4 can be restated as the claim that

$$d\left(T_{\hat{q}_\epsilon}\left(\Sigma_\epsilon \cdot T_\epsilon\right), T_{r_0}\left(W^u\left(J_0 \circ I\right)\right)\right) = O(\epsilon). \tag{5.4}$$

The Exchange Lemma tells us how the manifold is configured upon exit from the neighborhood of the slow manifold. Indeed, it gives precise information about the directions present in the tangent space at that point. At the risk of being repetitious, to understand the lemma, some thought should again be given to the dimensions of the manifolds involved. The shooting manifold is $k+1$-dimensional and the unstable manifold of M_0 is $k + l$-dimensional. As commented earlier, the Exchange Lemma addresses which of the (slow) l directions is picked out by the manifold at its exit point from D. The set $J_0 \circ I$ is the (slow) trajectory through the point J_0 and is thus 1-dimensional. Its unstable manifold, $W^u(J_0 \circ I)$ is $k + 1$-dimensional. The dimensions thus agree and the Exchange Lemma tells us that the extra direction picked out is exactly that determined by the slow flow itself. A picture is useful here. The basic Figure (11) that we have been using is deceptive as we can only represent there a 1- dimensional slow manifold. In Figure 16, the stable manifold is omitted and the case of a 2-dimensional M_0 is depicted with a 3-dimensional unstable manifold. The manifold $W^u(J_0 \circ I)$ is shown and the Exchange Lemma says that the shooting manifold Σ_ϵ will exit D with a nearby tangent space.

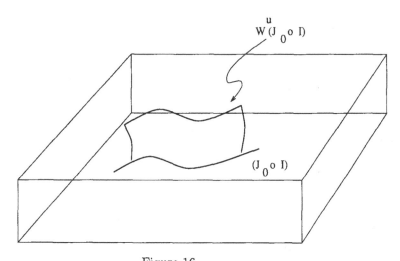

Figure 16
The unstable manifold to $J_0 \circ I$.

The application to the FitzHugh-Nagumo pulse will be given below, as well as the formulation of a general theorem concerning the construction of homoclinic orbits. However, we will first show why differential forms are relevant to the Exchange Lemma.

5.2   Differential forms and the Exchange Lemma

In order to show the C^1-closeness that is given in the Exchange Lemma, the linear spaces $T_{r_0}(W^u(J_0 \circ I))$ and $T_{\hat{q}_\epsilon}(\Sigma_\epsilon \cdot T_\epsilon)$ must be shown to be $O(\epsilon)$ close. This can be achieved by showing that the Plücker coordinates given by applying all the basis of $(k+1)$-forms to these tangent spaces are (projectively) $O(\epsilon)$ close. In other words it is not necessary that the length of the Plücker vectors are close, but only that the associated directions of these vectors are close. Amongst the Plücker coordinates there are two distinguished sets, we call block-1 forms and block-2 forms, as the resulting linear equations on these coordinates are a perturbation of a system in block form. The block-1 coordinates are those resulting from an application of the forms

$$da_1 \wedge da_2 \wedge \ldots \wedge da_k \wedge dy_i, \quad i = 1, \ldots, l, \tag{5.5}$$

and the block-2 forms are all the others. When the block-2 forms are applied to the tangent space $T_{r_0}(W^u(J_0 \circ I))$ the result is always 0. One part of the proof of the Exchange Lemma is thus to show that the block-2 forms applied to $T_{\hat{q}_\epsilon}(\Sigma_\epsilon \cdot T_\epsilon)$ are $O(\epsilon)$.

The equation for the block-1 forms must then be calculated. An approximation can be easily derived using the variational equation of the Fenichel Normal Form. Indeed, this calculation is instructive as it shows clearly why this normal form is so crucial. First, the variational equation in Fenichel coordinates can be obtained by differentiating (3.21)

$$
\begin{aligned}
da' &= \Lambda da + D_z(\Lambda dz)a \\
db' &= \Gamma db + D_z(\Gamma dz)b \\
dy' &= \epsilon\{D_y h(y,\epsilon) + H(da,b) + H(a,db) + D_z H(a,b)(a,b,dz)\},
\end{aligned}
\tag{5.6}
$$

where $z = (a,b,y)$ and the arguments of H, Λ and Γ have been suppressed. A simple approximation to (5.6) can be obtained if the underlying orbit actually lies on M_ϵ (which is not true for the orbit of interest but it does give the lowest order approximation). On M_ϵ we have that $a = b = 0$ and (5.6) simplifies to

$$
\begin{aligned}
da' &= A^\epsilon(y)da \\
db' &= B^\epsilon(y)db \\
dy' &= \epsilon\{D_y h(y,\epsilon)dy\},
\end{aligned}
\tag{5.7}
$$

where $A^\epsilon(y) = \Lambda(0,0,y,\epsilon)$ and $B^\epsilon(y) = \Gamma(0,0,y,\epsilon)$. Notice that the simple structure of (5.7) follows precisely from the elements of the Fenichel Normal Form. For instance, that the slow equation is not influenced by any fast (infinitesmal) variables comes directly from the bilinear form H. For this approximation, the equation for the block-1 forms can be easily calculated using the product rule.

$$
\begin{aligned}
(da_1 \wedge \ldots \wedge da_k \wedge dy_i)' &= \sum_{j=1}^k da_1 \wedge \ldots \wedge da_j' \wedge \ldots \wedge da_k \wedge dy_i \\
&\quad + da_1 \wedge \ldots \wedge da_k \wedge dy_i', \\
&= da_1 \wedge \ldots \wedge A_{jl}^\epsilon(y)da_l \wedge \ldots \wedge da_k \wedge dy_i \\
&\quad + da_1 \wedge \ldots \wedge da_k \wedge \epsilon[\tfrac{\partial}{\partial y_l}h_i(y,\epsilon)dy_l],
\end{aligned}
\tag{5.8}
$$

where the summation convention is being used, A_{jl}^ϵ is the jlth entry in A and h_i denotes the ith component of $h(y,\epsilon)$. In the first part, the only terms that will remain are those including $A_{jl}^\epsilon(y)da_l$, all the others will cancel. It follows that

$$
\begin{aligned}
(da_1 \wedge \ldots \wedge da_k \wedge dy_i)' &= \operatorname{Tr} A^\epsilon(y)[da_1 \wedge \ldots \wedge da_k \wedge dy_i] \\
&\quad + da_1 \wedge \ldots \wedge da_k \wedge \epsilon\tfrac{\partial}{\partial y_l}h_i(y,\epsilon)[da_1 \wedge \ldots \wedge da_k \wedge dy_l].
\end{aligned}
\tag{5.9}
$$

Now setting

$$
\Omega = \begin{pmatrix} da_1 \wedge \ldots \wedge da_k \wedge dy_1 \\ \vdots \\ da_1 \wedge \ldots \wedge da_k \wedge dy_l \end{pmatrix},
$$

(5.9) can be abbreviated as

$$\Omega' = \operatorname{Tr} A^\epsilon(y)\Omega + \epsilon D_y h(y,\epsilon)\Omega. \tag{5.10}$$

Since $\operatorname{Tr} A^\epsilon(y)$ is a scalar, an integrating factor can be introduced, which does not change the Plücker coordinates projectively. An equation for

$$\hat{\Omega} = \exp\left\{-\int_0^t \operatorname{Tr} A^\epsilon(y)\right\} \Omega$$

is easily computed as

$$\hat{\Omega}' = \epsilon D_y h(y,\epsilon)\hat{\Omega}, \tag{5.11}$$

which is exactly the variational equation for the slow flow! This fact is the key to the Exchange Lemma is used in the proof, sketched below, to show that the slow direction is that picked out for the remaining tangent vector for the shooting manifold after its passage near the slow manifold.

5.3 The FitzHugh-Nagumo pulse

In this section, we shall show how the Exchange Lemma can be used to complete the construction of the FitzHugh-Nagumo pulse. The shooting manifold Σ^ϵ in this case is W^u_-, the unstable manifold of the curve of rest states, which is 2-dimensional ($2 = k+1$ since $k = 1$ here). In order to emphasize the dependence on ϵ we shall write $W^{u,\epsilon}_-$. In order to apply the Exchange Lemma, we first need to see that

$$W^{u,0}_- \pitchfork_T W^s(M^R_0),$$

but this is a statement about the $\epsilon = 0$ flow, wherein $W^{u,0}_-$ lies in the plane $w = 0$. The flow in $w = 0$ is exactly given by the equation for the travelling waves of the scalar reaction-diffusion equation, namely (4.10). Furthermore, $W^s(M^R_0) \cap \{w = 0\}$ is exactly W^s_+ which is the stable manifold of the curve $\{(1,0,c) : c \text{ near } c^*\}$. It therefore follows that $W^{u,0}_-$ transversely intersects $W^s(M - 0^R)$ inside $w = 0$. Since the full space only adds the w-direction, which is in $W^s(M^R_0)$, it follows that

$$W^{u,0}_- \pitchfork_T W^s(M^R_0),$$

and the hypothesis of the Exchange Lemma holds.

Next, set $r_0 \in \mathcal{F}_2 \cap \partial D$, which is the intersection of the singular orbit with the boundary of D along the fast jump as it veers away from M^R_0. From Proposition 1, it follows that there is a point $\hat{q}_\epsilon = q_\epsilon \cdot T_\epsilon$ in $W^{u,0}_- \cap \partial D$ which lies at most $O(\epsilon)$ from r_0.

In order to carry the manifold $W^{u,\epsilon}_-$ over the fast jump away from M^R_0, we need to assess the configuration of $W^{u,\epsilon}_-$ at this exit point $\hat{q}_\epsilon$. This is precisely the information offered by the Exchange Lemma. Indeed, from Lemma 4 it follows that $W^{u,\epsilon}_-$ at $\hat{q}_\epsilon$ is $O(\epsilon)$ close to $W^u(J_0 \circ I)$.

The next step is then to show that

$$W^u(J_0 \circ I) \cap_T W^s(M_0^L),$$

which is a statement about the $\epsilon = 0$ flow with, moreover, c fixed at $c = c^*$, see Figure17.

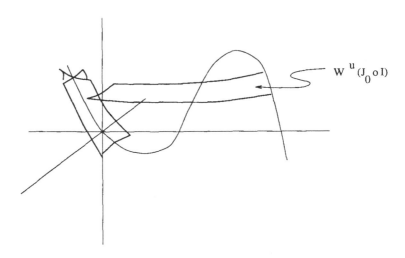

Figure 17
The transversality over the return jump.

The equation to be considered to verify this transversality is then

$$
\begin{aligned}
u' &= v \\
v' &= -c^*v - f(u) + w \\
w' &= 0.
\end{aligned}
\tag{5.12}
$$

The transversality argument is then very similar to that given, except that w is now the parameter and c is fixed, over the first fast piece and is omitted here.

Since $W_-^{u,\epsilon}$ is $O(\epsilon)$ from $W^u(J_0 \circ I)$ and $W^s(M_0^L)$ is $O(\epsilon)$ from $W^s(M_\epsilon^L)$, there will also be a transversal intersection between $W_-^{u,\epsilon}$ and $W^s(M_\epsilon^L)$. It can easily be checked that $W^s(M_\epsilon^L)$ is actually the stable manifold of the curve of critical points $\{(0,0,c) : c \text{ near } c^*\}$. Thus we have constructed a homoclinic orbit from this curve of critical points to itself. Since c is actually a parameter, the orbit must lie in a fixed c slice and hence we have a homoclinic orbit to the rest state for some c near c^*.

Remark It is instructive to see how the "exchange" of information occurs in this passage near the slow manifold. All information in the speed parameter c is lost as the shooting manifold veers near to M_ϵ. However new information, namely in the w-direction is acquired. This is exactly the direction of the slow flow in the manifold M_0^R. If multiple passages near slow manifolds occur then each passage is characterized by a loss of information acquired at the previous such passage and new information from the current slow flow is substituted.

5.4   General application

The Exchange Lemma will be used in this section to derive a very general theorem on the existence of homoclinic orbits to a 1-dimensional invariant set, due to Jones and Kopell, see [28]. We consider again the general singularly perturbed system (1.1) and suppose that there is a particular slow manifold M_ϵ^1 which contains an invariant curve P_ϵ for ϵ sufficiently small, and moreover that P_ϵ is attracting relative to the (slow) flow on M_ϵ. One should think of P_ϵ as being either a curve of critical points or a periodic orbit.

The singular orbit will be constructed with (arbitrarily) many fast jumps and intervening slow pieces. We shall assume that the following ingredients are given from which this will be put together. Let ρ be some fixed integer, which will be the number of critical manifolds visited by the singular orbit.

(A1) The following sets are assumed to exist for the equation (1.1):

M_0^j, $0 \le j \le \rho$: These are each normally hyperbolic critical manifolds, given, as usual, by the graphs of functions i.e., each satisfying (H1)-(H3). These are not necessarily distinct and the number of normal stable and unstable directions is independent of j. Moreover $M_0^\rho = M_0^0$.

$\mathcal{F}_0^j$, $1 \le j \le \rho$ Each being a heteroclinic orbit from M_0^{j-1} to M_0^j, including its α and ω-limit sets.

S_0^j, $0 \le j \le \rho$ If $j = 0$, $S_0^j = P_0$. If $0 < j < \rho$, S_0^j is a trajectory of the slow flow (1.7) which connects the end-point of $\mathcal{F}_0^j$ to the beginning point of $\mathcal{F}_0^{j+1}$. If $j = \rho$, it is a trajectory of (1.7) on M_0^0 connecting the end-point of $\mathcal{F}_0^\rho$ to P_0. The notation $\hat{S}_0^j$ will refer to the trajectories of (1.7) extended beyond the jump points.

The singular orbit can now be constructed in the obvious way by piecing together the fast jumps with in the intervening slow pieces. We call this singular homoclinic orbit $\mathcal{H}$, see Figure 18, so that

$$\mathcal{H} = \cup_{j=1}^{j=\rho} \mathcal{F}_0^j \cup S_0^j. \tag{5.13}$$

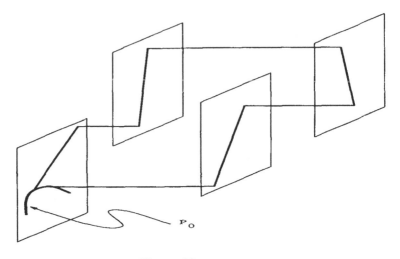

Figure 18
The singular homoclinic orbit.

A transversality assumption must also be made.

(A2) For every j so that $1 \leq j \leq \rho$,

$$W^u(\hat{\mathcal{S}}_0^{j-1}) \cap_T W^s(M_0^j). \tag{5.14}$$

Then we have the following theorem, see [28]

Theorem 8 *Under the assumptions (A1)-(A2), for $\epsilon > 0$, but sufficiently small, there is an orbit homoclinic to the set P_ϵ and $O(\epsilon)$ close to $\mathcal{H}$. Moreover, there is a neighborhood of $\mathcal{H}$ in which it is the unique such homoclinic orbit.*

Proof The proof is very simple, given the Exchange Lemma, and follows exactly the lines of the argument for the FitzHugh-Nagumo case. One takes $W^u(P_\epsilon)$, the existence of which is guaranteed by Theorem 6, and follows it around the phase space. The first transversality hypothesis $j = 1$ allows us to invoke the Exchange Lemma (Lemma 4) and conclude that we can make P_ϵ exit $O(\epsilon)$ close to the next fast jump where it exits the neighborhood of M_0^1. At that point, it is a consequence of Lemma 4 that it is $O(\epsilon)$ close to $W^u(\mathcal{S}_0^1)$ and the next transversality assumption can be used to get it by the next manifold. One continues in this fashion until $W^u(P_\epsilon)$ is seen to transversely intersect $W^s(M_0^\sigma)$. Since P_ϵ is attracting in M_0^σ, it is easily checked that the resulting intersection gives an orbit homoclinic to P_ϵ.

Theorem 8 can be applied to many problems. For instance, the reader can construct examples in which Silnikov orbits are easily seen to exist, see [57] for a discussion of the consequences of such orbits. Homoclinic orbits for the

Hodgkin-Huxley equations can be constructed using this Theorem. Bose [5] has constructed pulses for the Keener model [34] of two coupled nerve fibers each governed by the FitzHugh-Nagumo system.

5.5 Sketch of proof

The proof of the Exchange Lemma involves a fair amount of estimation and can only be sketched here. We will show, however, how the idea introduced above concerning the use of differential forms can be made into a proof. The proof given here is based on the approach developed by Tin, see [54]

Recall from the above that the forms applied to $T_{q_\epsilon} \cdot t$ are divided into two blocks. Let $Z(t)$ denote the vector of block-1 forms and $X(t)$ denote the vector of block-2 forms, the equations can then be calculated, with all the higher order terms appearing as

$$
\begin{aligned}
Z' &= (\text{ Tr } \Lambda + \phi + \epsilon[D_y h(y, \epsilon) + \Theta_1]) Z + \Theta_2 X \\
X' &= (G + \Psi_1) X + \Psi_2 Z,
\end{aligned} \tag{5.15}
$$

where ϕ, Θ_i and Ψ_i are higher order terms. To be precise,

$$
\begin{aligned}
&\|\Theta_1\| \le c_1 |a| |b|, \quad \|\Theta_2\| \le c_2 |a|, \quad \|\Psi_1\| \le c_3(|a| + |b|), \\
&\|\Psi_2\| \le c_4 |b|, \quad |\phi| \le c_5 |a|,
\end{aligned} \tag{5.16}
$$

where each of these estimates follows from the structure of the Fenichel Normal Form. Using an integrating factor as before, and keeping in mind that we only care about vectors being close in a projective sense, we can scale Z and X to obtain

$$
\begin{aligned}
\hat{Z}' &= \epsilon [D_y h(y, \epsilon) + \Theta_1] \hat{Z} + \Theta_2 \hat{X}, \\
\hat{X}' &= (G - \text{ Tr } \Lambda - \phi + \Psi_1) \hat{X} + \Psi_2 \hat{Z}.
\end{aligned} \tag{5.17}
$$

The dominant term in the second equation is $G - \text{ Tr } \Lambda$, and the idea is that its spectrum has negative real part, on account of the fact that $\text{ Tr } \Lambda$ is the sum of the real parts of the eigenvalues with positive real part (for the case that the underlying point lies in M_0). This can be used to show that $\hat{X}$ is forced to decay exponentially, although the actual estimates are quite difficult due to the coupling.

We need to determine $\hat{Z}(t)$ and this is achieved by a series of approximation. Consider the equation

$$
\hat{\Omega}' = \epsilon D_y h(\hat{y}, \epsilon) \hat{\Omega}, \tag{5.18}
$$

where $\hat{y}(t)$ is a solution of (3.21) chosen as follows: let p_ϵ be chosen in $\Sigma_\epsilon \cap W^s(M_\epsilon)$ exponentially close to q_ϵ, which is possible (why?). Next set $p_\epsilon \cdot t = (\hat{x}(t), \hat{y}(t), \epsilon)$, and check that there are $\kappa_1 > 0$ and $\beta_1 < 0$ so that

$$
|y(t) - \hat{y}(t)| \le \kappa_1 e^{-\beta_1 t}. \tag{5.19}
$$

Concerning $\hat{Z}(t)$ and $\hat{\Omega}(t)$ we then have the

Lemma 5 $\left|\hat{Z}(t) - \hat{\Omega}(t)\right| \leq \kappa_1 e^{-\beta_1 t}$.

The idea behind the proof of Lemma 5 is to estimate the terms in the equation satisfied by $\hat{Z}(t) - \hat{\Omega}(t)$, which is

$$\left(\hat{Z} - \hat{\Omega}\right)' = \epsilon\left[D_y h(y, \epsilon)(\hat{Z} - \hat{\Omega}) \\ +(D_y h(y, \epsilon) \quad -D_y h(\hat{y}, \epsilon))\hat{\Omega} + \Theta_1 \hat{Z}\right] + \Theta_2 \hat{X}. \tag{5.20}$$

One then uses the known estimates on the last three terms followed by the Gronwall inequality. Each of these terms is exponentially small, provided $\hat{Z}$ and $\hat{\Omega}$ are bounded. The first on account of (5.19), the second because $\|\Theta_1\| \leq c\,|a|\,|b|$ and ab is exponentially small as both are bounded and one of them is always exponentially small. Similarly for the last term as $\hat{X}$ is exponentially decaying.

It should be noted here that this is the key point at which the structure of the bilinear term, which encodes the Fenichel fibering, is used. Indeed, the bound on $\|\Theta_1\|$ in terms of ab is essential.

The next step in the proof of the Exchange Lemma is to introduce an approximation to $\hat{\Omega}(t)$. Let $\tilde{\Omega}(t)$ satisfy the same equation as $\hat{\Omega}(t)$, namely (5.18), but with $\tilde{\Omega}(0)$ being the block-1 forms of the space $T_{\pi^+(\hat{q}_\epsilon)} W^u(S_{\pi^+(\hat{q}_\epsilon)})$, where S_y is the slow trajectory through the point y on M_ϵ. Note that this space has block-2 forms all being 0. Note that $\pi^+(\hat{q}_\epsilon) = \hat{y}(0)$ and hence $\hat{\Omega}(t)$ will be the block-1 forms of

$$T_{(0,\hat{y}(t),\epsilon)} W^u(S_{\hat{y}(t)}),$$

which are $O(\epsilon)$ close to those of

$$T_{(0,\hat{y}(t),0)} W^u(J_0 \circ I),$$

for appropriate choice of the interval I and normalization. It thus remains to estimate $\hat{\Omega}(t) - \tilde{\Omega}(t)$. But it can be shown that $\left|\hat{\Omega}(0) - \tilde{\Omega}(0)\right|$ is exponentially small and, since they satisfy the same equation, we can conclude that, for appropriate values of t $\left|\hat{\Omega}(t) - \tilde{\Omega}(t)\right|$ is also exponentially small. The steps are now completed by invoking Lemma 5, which implies the Exchange Lemma provided that $\hat{Z}(0)$ is bounded away from 0, otherwise the estimates may be vacuous. But this follows from the transversality hypothesis.

Chapter 6

Generalizations and Future Directions

The Exchange Lemma given in the previous lecture is inadequate in two respects.

1. *The $(k+1)$ Exchange Lemma will be inadequate for many applications.* For instance suppose that the periodic orbit or curve of critical points in the example above were *not attracting* in the *slow manifold*, we would then have to follow an invariant manifold that was larger than $k+1$ as it would have more slow directions. The same would be true if the invariant sets, to which we sought homoclinic orbits, were higher-dimensional.

2. *$O(\epsilon)$ is not good enough.* If, for instance, we are considering a perturbed Hamiltonian system the transversality of the shooting manifold Σ_ϵ with the stable manifold to the slow manifold, namely $W^s(M_\epsilon)$ may only be of $O(\epsilon)$ and not $O(1)$, as we have used above. This turns out not to be a problem in reaching the conclusion of the Exchange Lemma, but that conclusion will itself be useless in this case as we would be following an invariant manifold by an $O(\epsilon)$ approximate version and the approximate version would itself intersect the next stable manifold only at $O(\epsilon)$, since these might cancel a more accurate estimate must be found in the Exchange Lemma.

In the first few sections of this last lecture, I will give the generalizations of the Exchange Lemma that address these points. In the last sections, I will discuss various applications and the directions for further study that they suggest.

6.1 The $(k+\sigma)$ Exchange Lemma

We suppose now that the shooting manifold Σ_ϵ is a $(k+\sigma)$-dimensional locally invariant manifold, again smooth in ϵ. Recall that k is the number of unstable directions for the slow manifold. The cases of interest are when $1 < \sigma < l$, for then the Exchange Lemma of the last lecture does not apply and yet the dimension of the shooting manifold is still not equal to the full dimension of the unstable manifold of the slow manifold. However the Exchange Lemma I shall state in fact applies for $0 \leq \sigma \leq l$.

An assumption must again be made as to how the shooting manifold enters a neighborhood of the slow manifold. Indeed, we assume

(H5) There is a $(k + \sigma)$-dimensional, locally invariant manifold Σ_ϵ, defined for $0 < \epsilon \ll 1$, and smooth in ϵ, so that

$$\Sigma_0 \cap_T W^s(M_0) \neq \emptyset,$$

at a point $q \in \partial D$.

Note that in this case the intersection will be more than a trajectory indeed, by a dimension count, we expect it to be σ-dimensional. Let V be some suitably chosen neighborhood of the point $q \in \Sigma_0 \cap W^s(M_0) \cap \partial D$ and consider the set

$$
\begin{aligned}
J_0 &= \omega\left(\Sigma_0 \cap W^s(M_0) \cap V\right) \\
&= \pi^-\left(\Sigma_0 \cap W^s(M_0) \cap V\right),
\end{aligned}
\tag{6.1}
$$

in M_0. By the transversality hypothesis (H5) the set J_0 will be a $\sigma - 1$ dimensional submanifold of M_0. We need a transversality hypothesis on the slow flow.

(H6) The set J_0 is a $(\sigma - 1)$-dimensional manifold and the slow flow i.e., that associated with (1.7), is *not tangent* to J_0, in other words

$$g(\pi^-(q), 0) \notin T_{\pi^-(q)} J_0. \tag{6.2}$$

With the same notation as in the previous section, I can state the $(k + \sigma)$-Exchange Lemma, which is due to Tin and the author, see [30].

Lemma 6 $(k + \sigma)$-**Exchange Lemma** *The manifold $\Sigma_\epsilon \cdot T_\epsilon$ is C^1 $O(\epsilon)$ close to $W^u(J_0 \circ I)$ in a neighborhood of the point $\hat{q}_\epsilon = q_\epsilon \cdot T_\epsilon$.*

The proof follows the same lines as the sketch of the $k + 1$ case given in the previous lecture, see [30] or [54]. In his thesis, Tin [54], formulated a general theorem concerning the existence of homoclinic orbits to invariant subsets of a slow manifold. Let P_0 be a γ-dimensional invariant, compact submanifold of M_0^1 that is normally hyperbolic under the slow flow on M_0^1. Further, let U_0 be the $\gamma + \eta$-dimensional unstable manifold of P_0 in M_0^1, in the slow flow on M_0^1, given as usual by (1.7). In a similar fashion to the application given in the previous lecture, we assume the presence of the objects given in the next hypothesis. (A3) The following sets exist as stated:

M_0^j, $0 \leq j \leq \rho$: These are each normally hyperbolic critical manifolds, given, as usual, by the graphs of functions i.e., satisfying (H1)-(H3). These are not necessarily distinct and the number of normal stable and unstable directions is independent of j. Moreover $M_0^\rho = M_0^0$.

$\mathcal{F}_0^j$, $1 \leq j \leq \rho$: Each being a heteroclinic orbit from M_0^{j-1} to M_0^j.

S_0^j, $0 \leq j \leq \rho$: If $j = 0$, S_0^j is P_0 together with a curve in U_0^j that connects P_0 to the beginning point of $\mathcal{F}_0^1$. If $0 < j < \rho$, S_0^j is a trajectory of the slow flow (1.7) which connects the end-point of $\mathcal{F}_0^j$ to the beginning point of $\mathcal{F}_0^{j+1}$. If $j = \rho$, it is a trajectory of (1.7) on M_0^0 connecting the end-point of $\mathcal{F}_0^\rho$ to P_0.

The singular orbit is given similarly by the expression (5.13).

Since P_0 is assumed to be normally hyperbolic in M_0^0, see Fenichel [15], it will perturb to a (locally) invariant manifold in M_ϵ^0, say P_ϵ. The next theorem will give conditions under which the singular orbit $\mathcal{H}$ perturbs to an actual homoclinic orbit to P_ϵ when $\epsilon > 0$ but sufficiently small.

Needless to say, there are transversality hypotheses to be satisfied along each of the fast jumps. However, another collection of sets need to be determined in order to express these transversality conditions.

(A4) The following sets exist as stated:

U_0^j, $0 \leq j \leq \rho - 1$: Subsets of M_0^j. If $j = 0$, U_0^j is the unstable manifold in M_0^1 of P_0 under the flow of (1.7). They are then defined inductively along with the sets below by

$$U_0^j = J_0^j \circ I, \tag{6.3}$$

where $\circ$ refers to the slow flow on M_0^j, and $I = (\hat{\tau}_j - \eta_j, \hat{\tau}_j + \eta_j)$, with $\hat{\tau}_j$ chosen so that the beginning point of $\mathcal{F}_0^{j+1}$ is contained in U_0^j.

J_0^j, $1 \leq j \leq \rho - 1$: Subsets of M_0^j, that are defined by

$$\omega \left(W^u(U_0^{j-1}) \cap W^s(M_0^j) \right), \tag{6.4}$$

inductively along with the above.

For these sets to be well-defined and to be able to apply the Exchange Lemma, the following transversality hypotheses must hold.
(A5) If $1 \leq j \leq \rho - 2$ then

$$W^u(U_0^j) \cap_T W^s(M_0^{j+1}) \neq \emptyset, \tag{6.5}$$

along $\mathcal{F}_0^j$.

(A6) If $1 \leq j \leq \rho - 2$ then the slow vector field is not tangent to the set J_0^{j+1}, which is assumed to be a $(\sigma - 1)$-dimensional manifold.

A final condition is needed for the last jump. The set V_0 is the stable manifold of P_0 in M_0^0, relative to the slow flow.

(A7) Along $\mathcal{F}_0^\rho$

$$W^u(U^{\rho-1})_0 \cap_T W^s(V_0) \neq \emptyset.$$

The following theorem has been proved by Tin, see [54].

Theorem 9 *Under the assumptions (A3)-(A7), if $\epsilon > 0$, but sufficiently small, then there is an orbit homoclinic to P_ϵ that is within $O(\epsilon)$ of the singular homoclinic orbit $\mathcal{H}$.*

6.2 Exponentially small Exchange Lemma

The estimate in the Exchange Lemma can be significantly tightened if we are willing to drop the comparison with the $\epsilon = 0$ case. Indeed, a perusal of the proof shows that the $O(\epsilon)$ estimate comes in when going from the tangent space to the unstable manifold of the appropriate subset of M_ϵ to that of the slow trajectory in M_0. Since the Fenichel theory does supply us with a full structure when $\epsilon \neq 0$, we can easily make the comparison with the $\epsilon \neq 0$ object.

Furthermore, the transversality at entry to D can be significantly weakened, as we only need to assume that the transversality occurs when $\epsilon \neq 0$. The transversality can be measured by taking bases for the tangent spaces and wedging the entire set of vectors (taking care not to repeat the vector field and keeping the basis bounded away from 0 and ∞ as $\epsilon \to 0$) to make a volume form. For the Exchange Lemma with Exponentially Small Error we make the following hypothesis.

(H6) There is a $(k + \sigma)$-dimensional, locally invariant manifold Σ_ϵ, defined for $0 < \epsilon \ll 1$, and smooth in ϵ, so that

$$\Sigma_\epsilon \cap_T W^s(M_\epsilon) \neq \emptyset,$$

at a point $q \in \partial D$ and the transversality is of $O(\epsilon^\rho)$ for some ρ.

Set

$$J_\epsilon = \pi^- \left(\Sigma_\epsilon \cap W^s(M_\epsilon) \right), \tag{6.6}$$

where again π^- is the Fenichel map sending points to the base points of their fiber. Notice that, in this case, the Fenichel map cannot be replaced by an ω-limit set as there is a genuine flow on the manifold M_ϵ. By the same token, the set J_ϵ contains directions from the slow flow and thus J_ϵ has one higher dimension than J_0. We can then replace $J_0 \circ I$ in the statement of the Exchange Lemma by $J_\epsilon \cdot T$ for some appropriately chosen T.

Lemma 7 Exchange Lemma with Exponentially Small Error *The manifold $\Sigma_\epsilon \cdot T_\epsilon$ is C^1 $O(\exp\{\frac{-c}{\epsilon}\})$ close to $W^u(J_\epsilon \cdot T)$ at $q_\epsilon \cdot T_\epsilon$, for some $c > 0$.*

This Lemma was first proved by Jones, Kaper and Kopell [27] in the $\sigma = 1$ case and later by Jones and Tin [30] for the general case. It is crucial that the full structure of the Fenichel Normal Form be used in order to obtain this result. In particular, the decomposition of the slow vector field as a bilinear form, which comes from the fibering, is crucial. The exponential closeness, and its relation to the bilinear form, was noticed independently by Sandstede (private communication).

6.3 Pendulum forced by two frequencies

In this section, I will outline an application that uses the Exchange Lemma with Exponentially Small Error. Consider the pendulum forced by two separate frequencies

$$
\begin{aligned}
q'' + \sin q &= \epsilon\left(\delta q' + \gamma(\tau_1, \tau_2)\right) \\
\tau_1' &= \epsilon\omega_1(\tau_1, \tau_2) \\
\tau_2' &= \epsilon\omega_2(\tau_1, \tau_2),
\end{aligned}
\tag{6.7}
$$

where γ and ω_i, $i = 1, 2$ are all periodic, with period 2π in each of their arguments and also assumed to be smooth. Equation (6.7) is rewritten as a system

$$
\begin{aligned}
q' &= p \\
p' &= -\sin q + \epsilon\left(\delta p + \gamma(\tau_1, \tau_2)\right) \\
\tau_1' &= \epsilon\omega_1(\tau_1, \tau_2) \\
\tau_2' &= \epsilon\omega_2(\tau_1, \tau_2).
\end{aligned}
\tag{6.8}
$$

The fast flow is obtained, as usual, by setting $\epsilon = 0$. There are critical manifolds given by any of the critical points of the pendulum. We set then

$$
M_0^\pm \subset \{p = 0, \ q = \pm\pi\}.
\tag{6.9}
$$

Since both of these points are saddles of the pendulum, the manifolds $M_0^\pm$ are normally hyperbolic and hence the Fenichel Theorems apply. Various identifications can be made in the phase space. Since the vector field is periodic in τ_1 and τ_2, both τ_1 and τ_2 can be identified modulo 2π. The critical manifolds would then become tori. We have not set the theory up to cover such manifolds and hence we shall take this point of view. I shall assume that $M_0^\pm$ is defined on a sufficiently large region that it will contain a fundamental domain of the torus. Thus $M_\epsilon^\pm$ can be viewed as a torus after the fact by identifying τ_1 and τ_2 after the results have been obtained.

The slow flow on $M_0^\pm$ is given by the equations:

$$
\begin{aligned}
\dot{\tau}_1 &= \omega_1(\tau_1, \tau_2) \\
\dot{\tau}_2 &= \omega_2(\tau_1, \tau_2),
\end{aligned}
\tag{6.10}
$$

which are also the exact equations on $M_\epsilon^\pm$ as the fast variables do not appear in the slow equations at all. I shall assume that, on $M_0^\pm$ there are two periodic orbits, one attracting, say P_a, and the other repelling P_r, see Figure 19. Of interest will be the construction of orbits homoclinic to P_a, which will require use of the $(k+1)$- Exchange Lemma with the exponentially small error. From the point of view of applying the Exchange Lemma, the construction of orbits homoclinic to P_r, although possible, is not as interesting (why?).

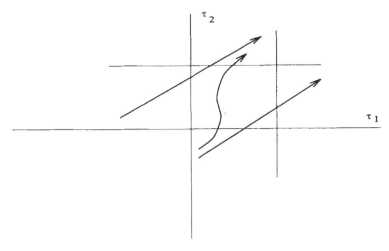

Figure19
The slow flow.

Note that the fast flow decouples completely from the slow variables, when $\epsilon = 0$. Since there is a heteroclinic orbit from $(0, -\pi)$ to $(0, \pi)$ every point on M_0^- is connected to its partner on M_0^+. There is therefore a problem in even constructing the singular orbit as it has to be decided which heteroclinic orbit is picked out. One can formulate a Melnikov function $\Gamma^+(\gamma, \tau_1, \tau_2)$ whose zeroes indicate the location of heteroclinics that exist for small values of ϵ from M_ϵ^- to M_ϵ^+. Another Melnikov function $\Gamma^-(\gamma, \tau_1, \tau_2)$ indicates the potential heteroclinic orbits from M_ϵ^+ back to M_ϵ^-. I assume that the function γ and the quantity δ are chosen so that the zero sets $\Gamma^- = 0$ and $\Gamma^+ = 0$ cross P_a transversely, for an example see [27]. The quantity δ is then fixed.

A singular orbit is constructed with ρ jumps as follows. If j is odd then $M_0^j = M_0^-$ and if j is even $M_0^j = M_0^+$. The singular orbit is then composed of the following pieces

$\mathcal{F}_0^j$ $1 \leq j \leq \rho$, being each the heteroclinic jump from M_0^j to M_0^{j+1} from a point (τ_1, τ_2) where $\Gamma^+ = 0$, if j is odd and from a point where $\Gamma^- = 0$ if j is even.

$\mathcal{S}_0^j$ $1 \leq j \leq \rho$ are pieces of P_a between end-points of $\mathcal{F}_0^j$ and the beginning point of $\mathcal{F}_0^{j+1}$.

The singular orbit is constructed as usual

$$\mathcal{H} = \cup_{i=1}^{\rho} \mathcal{F}_0^j \cup \mathcal{S}_0^j.$$

It is shown in [27] that there is a (real) homoclinic orbit nearby this singular orbit. The strategy is, by now, standard. One follows the unstable manifold of P_a around the phase space using the singular orbit as a template. The Melnikov calculation gives the required transversality upon entering each slow manifold and the exponentially small error in the Exchange Lemma Lemma 7 allows the passage of the unstable manifold near each slow manifold and sets the shooting manifold up for its next jump. The exponentially small error is needed as the transversality over the jumps is only $O(\epsilon)$. If the unstable manifold were only tracked up to $O(\epsilon)$ during its passage near the slow manifold, a transversal intersection with the next manifold could not be guaranteed. The details are given in [27]

6.4   Recent results and new directions

In this final section, I shall highlight some recent pieces of work that indicate directions in which the theory described above needs to be extended.

6.4.1   Orbits homoclinic to resonance

The problem of orbits homoclinic to resonance has been considered recently by many authors. The equations are as follows

$$
\begin{aligned}
\dot{x} &= JD_x H(x, I) + \delta f_1(x, I, \theta, \lambda) \\
\dot{I} &= \delta g_1(x, I, \theta, \lambda) \\
\dot{\theta} &= D_I H(x, I) + \delta g_2(x, I, \theta, \lambda),
\end{aligned}
\tag{6.11}
$$

where all the functions are smooth and periodic of period 2π in θ The resonance occurs at a point $(\hat{x}, \hat{I})$ which is a critical point of the $\epsilon = 0$ system, that is assumed to be a saddle point for the x equation. This is only a circle of critical points, but, as is typical in such resonance problems, can be blown up to expose some interesting structure. Setting $I = \hat{I} + \sqrt{\delta} h$, one arrives at the system, with $\epsilon = \sqrt{\delta}$

$$
\begin{aligned}
\dot{x} &= JD_x H(x, h, \epsilon) + \epsilon^2 f_1(x, h, \theta, \lambda, \epsilon) \\
\dot{h} &= \epsilon g_1(x, h, \theta, \lambda, \epsilon) \\
\dot{\theta} &= D_I H(x, \hat{I}) + \epsilon g_3(x, h, \theta, \lambda, \epsilon),
\end{aligned}
\tag{6.12}
$$

which has a manifold of critical points M_0 given by $x = \hat{x}$, this is now an annulus. The system (6.12) is not in the usual singular perturbation form due to the term $D_I H(x, \hat{I})$, but it does possess a manifold of critical points parametrized by θ and I. In fact, this is all that is used by the results given in these lectures and everything applies to (6.12) as if both h and θ were slow variables. The normal hyperbolicity of $M_0 \subset \{x = \hat{x}\}$ follows from the fact that $\hat{x}$ is a saddle of the x-equation.

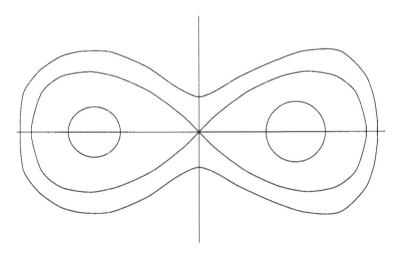

Figure 20
The scaled slow flow.

The original motivation for the study of (6.11) is a 2-mode truncation model
of the nonlinear Schrödinger equation. It is desired to find orbits that are observed in certain spatio-temporal chaotic patterns, see McLaughlin et al. [4].
The slow flow will have the form of a "bow-tie" as shown in Figure 20 and the
fast flow will be a "fish", see Figure 21. Of interest are orbits that are homoclinic to various different invariant sets in the slow flow, in particular: the
saddle, the center(s) and the periodic orbits surrounding the centers. This has
been treated by many authors using a combination of Hamiltonian techniques
and those of singular perturbation theory. In the earliest work, Kovacic and
Wiggins [35] found orbits connecting the center to itself involving one fast jump,
however they needed some negative damping to get the orbit. Mclaughlin et
al. [43] found saddle to saddle connections, with one fast jump in the case of
dissipative perturbations. Simultaneously, Kovačič also solved this problem [36].
For Hamiltonian perturbations, Kovačič [37] and Haller and Wiggins [22] found
periodic to periodic orbits, amongst others, with one fast jump. Haller and
Wiggins, in further work [23], found multi-jump orbits for the problem with
Hamiltonian perturbations. These are different, however, from those that one
would construct using an Exchange Lemma argument as the time spent near the
slow manifold is not great. Tin and Camassa [9] have found multi-jump orbits in
a problem that is very close to the above but in the non-resonant case. In some
ways, this case is harder as it is not a singular perturbation problem. This, and
related work [8], [55], have made an important contribution to the problem of
whether slow manifolds in atmospheric flows exist, see [40]. Two aspects of this
problem are worthy of mention in the current context. First, Tin [55] developed
a new strategy for proving transversality as the traditional Melnikov method was
not applicable. Second, an Exchange Lemma that works for perturbed invariant
manifolds in more general situations than found in singular perturbation theory

was needed. Tin [54] developed such an Exchange Lemma for passages near "non-slow" manifolds.

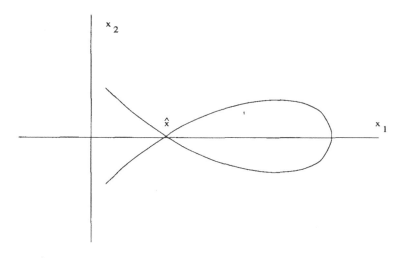

Figure 21
The fast flow.

In the most recent contribution to the orbits homoclinic to resonance problem, Kaper and Kovacic [32] have applied the Exchange Lemma to build on the single jump pulses and found multi-jump orbits homoclinic to all the available different invariant sets, for the details see [32].

The problems considered here only apply to an approximation to the real problem, namely the 2-mode truncation. The real problem is in infinite dimensions and this needs to be resolved. In particular, the Fenichel structure are not clearly understood in infinite dimensions. The Exchange Lemma also appears to be hard to generalize.There are many problems and issues in this area that, I believe, will receive much attention over the coming years.

6.4.2 Stability of travelling waves

Much of the motivation for the Exchange Lemma came from the need to compute directions of transversality for assessing the stability of travelling waves. As seen in many of the examples above, travelling waves are constructed as homoclinic orbits for particular systems of ODE's. These will be constructed then as intersections of stable and unstable manifolds of the relevant critical point. A parameter is obviously needed to make this situation robust. In most cases of travelling waves the parameter is supplied by the speed. It was shown by Evans [14] that the direction in which the unstable manifold crosses the stable manifold as the speed parameter varies renders some crucial information about the stability of the wave. Indeed, it determines the parity of the number of eigenvalues of the linearization of the PDE at the wave that lie in the right half-plane. Determining the nature of the transversal intersection in many of these travelling wave problems then becomes an important issue.

This direction is implicit in the Exchange Lemma analysis but has not, up to this point, been explicitly incorporated into the theory. In recent work, Bose and Jones [6] have studied the example of travelling pulses in coupled nerve fibers. Keener [34] formulated a model for this situation in terms of a pair of coupled FitzHugh-Nagumo equations. The fibers are coupled by reciprocal diffusive coupling. The PDE's are

$$
\begin{aligned}
u_{1t} &= u_{1xx} + f(u_1) - w_1 + d(u_2 - u_1) \\
w_{1t} &= \epsilon(u_1 - \gamma w_1) \\
u_{2t} &= u_{2xx} + f(u_2) - w_2 + d(u_1 - u_2) \\
w_{2t} &= \epsilon(u_2 - \gamma w_2).
\end{aligned}
\tag{6.13}
$$

It is clear that there will be an "in-phase" wave, when ϵ is sufficiently small, for which $u_2 = u_1$, which is just the individual FitzHugh-Nagumo pulse constructed above on each fiber. However, its stability is not obvious. Bose and Jones [6] have proved the stability using the appropriate generalization of Evans' idea due to Alexander, Gardner and Jones [1]. The Exchange Lemma is crucial in tracking the relevant unstable manifold and studying its intersection with the stable manifold. This contains much more information than is in the individual travelling pulses. However, the transversality given by the Exchange Lemma does not readily give the "direction" information required in applying the Evans stability idea. Bose [5] developed a way of keeping track of this sign information in the proof of the Exchange Lemma and this is a key ingredient in the stability analysis of these "in-phase" waves.

A general theory is needed for tracking the invariant manifolds in the Exchange Lemma including the "sign" of a relevant basis. I anticipate that the Exchange Lemma will have many further applications in the stability analysis of travelling waves. Indeed, the eigenvalue equations that occur in the stability analysis are ODE's living naturally in the tangent bundle to the travelling wave phase space. If the underlying problem is singularly perturbed then so are the eigenvalue equations. The eigenvalues occur at values of the eigenvalue parameter at which certain heteroclinic orbits exist. The structure of these orbits, and consequently the distribution of eigenvalues, promises to be significantly illuminated by direct application of the Exchange Lemma.

Bibliography

[1] J.Alexander, R.Gardner & C.Jones, A topological invariant arising in the stability analysis of travelling waves, *J. reine angew. Math.* 410 (1990) 167-212.

[2] D.Aronson & H.Weinberger, Nonlinear diffusion in population genetics, combustion and nerve propagation, in *Partial Differential Equations and Related Topics*, J.Goldstein ed., Lecture Notes in Mathematics, 446 (1975), Springer-Verlag, New York.

[3] P.Bates & C.Jones, Invariant manifolds for semilinear partial differential equations, *Dynamics Reported vol.2*, Wiley (1989), 1-38.

[4] A.R.Bishop, M.G.Forest, D.W.McLaughlin and E.A.Overman II, Correlations between chaos in a perturbed Sine-Gordon equation and a truncated model system, *SIAM J. Math. Anal.* (1990) 1511-1536.

[5] A.Bose, Existence and stability of travelling waves for coupled nerve axon equations, Ph.D. thesis, Brown U., 1993.

[6] A.Bose & C.Jones, Stability of the in-phase travelling wave solution in a pair of coupled nerve fibers, preprint (1994).

[7] G.Caginalp & P.Fife, Higher order phase field models and detailed anisotropy, *Phys. Rev. B*, 34 (1986) 4940-4943.

[8] R.Camassa, On the geometry of a slow manifold, preprint (1994).

[9] R.Camassa & S.-K.Tin, The global geometry of the slow manifold in Lorenz-Krishnamurthy model, preprint (1994).

[10] G.Carpenter, A geometric approach to singular perturbation problems with applications to nerve impulse equations, *JDE* 23 (1977) 335-367.

[11] K.W.Chang & F.A.Howes, *Nonlinear Singular Perturbation Phenomena: Theory and Applications*, Springer-Verlag, New York, 1984.

[12] B.Deng, The existence of infinitely many traveling front and back waves in the FitzHugh-Nagumo equations, *SIAM J. Math. Anal.* 6 (1991).

[13] N.Ercolani, D.McLaughlin & H.Roitner, Attractors and transients for a perturbed perioidc KdV equation: a nonlinear spectral analysis, *J. Nonlin. Sci.* in press.

[14] J.W.Evans, Nerve impulse equations I-IV, *Indiana Univ. Math. Journal*, 21,22,24 (1972-5).

[15] N.Fenichel, Persistence and smoothness of invariant manifolds for flows, *Indiana Univ. Math. Journal*, 21 (1971) 193-226.

[16] N.Fenichel, Asymptotic stability with rate conditions, *Indiana Univ. Math. Journal*, 23 (1974) 1109-1137.

[17] N.Fenichel, Asymptotic stability with rate conditions II, *Indiana Univ. Math. Journal*, 26 (1977) 81-93.

[18] N.Fenichel, Geometric singular perturbation theory for ordinary differential equations, *J. Diff. Eq.* 31 (1979) 53-98.

[19] R.Gardner & C.Jones, Traveling waves of a perturbed diffusion equation arising in a pahse field model, *Indiana U. Math. J.* 38 (1989) 1197-1222.

[20] J.Guckenheimer & P.Holmes, *Nonlinear Oscillations, Dynamical Systems, and Bifurcations of Vector Fields*, Springer-Verlag, New York, 1983.

[21] J.Hadamard, Sur l'iteration et les solutions asymptotiques des equations differentielles, *Bull. Soc. Math. France* 29 (1901) 224-228.

[22] G.Haller & S.Wiggins, Orbits homoclinic to resonances: the Hamiltonian case, *Physica D* 66 (1993) 298-346.

[23] G.Haller & S.Wiggins, N-pulse homoclinic orbits in perturbations of resonant Hamiltonian systems, to appear in *Arch. Rat. Mech. Anal.*(1995).

[24] S.Hastings, On travelling wave solutions of the Hodgkin-Huxley equations, *Arch. Rat. Mech. Anal.* 60 (1976) 229-257.

[25] M.Hirsch, C.Pugh & M.Shub, *Invariant Manifolds*, Lecture Notes in Mathematics, 583 (1977), Springer-Verlag, New York.

[26] C.Jones, Stability of the travelling pulse of the FitzHugh-Nagumo system, *Trans. AMS* 286 (1984) 431-469.

[27] C.Jones, T.Kaper & N.Kopell, Tracking invaraint manifolds up to exponentially small errors, *SIAM J. Math. Anal.*, to appear.

[28] C.Jones & N.Kopell, Tracking invariant manifolds with differential forms in singularly perturbed systems, *J. Diff. Eq.*, in press.

[29] C.Jones, N.Kopell & R.Langer, Construction of the FitzHugh- Nagumo pulse using differential forms, in *Patterns and Dynamics in Reactive Media*, H.Swinney, G.Aris, and D.Aronson eds., IMA Volumes in Mathematics and its Applications, 37 (1991), Springer-Verlag, New York.

[30] C.Jones & S.-K.Tin, On the dynamics of tangent spaces near a normally hyperbolic manifold, in preparation.

[31] C.Jones, J.Rubin & M.Maxey, Settling and asymptotic motion of aerosol particles in a cellular flow field, preprint (1994).

[32] T.J.Kaper & G.Kovačič, Multi-bump orbits homoclinic to resonance bands, preprint (1993).

[33] T.J.Kaper & S.Wiggins, On the structure of separatrix swept regions in singularly perturbed Hamiltonian systems, *J. Diff. and Int. Eq.* 5 (1992) 1363- 1381.

[34] J.Keener, Frequency decoupling of parallel excitable fibers, *SIAM J. Appl. Math.* 49 (1989) 211-230.

[35] G.Kovačič & S.Wiggins, Orbits homoclinic to resonance with an application to chaos in a model of the forced damped sine-Gordon equation, *Physica D* 57 (1992) 185-225.

[36] G.Kovačič, Singular perturbation theory for homoclinic orbits in a class of near-integrable dissipative systems, to appear in *SIAM J. Math. Anal.*

[37] G.Kovačič, Singular perturbation theory for homoclinic orbits in a class of near-integrable Hamiltonian systems, *J. Dyn. Diff. Eq.* 5 (1993) 559-597.

[38] R.Langer, Existence and Uniqueness of Pulse Solutions to the FitzHugh-Nagumo Equations, Ph.D. Thesis, Northeastern University, 1980.

[39] M.Levi, *Qualitative Analysis of the periodically forced relaxation oscillations*, Vol.244 in Memoirs AMS, AMS, Providence, RI, 1981.

[40] E.N.Lorenz, The slow manifold-What is it?, *J. Atmos. Sci.* 49 (1992) 2449-2451.

[41] W.S.Massey, *Singular Homology Theory*, Graduate Texts in Mathematics 70, Springer-Verlag, New York, 1980.

[42] R.McGehee, The stable manifold theorem via an isolating block, Symposium on ODE, W.Harris & Y.Sibuya eds., Springer-Verlag,New York (1973) 135-44.

[43] D.W.McLaughlin, E.A.Overman II, S.Wiggins & C.Xiong, Homoclinic orbits in a four dimensional model of a perturbed NLS equation: a geometric singular perturbation study, to appear in *Dynamics Reported*.

[44] E.F.Mischenko & N.Rozov, *Differential Equations with small parameters and realaxtion oscillations*, Plenum Press, New York, 1980.

[45] Y.Nishiura & H.Fujii, Stability of singularly perturbed solutions to a system of reaction-diffusion equations, *SIAM J. Math. Anal.*, 18 (1987) 1726- 1770.

[46] Y.Nishiura & M.Mimura, Layer oscillations in reaction-diffusion systems, *SIAM J. Appl. Math.* 49 (1989) 481-514.

[47] T.Ogawa, Travelling wave solutions to perturbed Korteweg-de Vries equations, preprint (1993), to appear in Hiroshima J. Math..

[48] O.Perron, DieStabilitätsfrage bei Diffferentialgleichungsysteme, *Math. Zeit.* 32 (1930) 703-728.

[49] D.Ruelle, *Elements of Diffrentiable Dynamics and Bifurcation Theory*, Academic Press, San Diego, 1989.

[50] Yu.Rzhanov, H.Richardson, A.Hagberg & J.Moloney, Spatio- temporal oscillations in a semiconductor etalon, preprint (1992).

[51] K.Sakamoto, Invariant manifolds in singular perturbation problems for ordinary differential equations, *Proc. Roy. Soc. Ed.*, 116A (1990) 45-78.

[52] H.Stommel, Trajectories of small bodies sinking slowly through convection cells, *J. Mar. Res.*, 8 (1949) 24-29.

[53] P.Szmolyan, Analysis of a singularly perturbed travelling wave problem, *SIAM J. Appl. Math.* (1992)

[54] S.-K.Tin, On the dynamics of tangent spaces near a normally hyperbolic manifold, Ph.D. Thesis, Brown University, 1994.

[55] S.-K.Tin, Transversality of double-pulse homoclinic orbits in some atmospheric equations, preprint (1994).

[56] J.Topper & T.Kawahara, Approximate equations for long nonlinear waves on a viscous fluid, *J. Phys. Soc. Japan* 44 (1978) 663-666.

[57] S.Wiggins, *Global Bifurcations and Chaos*, Springer-Verlag, New York, 1988.

[58] S.Wiggins, *Normally Hyperbolic Invariant Manifolds in Dynamical Systems*, Springer-Verlag, New York, 1994.

Conley Index Theory

Konstantin Mischaikow*
Center for Dynamical Systems and Nonlinear Studies
Georgia Institute of Technology
Atlanta. GA 30332
mischaik@math.edu.gatech

Contents

*Research was supported in part by a grant from the NSF.

Chapter 1

Introduction

Perhaps the best way to begin these set of notes is with a description of what they are and are not meant to be. The standard reference for Charles Conley's view of dynamical systems is his monograph, "Isolated Invariant Sets and the Morse index" [4]. However, in the intervening 15 years considerable progress and developments have occured. These results are scattered throughout the literature, and no single text covers all this material; this article won't either. Rather, the goal is to provide the reader with an intuition of and motivation for some of these recent developments. It is hoped that at the very least these notes will aid the reader in understanding the literature. We begin therefore, with three sections which cover the by now classical aspects of the theory: the Conley index, decompositions of invariant sets, and Conley's connection matrix. Since most of these topics have appear in expository form, these sections make no pretense of providing a detailed or complete description of these results. Rather, they are presented with the goal of providing a common notation and a self contained set of definitions to be used in the later chapters.

To better describe the philosophy which these notes are intended to convey it is best to become somewhat more technical. Recall that a *flow* $\varphi : \mathbf{R} \times X \to X$ on a locally compact metric space X with metric $\mathbf{d}$ is a continuous map satisfying:

$$\varphi(0, x) = x$$
$$\varphi(t, \varphi(s, x)) = \varphi(t + s, x)$$

A set S is an *invariant set* for the flow φ if

$$\varphi(\mathbf{R}, S) := \bigcup_{t \in \mathbf{R}} \varphi(t, S) = S.$$

The objects of primary interest in Conley's approach to dynamical systems are isolating neighborhoods and their associated invariant sets.

Definition 1.0.1 An *isolating neighborhood* is a compact set N such that its maximal invariant set lies in its interior, i.e.

$$\text{Inv}(N, \varphi) := \{x \in N \mid \varphi(\mathbf{R}, x) \subset N\} \subset \text{int}(N).$$

S is an *isolated invariant set* if $S = \mathrm{Inv}(N)$ for some isolating neighborhood N.

Why the emphasis on isolating neighborhoods? As is well known by now invariant sets, one of the prime objects of interest in dynamical systems, are notoriously fickle with respect to perturbation. In particular, they can disappear (saddle-node bifurcations), change their topological type (heteroclinic bifurcations), and change their stability (Hopf bifurcation). In contrast, isolating neighborhoods are *robust*, i.e. if N an isolating neighborhood for a particular flow φ, then it is an isolating neighborhood for all nearby flows ψ. This is easily seen by first observing that assuming N is an isolating neighborhood for φ is equivalent to assuming that if $x \in \partial N$, then there exists $t_x \in \mathbf{R}$ such that $\varphi(t_x, x) \notin N$. By compactness, there exists $T > 0$ such that $|t_x| < T$ for all $x \in \partial N$. Now by continuity, for ψ sufficiently close (in the C^0 topology) to φ, $\psi([-T, T], x) \not\subset N$ for all $x \in \partial N$. Furthermore, isolating neighborhoods are abundant. In fact, it can be shown [55] that they are prevalent [25].

Thus isolating neighborhoods are nice objects; one expects to be able to find them and once found they have a tendency to stay put. Unfortunately, as is remarked above, the object of ultimate interest is really the invariant set. Thus, the fundamental problem with Conley's approach is:

N *is computable, but* $\mathrm{Inv}(N)$ *is the object of interest.*

The goal of these notes is to indicate that via the Conley index, knowledge concerning N can be translated into information about the dynamics of $\mathrm{Inv}(N)$.

Since the Conley index plays such an important role in our analysis we spend all of Chapter 2 defining it and describing it properties. For the moment, it is enough just to mention its three most important qualities.

1. It is an index of isolating neighborhoods with the following property. If N and N' are isolating neighborhoods and $\mathrm{Inv}\, N = \mathrm{Inv}\, N'$, then the Conley index of N is the same as the Conley index of N'. Observe, that this implies that we can also consider the Conley index as an index of isolated invariant sets. We adopt the first point of view for applications, but the latter when making definitions.

2. If the Conley index of N is not trivial, then $\mathrm{Inv}\, N \neq \emptyset$.

3. If N is an isolating neighborhood for a continuous parameterized family of flows φ^λ, $\lambda \in [0, 1]$, that is

$$\mathrm{Inv}(N, \varphi^\lambda) \subset \mathrm{int}\, N \quad \text{for } \lambda \in [0, 1]$$

then the Conley index of N under φ^0 is the same as the Conley index of N under φ^1.

The reader who is familiar with degree theory may notice the similarities shared by the Conley index. Recall that the degree of a map is defined in terms of a neighborhood such that the map has no fixed points on the boundary. Furthermore, if this property is satisfied for a continuous family of maps then the degree of these maps are the same. And finally, if the degree is not zero, then the set of fixed points is nonempty. We emphasize this similarity, because fixed point theory is the single most successful tool in nonlinear analysis. Our goal is to show that the Conley index theory has the same potential for problems in global nonlinear dynamics. With this in mind we make the simple observation that degree theory is used to describe the existence of fixed points. Thus, we should strive to use the Conley index theory to describe isolated invariant sets. However, nonlinear systems exhibit extremely complicated and wide variety of isolated invariant sets. Therefore, the first point to consider is what should be meant by describing isolated invariants sets.

The first, and simplest, possibility would be to directly mimic degree theory and state that the goal is to prove the existence of specific invariant sets such as fixed points or periodic orbits. As will be discussed in Chapters 2 and 5, this can be done. In particular, in Chapter 5 we shall prove a theorem to the effect that given the appropriate hypothesis on the Conley index and the existence of a Poincaré section for the isolating neighborhood, then the isolated invariant set contains a periodic orbit.

The second, and perhaps more interesting approach, is to try to describe the structure of the isolated invariant set as a whole. A generally accepted means of describing these structures is to prove the existence of a conjugacy with respect to a well understood or more easily computable system. To be more precise, two flows $\varphi : \mathbf{R} \times X \to X$ and $\psi : \mathbf{R} \times Y \to Y$ are said to be conjugate if there exists a homeomorphism $h : Y \to Y$ which carries orbits of φ to orbits of ψ while preserving the direction of time. Another way of stating this property is to assume the existence of a flow $\widetilde{\varphi}$ obtained from a time reparameterization of φ such that the following diagram commutes

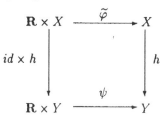

In the setting of discrete dynamics, the classical example is Smale's homeomorphism from the invariant set of the horseshoe map to the shift dynamics on bi-infinite sequences of two symbols. While from a qualitative and theoretical point of view such a description is quite desirable, in practice there are at least two serious drawbacks to this approach. First, a given system of, say, ordinary differential equations, obtaining a rigorous proof of the desired conjugacy can be extremely difficult. The Lorenz equations are an excellent example of this. Most of our mathematical understand of the Lorenz system comes from analysis of geometric Lorenz systems. However, a rigorous relationship between the Lorenz equations and the geometric model has, as of yet, not been obtained. Second, given a conjugacy for a fixed parameter value determining the range for which the conjugacy is valid or how the conjugacy changes as a function of the parameter is an equally daunting task. To see this, one needs only note that there exist conjugacies between structurally stable and non-structurally stable flows, e.g., $\dot{x} = -x$ and $\dot{x} = -x^3$. These drawbacks can be a serious handicap. For many models arising from the sciences and engineering, not only are the parameters not known to great precision, but often the class of potential non-linearities is large. Therefore one is faced with the task of providing a rigorous description of the dynamics for classes of equations for which one expects (or for which the numerics indicates) a wide range of behaviors.

These difficulties suggest that we weaken our concept of describe. Returning to the theme of degree theory, it is worth remarking that the question degree theory is intended to answer is, does there exist a fixed point; not how many fixed points exist or is there a unique fixed point. These latter questions are almost always resolved by different techniques. Thus, with degree theory one seeks a lower bound (one) on the number of fixed points. The analogy for dynamical systems which we shall adopt is that of a *semi–conjugacy*. This means given an invariant set S and its dynamics $\varphi : \mathbf{R} \times S \to S$, choosing a well understood flow $\psi : \mathbf{R} \times Y \to Y$ and showing that there exists a continuous *surjection* $\rho : S \to Y$ such that the following diagram commutes

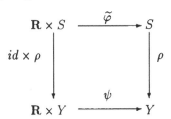

where once again $\widetilde{\varphi}$ represents a time reparameterization of φ. In Chapters 3, 6, and 7, examples of semi–conjugacies will be presented.

Before leaving this topic, however, let us remark on what information can be obtained from the existence of a semi–conjugacy. Observe that for any $y \in Y$ the pre-image of its trajectory $\rho^{-1}(\psi(\mathbf{R}, y))$ corresponds to a union of trajectories in S under φ. A more global statement is that the entropy of φ on S is greater than or equal to the entropy of ψ. Thus, the dynamics of ψ on Y acts as a lower bound on the dynamics of S, the system of interest. At the same time it must, also, be pointed out that orbits in the pre–image need not have the same qualitative behavior as those of the model flow. As an example consider the projection of an irrational flow on a torus onto a periodic orbit. Of course, with additional hypothesis it is possible to assure that the qualitative dynamics of the model orbits corresponds to that of its pre-images. On the most basic level, the compactness of S implies that recurrent dynamics in the model flow lifts to reccurent dynamics on S.

As has been stated repeatedly, isolated invariant sets can be extremely complicated objects, thus it is reasonable to ask whether they posses a natural decomposition which preserves the perturbation properties of the corresponding isolating neighborhoods. This leads to Conley's idea of a *Morse decomposition* of an invariant set S, which is denoted by

$$\mathcal{M}(S) = \{M(p) \mid p \in \mathcal{P}\}.$$

A Morse decomposition consists of at most a finite number of disjoint compact invariant subsets of S which contain the chain recurrent set of S. These individual isolated invariant subsets $M(p)$ are called *Morse sets*, and the remaining portion $S \setminus \bigcup M(p)$ is referred to as the set of connecting orbits. A complete description of Morse decompositions can be found in Chapter 3.

As a decomposition $\mathcal{M}(S)$ is very robust with regards to perturbations. This does not mean that the Morse sets as sets are preserved under perturbation, but rather that the decomposition and the associated isolating neighborhoods are preserved. Given this decomposition, providing a description of the invariant set S is reduced to the following two problems.

- *Describe the structure of the Morse sets.*

- *Describe the structure of the set of connecting orbits.*

In this setting it is natural to view the first as a "local" question and the second as a "global" question.

Since Morse sets are isolated invariant sets, one can compute their Conley index and then hope to use this index information to describe their structure. This will be done in Chapter 7 and shows the potential for the further development of techniques for answering the local question. To describe the global structures requires being able to compare the Conley indices for Morse sets with the Conley index for the global invariant set S. This is the purpose of the connection matrix which is discussed in Chapter 4, which is then applied to an example in Chapter 6.

Throughout this discussion, it has been assumed that isolating neighborhoods are known and that their corresponding Conley indices can be computed. This is a rather significant assumption, since this is exactly the point where the nonlinear analysis needs to be performed. Therefore, it is essential that we have techniques for finding isolating neighborhoods and then for computing their indices. Though out of the scope of these lectures, it should be remarked that it is possible to perform these computations using the computer and even to obtain rigorous results in this manner [38, 13]. What will be discussed in the final Chapter is a method for constructing isolating neighborhoods and determining their indices in the setting of singular perturbations. These techniques for computing the indices are new, and hence, it will take time to see if they can successfully be applied to interesting problems.

Chapter 2

The Conley Index

This chapter provides a brief introduction to the Conley index both for flows and maps. As was mentioned in the introduction the goal is to provide an intuitive understanding of the index rather than a rigorous development. We begin with a brief collection of preliminary definitions from topology and algebra. Next is a discussion of the index in the setting of flows on locally compact spaces. For a complete description of this topic the reader is referred to [4, 50, 52]. This is followed by a look at the index for homeomorphisms. Here the best reference is [41]

2.1 Some Preliminaries

Recall that a *pointed space* (Y, y_0) is a topological space Y with a distinguished point y_0. Given a pair (N, L) of spaces with $L \subset N$,

$$(N/L, [L]) := (N \setminus L) \cup [L]$$

where $[L]$ denotes the equivalence class: $x \sim y$ if an only if $x, y \in L$. The topology on $(N/L, [L])$ is defined as follows: a set $U \subset N \setminus L$ is open if U is open in N and $U \cap L = \emptyset$, or the set $(U \cap (N \setminus L)) \cup L$ is open in N. If $L = \emptyset$, then

$$(N/L, [L]) := (N \cup \{*\}, \{*\})$$

where $*$ denotes the equivalence class consisting of the empty set.

Let (X, x_0) and (Y, y_0) be pointed topological spaces and let

$$f, g : (X, x_0) \to (Y, y_0)$$

be continuous functions. Implicit in this notation is the assumption that $f(x_0) = g(x_0) = y_0$.

Definition 2.1.1 f is *homotopic* to g

$$f \sim g$$

if there exists a continuous function $F : X \times [0,1] \to Y$ such that

$$
\begin{aligned}
F(x,0) &= f(x) \\
F(x,1) &= g(x) \\
F(x_0, s) &= y_0, \quad 0 \le s \le 1.
\end{aligned}
$$

Definition 2.1.2 Two pointed topological spaces (X, x_0) and (Y, y_0) are *homotopic*

$$(X, x_0) \sim (Y, y_0)$$

if there exists $f : (X, x_0) \to (Y, y_0)$ and $g : (Y, y_0) \to (X, x_0)$ such that

$$f \circ g \sim \mathrm{id}_Y \quad \text{and} \quad g \circ f \sim \mathrm{id}_X$$

Observe that homotopy defines an equivalence class on the set of topological spaces. The reader for whom homotopy is a new concept may wish to check that $\mathbf{R}^2 \setminus \{0\} \sim S^1$.

We shall also need to make use of direct limits, but only in a very simple context which we present here. Let G denote an abelian group and $f : G \to G$ a group homomorphism. This can be viewed as direct system of abelian groups and homomorphisms by setting $G_n = G$ and $f_n = f$ for $n = 0, 1, 2, \dots$. With this notation we obtain the sequence

$$G_0 \xrightarrow{f_0} G_1 \xrightarrow{f_1} G_2 \xrightarrow{f_2} G_3 \dots$$

Let $\widetilde{G} := \prod_{n=0}^{\infty} G_n$ and for $m > n$ define an equivalence relation $g_n \sim g_m$ if $f^{m-n}(g_n) = g_m$ where $g_n \in G_n$. The *direct limit* is the resulting quotient space, denoted by

$$\varinjlim G := \widetilde{G}/\sim,$$

along with the group structure

$$[g_n] + [g_m] = [f^{m-n}(g_n) + g_m]$$

where, again, it is assumed that $m > n$. Observe that an element of $\widetilde{G}$ can be written in the form (n, g). It is left to the reader to check that the map

$$(n, g) \to (n, f(g))$$

induces an automorphism on the direct limit $\varinjlim G$. The reader is referred to [27] for a more complete discussion of direct limits. To simplify the notation we will usually write

$$
\begin{aligned}
\mathcal{L}(G) &:= \varinjlim G \\
\mathcal{L}(f) &:= \varinjlim f
\end{aligned}
$$

A final comment; readers familiar with homotopy theory are, of course, well aware of the fact that determining equivalance classes up to homotopy is extremely difficult. Therefore, we shall make use of algebraic topology, in particular the Alexander–Spanier cohomology [27, 53], to distinguish between different homotopy classes. Lack of familiarity with this particular cohomology theory is no reason for concern. It is used for two purely technical reasons. The first is that the supports of Alexander–Spanier cochains are particularly easy to work with and the second is the weak continuity property stated in the following theorem (see [53, Theorem 6.6.6]).

Theorem 2.1.3 *Let* $\{(N_\alpha, L_\alpha)\}$ *be a family of compact Hausdorff pairs directed downwards by inclusion, and let* $(N, L) = (\cap N_\alpha, \cap L_\alpha)$. *The inclusion maps* $i_\alpha : (N, L) \to (N_\alpha, L_\alpha)$ *induce an isomphism*

$$i^* : \varinjlim H^*(N_\alpha, L_\alpha) \to H^*(N, L).$$

2.2 Conley Index for Flows

Throughout these notes X will denote a locally compact metric space with metric d and $\varphi : \mathbf{R} \times X \to X$ a continuous flow on X. Let $N \subset X$. $L \subset N$ is *positively invariant* in N if given $x \in L$ and $\varphi([0, t], x) \subset N$, then $\varphi([0, t], x) \subset L$. L is an *exit set* for N if given $x \in N$ and $t_1 > 0$ such that $\varphi(t_1, x) \notin N$, then there exists $t_0 \in [0, t_1]$ for which $\varphi([0, t_0], x) \subset N$ and $\varphi(t_0, x) \in L$.

Definition 2.2.1 Let S be an isolated invariant set. A pair of compact sets (N, L) where $L \subset N$ is called an *index pair* for S if:

1. $S = \text{Inv}(\text{cl}(N \setminus L))$ and $N \setminus L$ is a neighborhood of S;

2. L is positively invariant in N;

3. L is an exit set for N.

Definition 2.2.2 The *homotopy Conley index* of S is

$$h(S) = h(S, \varphi) \sim (N/L, [L]).$$

The index has been defined in terms of an isolated invariant set, but it can be extended to an index of isolating neighborhoods as follows. Let N be an isolating neighborhood. The Conley index of N is defined to be

$$h(N) = h(N, \varphi) \sim h(\text{Inv}(N, \varphi)).$$

Observe from the definition that the homotopy Conley index is the homotopy type of a topological space. Unfortunately, as was mentioned earlier working with homotopy classes of spaces is extremely difficult. To get around this, it is useful to consider the cohomological Conley index defined by

$$CH^*(S) := H^*(N/L, [L]) \approx H^*(N, L)$$

where H^* denotes the Alexander–Spanier cohomology [27, 53] with integer coefficients. .

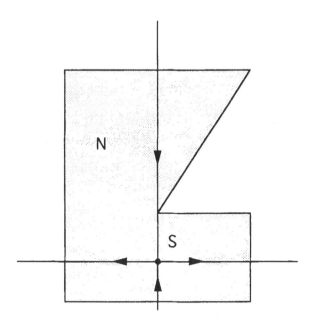

Figure 2.1: An isolating neighborhood which cannot be used in an index pair.

Theorem 2.2.3 (The Conley index is well defined) *Let* (N, L) *and* (N', L') *be index pairs for an isolated invariant set* S. *Then*

$$(N/L, [L]) \sim (N'/L', [L']).$$

Since the cohomology of two homotopic spaces is the same, the cohomology index is well defined.

Theorem 2.2.4 (The Conley index exists) *Let* S *be an isolated invariant set. Then there exists an index pair for* S.

Remark 2.2.5 With regard to index pairs an important point is the following. Let N be an isolating neighborhood for S. While Theorem 2.2.4 guarantees that S has an index pair it does not guarantee that the index pair will take the form (N, L). It is left to the reader to check that Figure 2.1 provides an example of such an isolating neighborhood N.

Let us now consider some examples of the Conley index.

Example 2.2.6 Observe that the empty set $\emptyset$ is vacuously an isolated invariant set. Furthermore, $(\emptyset, \emptyset)$ is an index pair for the empty set. Thus

$$CH^*(\emptyset) \approx 0$$

The converse of this example though trivial has significant enough implications to be designated as a theorem.

Theorem 2.2.7 (Ważewski Property) *Let N be an isolating neighborhood and assume that*

$$CH^*(N) \not\approx 0.$$

Then,

$$\text{Inv } N \neq \emptyset.$$

This result provides the simplest example of an existence result which can be obtained via the Conley index. It also demonstrates an important point concerning the way one wishes to view the Conley index. The computation of the index was done with knowledge of the invariant set. The theorem of interest was stated in terms of the isolating neighborhood.

To continue building our intuition of the Conley index consider the following series of simple examples.

Example 2.2.8 Let S be an attracting hyperbolic fixed point as in Figure 2.2. In this case N is an isolating neighborhood of S. Observe that every point on the boundary of N enters N immediately. Thus, the exit set is the empty set, which implies that $(N, \emptyset)$ is an index pair for S. Since $N \sim \{point\}$

$$\begin{aligned} CH^k(S) &\approx H^k(N, \emptyset) \\ &\approx H^k(N) \\ &\approx H^k(\{\text{point}\}) \\ &\approx \begin{cases} \mathbf{Z} & \text{if } k = 0 \\ 0 & \text{otherwise.} \end{cases} \end{aligned}$$

Example 2.2.9 Let S be a hyperbolic fixed point with a one dimensional unstable manifold as in Figure 2.3. The pair (N, L) is an index pair for S and hence can be used to compute the Conley index. In Figure 2.4 it is explicitly shown that $h(S) \sim (S^1, *)$. Hence,

$$\begin{aligned} CH^k(S) &\approx H^k(S^1, *) \\ &\approx \begin{cases} \mathbf{Z} & \text{if } k = 1 \\ 0 & \text{otherwise.} \end{cases} \end{aligned}$$

The following theorem, which follows from the Morse lemma [24, 33], generalizes the remarks of the last two examples.

Theorem 2.2.10 *Let S be a hyperbolic fixed point with an unstable manifold of dimension n. Then*

$$CH^k(S) \approx \begin{cases} \mathbf{Z} & \text{if } k = n \\ 0 & \text{otherwise} \end{cases}$$

Returning to one of the themes of these notes, we now ask the question what can one conclude from knowing that the Conley index of an isolating neighborhood is that of a hyperbolic fixed point. The ideal conclusion would be the following. Assume N is an isolating neighborhood and that

$$CH^k(N) \approx \begin{cases} \mathbf{Z} & \text{if } k = n \\ 0 & \text{otherwise,} \end{cases}$$

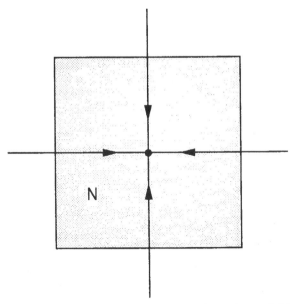

Figure 2.2: An attracting fixed point and isolating neighborhood

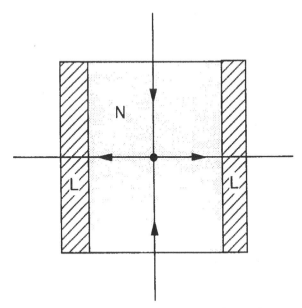

Figure 2.3: A saddle with an index pair

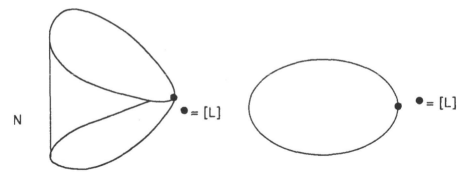

Figure 2.4: Computing the homotopy Conley index of a saddle

then Inv N contains a hyperbolic fixed point with an unstable manifold of dimension n. A result like this would provide information about both existence and stability. Unfortunately, as the simple example in Figure 2.5 indicates, such a theorem is too ambitious. However, the following existence result due to C. McCord [28] holds.

Theorem 2.2.11 *If*
$$CH^*(\text{Inv}(N)) \approx \mathbf{Z}$$
then $\text{Inv}(N)$ *contains a fixed point.*

The proof of this theorem follows from the Lefschetz fixed point theorem. While this theorem is probably of limited interest in terms of applications (in almost all cases one of the various fixed point indices will be more generally applicable), it does point out the possibility of using the index theory to obtain information concerning the structure of the dynamics of the invariant set from the isolating neighborhood.

Example 2.2.12 In Figure 2.6 the isolated invariant set S consists of a periodic orbit with a two dimensional unstable manifold. In Figure 2.7 it is shown how one can homotope the Conley index to the wedge of a two sphere and a circle. Thus
$$CH^k(S) \approx \begin{cases} \mathbf{Z} & \text{if } k = 1,2 \\ 0 & \text{otherwise.} \end{cases}$$

The Thom isomorphism theorem [3, 34, 53] provides a general means of computing the Conley index for normally hyperbolic invariant sets.

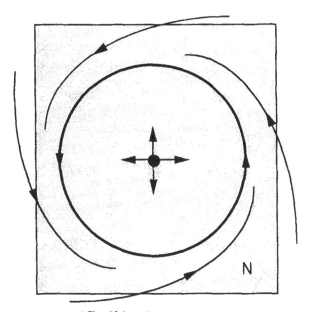

Figure 2.5: $CH^k(N) \approx \begin{cases} \mathbf{Z} & \text{if } k = 0 \\ 0 & \text{otherwise} \end{cases}$ but the only fixed point has a two dimensional unstable manifold.

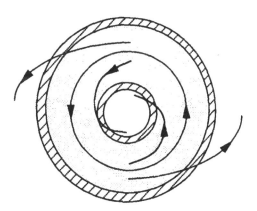

Figure 2.6: An unstable periodic orbit.

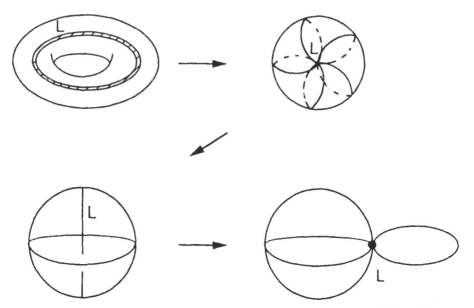

Figure 2.7: Computing the homotopy Conley index of a periodic orbit.

Theorem 2.2.13 *Assume that a manifold S is a normally hyperbolic invariant set. Let E be the vector bundle over S defined by the local unstable manifold of S. If E is a rank n orientable bundle, then*

$$CH^k(S) \approx H^{k+n}(S).$$

As a corollary we obtain the following useful result which generalizes Example 2.2.12.

Corollary 2.2.14 *Let S be a hyperbolic periodic orbit with an oriented unstable manifold of dimension $n + 1$. Then*

$$CH^k(S) \approx \begin{cases} \mathbf{Z} & \text{if } k = n, n+1 \\ 0 & \text{otherwise.} \end{cases}$$

Determining the orientation of an invariant manifold which is defined by a differential equation is often extremely difficult. For this reason it is often useful to use $\mathbf{Z}_2$ coefficients. In particular, the Thom isomorphism theorem implies the following theorem.

Theorem 2.2.15 *Assume that a manifold S is a normally hyperbolic invariant set. Let E be the vector bundle over S defined by the local unstable manifold of S. If E is a rank n bundle, then*

$$CH^k(S; \mathbf{Z}_2) \approx H^{k+n}(S, \mathbf{Z}_2).$$

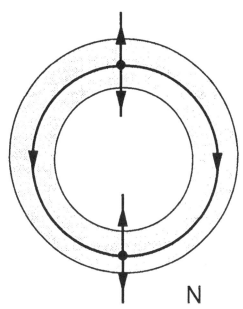

N

Figure 2.8: N has the Conley index of a periodic orbit but $\text{Inv}\, N$ contains no periodic orbit.

The reader might wish to compute the homotopy Conley index for a hyperbolic periodic orbit where the unstable manifold is two dimensional and locally forms a Möbius band.

Example 2.2.16 Given Theorem 2.2.11 it is reasonable to ask whether an equivalent theorem holds for periodic orbits. The answer is no as can be seen from Figure 2.8. However, with the additional assumption of the existence of a Poincaré section a positive answer can be given and is the subject of Section 5. This example points out the fact that knowledge of the index alone is seldom sufficient to make conclusions concerning the structure of the dynamics on the isolated invariant set. As has been repeated often, this is one of the major difficulties one faces when attempting to apply the Conley index theory.

The following theorem is fundamental to many of the most significant applications of the index theory to date.

Theorem 2.2.17 (Summation Property) *Assume $S = S_0 \cup S_1$ is an isolated invariant set where S_0 and S_1 are disjoint invariant sets. Then*

$$CH^*(S) \approx CH^*(S_0) \oplus CH^*(S_1).$$

Proof. Since S_0 and S_1 are disjoint invariant sets, there exist disjoint isolating neighborhoods N_0 and N_1 such that (N_0, L_0) and (N_1, L_1) are index pairs for S_0 and S_1 respectively. Thus,

$$CH^*(S) \;=\; H^*(N_0 \cup N_1, L_0 \cup L_1)$$

$$= H^*(N_0, L_0) \oplus H^*(N_1, L_1)$$
$$= CH^*(S_0) \oplus CH^*(S_1).$$

■

As will become clear in the later sections, in applications it is the converse of Theorem 2.2.17 which is of greatest use.

2.3 Continuation of the Conley Index

One of the most important properties of the Conley index is that it remains constant under appropriate homotopies of the flow. To be more precise let $\varphi^\lambda : \mathbf{R} \times X \to X$, $\lambda \in \Lambda$, be a continuously parameterized family of flows where the parameter space Λ is a compact, locally contractible, connected, metric space. Let N be an isolating neighborhood for the flow φ^{λ_0}. The proof of the following proposition was sketched in the introduction.

Proposition 2.3.1 *There exists $\epsilon > 0$ such that if $\mathbf{d}(\lambda, \lambda_0) < \epsilon$ then N is an isolating neighborhood for φ^λ.*

Unfortunately, the corresponding statement for index pairs need not be true. Thus, it is nontrivial to prove the following theorem which guarantees that the Conley index remains constant under perturbation.

Theorem 2.3.2 *Let N be an isolating neighborhood for φ^{λ_0}. Choose $\epsilon > 0$ such that if $\mathbf{d}(\lambda, \lambda_0) < \epsilon$, then N is an isolating neighborhood for φ^λ. Then*

$$CH^*(N, \varphi^\lambda) \approx CH^*(N, \varphi^{\lambda_0}).$$

To go beyond the level of perturbation it is useful to introduce the following notation. The *parameterized flow* corresponding to the family φ^λ is the continuous flow

$$\begin{aligned} \Phi : \mathbf{R} \times X \times \Lambda &\to X \times \Lambda \\ (t, x, \lambda) &\mapsto \Phi(t, x, \lambda) := (\varphi^\lambda(t, x), \lambda) \end{aligned}$$

Let $N \subset X \times \Lambda$. Let

$$N^\lambda := N \cap (X \times \{\lambda\})$$

Definition 2.3.3 *Let $\lambda_i \in \Lambda$, $i = 0, 1$, and let S^i be isolated invariant set for φ^{λ_i}. S^0 and S^1 are related by continuation if there exists an isolating neighborhood $N \subset X \times \Lambda$ of the parameterized flow Φ such that $\mathrm{Inv}(N^{\lambda_0}, \varphi^{\lambda_0}) = S^0$ and $\mathrm{Inv}(N^{\lambda_1}, \varphi^{\lambda_1}) = S^1$.*

Theorem 2.3.4 *If S^0 and S^1 are related by continuation, then*

$$h(S^0) \sim h(S^1)$$

and hence

$$CH^*(S^0) \approx CH^*(S^1).$$

2.4 The Discrete Conley Index

We now turn to the setting of discrete dynamics. In order to simplify the discussion we shall assume that $f : X \to X$ is a homeomorphism. Given $N \subset X$, let

$$\text{Inv}^+ N := \{x \in X \mid \forall n \in \mathbf{Z}^+ \; f^n(x) \in N\}$$
$$\text{Inv}^- N := \{x \in X \mid \forall n \in \mathbf{Z}^- \; f^n(x) \in N\}$$
$$\text{Inv}(N, f) := \bigcap_{n \in \mathbf{Z}} f^n(N)$$
$$= \text{Inv}^+ N \cap \text{Inv}^- N$$

N is an *isolating neighborhood* if

$$\text{Inv}(N, f) \subset \text{int } N.$$

As in the case of a flow one begins defining the Conley index in terms of an index pair.

Definition 2.4.1 A pair (N, L) of compact subsets of $W \subset X$ is an *index pair* for an isolated invariant set $K \subset W$ if:

1. N and L are positively invariant with respect to N;

2. $\text{Inv}^- W \subset \text{int}_W N$, $\text{Inv}^+ W \subset W \setminus L$;

3. $N \setminus L \subset \text{int } W \cap f^{-1}(\text{int } W)$.

One is tempted at this point to define the index in terms of the relative cohomology as was done in the flow case. However, as the following example shows this does not lead to an index which is an invariant of the invariant set.

Example 2.4.2 Consider the simple linear map $f : \mathbf{R}^2 \to \mathbf{R}^2$ given by the matrix

$$f = \begin{pmatrix} 2 & 0 \\ 0 & \frac{1}{2} \end{pmatrix}.$$

Then, the pair (N, L) shown in Figure 2.9 is an index pair and

$$H^k(N, L) \approx \begin{cases} \mathbf{Z} & \text{if } k = 1 \\ 0 & \text{otherwise} \end{cases}.$$

Observe that (N', L) as indicated in Figure 2.10 is also an index pair, but in this case

$$H^k(N, L) \approx \begin{cases} \mathbf{Z} & \text{if } k = 0, 1 \\ 0 & \text{otherwise} \end{cases}.$$

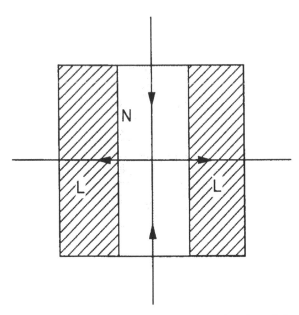

Figure 2.9: An index pair for the fixed point of f.

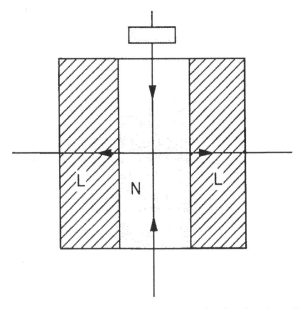

Figure 2.10: Another index pair for the fixed point of f.

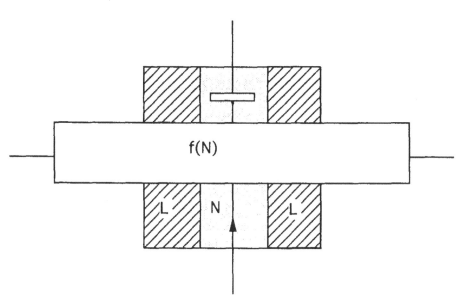

Figure 2.11: The image of the index pair under f.

One's impression is that the problem exhibited in Example 2.4.2 is due to the fact that the extra piece added onto N is a "bad" piece and the cohomology it induces should be ignored. To see how to do this consider the image of (N', L) under f shown in Figure 2.11. In particular, we can consider

$$f : (N', L) \to (N \cup f(N), L \cup f(L))$$

and correspondingly

$$f^* : H^*(N \cup f(N), L \cup f(L)) \to H^*(N', L).$$

Observe that f^1 is an isomorphism while $f^0 = 0$. Thus, the image of f^* identifies the "correct" cohomology generators in $H^*(N', L)$. It is easy to imagine how to further complicate the problem of Example 2.4.2. Thus, at the very least one needs to be able to determine the image of an arbitrarily large number of iterations of f^*. A minor technicality is the fact that spaces on which f is defined changes under each iteration. To overcome this, observe that for any index pair (N, L) excision defines an isomorphism

$$\eta : H^*(N \cup f(N), L \cup f(L)) \to H^*(N, L).$$

Definition 2.4.3 The *index map*

$$F_{N,L}^* := H^*(N, L) \to H^*(N, L)$$

associated to the index pair (N, L) is defined by

$$F_{N,L}^* := f^* \circ \eta^{-1}.$$

On occasion, where the index pair is unambiguous, but the particular dynamical system is not, we shall denote the index map by

$$F_f^*$$

where f generates the dynamics.

This definition allows us to construct for any index pair (N, L) the directed system $\{H^*(N, L), F_{N,L}^*\}$.

Definition 2.4.4 Let S be an isolated invariant set for the discrete dynamical system $f : X \rightarrow X$. Let (N, L) be an index pair for S. The *Conley index* for S is defined to be

$$\mathrm{Con}^*(S) = (CH^*(S), \chi^*(S)) := \varinjlim\{H^*(N, L), F_{N,L}^*\}$$

As before, we extend this index to an index of isolating neighborhoods as follows. If N is an isolating neighborhood then

$$\mathrm{Con}^*(N) := \mathrm{Con}^*(\mathrm{Inv}\, N).$$

The following two theorems state that the index theory exists and is well defined.

Theorem 2.4.5 (The Conley index is well defined.) *Let (N, L) and (N', L') be index pairs for an isolated invariant set S. Then,*

$$\varinjlim\{H^*(N, L), F_{N,L}^*\} \approx \varinjlim\{H^*(N', L'), F_{N',L'}^*\}$$

Theorem 2.4.6 (The Conley index exists.) *Let S be an isolated invariant set. Then, there exists an index pair for S.*

As in the case of the index theory for flows, the crucial continuation property, Wazewski's property, and the summation property hold. In addition we have the following theorem.

Theorem 2.4.7 *Let Y be another locally compact metric space and $g : Y \rightarrow Y$ be a homeomorphism. Let $\alpha : X \rightarrow Y$ and $\beta : Y \rightarrow X$ be continuous maps such that $f = \beta \circ \alpha$ and $g = \alpha \circ \beta$. If N is an isolating neighborhood for f, then $\alpha(N)$ is an isolating neighborhood of g and*

$$\mathrm{Con}(N, f) \approx \mathrm{Con}(\alpha(N), g).$$

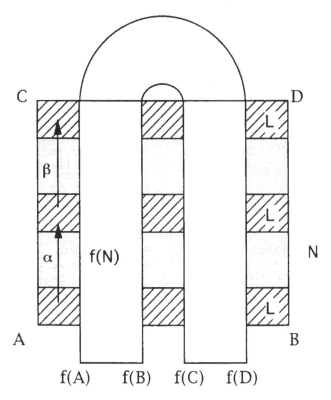

Figure 2.12: Smale's horseshoe.

Example 2.4.8 Smale's horseshoe map is defined as follows. Let $f : \mathbf{R}^2 \to \mathbf{R}^2$ be a plane homeomorphism which transforms the square Q with vertices A, B, C, D into the horseshoe-shaped region $A'B'C'D'$ as indicated in Figure 2.12. Observe that the shaded regions L represent the exit set for the isolating neighborhood N. The induced index map is $F^*_{N,L} : H^*(N, L) \to H^*(N, L)$ and can be represented by the matrix

$$F^1_{N,L} = \begin{bmatrix} 1 & -1 \\ 1 & -1 \end{bmatrix}$$

(That this is in fact the correct map can be most easily seen by looking at the 1-chains α and β indicated in Figure 2.12 and then examining their image under f.) Observe that $F^{*2}_{N,L} = 0$. Therefore, the index of the horseshoe is

$$\mathrm{Con}^*(\mathrm{Inv}\, N) \approx (0, 0).$$

Example 2.4.9 The G–horseshoe is defined like Smale's horseshoe except that one branch is brought back with the orientation preserved (see Figure 2.13.

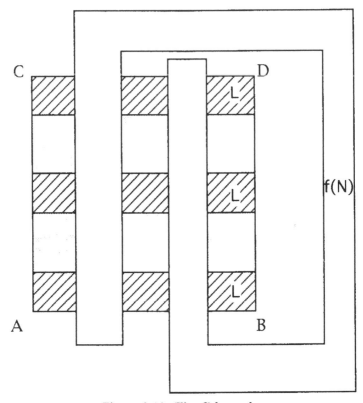

Figure 2.13: The G horseshoe.

Again, the shaded regions L represent the exit set for the isolating neighborhood N, but this time the induced index map takes the form

$$F^1_{N,L} = \begin{bmatrix} 1 & 1 \\ 1 & 1 \end{bmatrix}$$

If one now takes the direct limit (which in this case is equivalent to quotienting out by the kernel of $F_{N,L}$) one determines that the index using rational coefficients is

$$\mathrm{Con}^*(\mathrm{Inv}\, N) \approx (\mathbf{Q}, 2).$$

2.5 Appendix

This chapter cannot be ended without remarking that the Conley index is an extremely general object and applies to a much wider range of dynamics than has been described here. In particular, in the context of semi-flows on metric spaces one should consult the work of K. Rybakowski [49], V. Benci [1] and V. Benci and M. Degiovanni [2]. The prime obstacle to a meaningful theory in these general settings is the question of compactness. Rybakowski imposes a Palais–Smale type condition on the semi–flow which has the effect of insuring that the isolated invariant sets and their unstable sets are compact. Benci's approach is to view the Conley index from the start as an index for isolating neighborhoods and only worry about compactness after the theory has been developed. This leads to a very general theory, however, for applications one still needs to apply some form of compactness in order to draw conclusions concerning the existence and structure of the invariant set. The advantage is that it provides one with the flexibility of imposing whatever level of compactness is natural to the problem.

One can also extend the index to maps on non-compact spaces as was done by M. Mrozek and Rybakowski [46] and M. Degiovanni and M. Mrozek [9].

Another direction, is that of multi–valued dynamical systems which has been treated by Mrozek [42] and T. Kaczynski and Mrozek [26]. This latter theory is for multi–valued maps and leads to interesting possibilities for numerical applications [45].

The division of the Conley index into an index for flows and maps as presented in this chapter is artificial and was done only for pedagogical reasons. That the two theories are the same in the locally compact setting is explicitly stated in Mrozek's original work [41]. The extension of the discrete index to Rybakowski's setting and then Benci's appears in [46] and [9], respectively. On an even more abstract level, functorial descriptions of the Conley index have been developed by M. Mrozek [44] and A. Szymczak [54].

Chapter 3

Decomposing Invariant Sets

As was mentioned in the introduction, our goal is to use the Conley index to make statements concerning the dynamics of isolated invariant sets. Obviously, to do this we need to have a natural way of decomposing the invariant sets into "minimal" pieces. The key word, however, is natural; the decomposition must be compatible with the coarse nature of isolating neighborhoods and the index. Thus, we begin with the idea of chain recurrence and state Conley's decomposition theorem which shows that all invariant sets are made up of chain recurrent dynamics and gradient like dynamics. Unfortunately, the decomposition into components of the chain recurrent set is not robust, and hence, for applications we need to consider coarser decompositions. Thus, in Section 3.2 we discuss the coarsest of decompositions, namely attractor–repeller pairs. In Section 3.3 we show how knowledge of this decomposition in conjunction with the algebra of the index theory can be used to describe the structure of the isolated invariant set. This extremely simple example is meant to be a preview of the results described in Chapters 6 and 7. Finally, in Section 3.4 we review the concept of a Morse decomposition; which, while generalizing attractor–repeller pair decomposition, is still coarse enough to be robust with respect to perturbations.

3.1   Conley's Decomposition Theorem

An (ϵ, τ) *chain* from x to y is a finite sequence

$$\{(x_i, t_i)\} \subset X \times [0, \infty), \quad i = 1, \ldots n$$

such that $x = x_1$, $t_i \geq \tau$, $\mathbf{d}(\varphi(t_i, x_i), x_{i+1}) \leq \epsilon$ and $\mathbf{d}(\varphi(t_n, x_n), y) \leq \epsilon\}$. If there exists an (ϵ, τ) chain from x to y then we write $x \succeq_{(\epsilon, \tau)} y$. If $x \succeq_{(\epsilon, \tau)} y$ for all (ϵ, τ) then $x \succeq y$.

Definition 3.1.1 The *chain recurrent set* of X under the local semi-flow ϕ is defined by
$$\mathcal{R}(X, \phi) = \{x \in X \mid x \succeq x\}.$$

The chain recurrent set is fundamental for two reasons. First, it is minimal in the sense of the following theorem.

Theorem 3.1.2 (Conley [4]) *If S is a compact invariant set, then $\mathcal{R}(\mathcal{R}(S)) = \mathcal{R}(S)$.*

Second, it captures all the recurrent dynamics as is indicated by the following theorem.

Theorem 3.1.3 (Conley [4]) *Let $\mathcal{R}_i(S)$, $i = 1, 2, \ldots$, denote the connected components of $\mathcal{R}(S)$. Let S be a compact invariant set. Then there exists a continuous function*
$$V : S \to [0, 1]$$
such that:

(i) if $x \notin \mathcal{R}(S)$ and $t > 0$, then $V(x) > V(\varphi(t, x))$;

(ii) for each $i = 1, 2, \ldots$, there exists $\sigma_i \in [0, 1]$ such that $\mathcal{R}_i \subset V^{-1}(\sigma_i)$ and, furthermore, the $\{\sigma_i\}$ can be chosen such that $\sigma_i \neq \sigma_j$ if $i \neq j$.

Unfortunately, for direct applications the components of the chain recurrent set are difficult to work with since they are not robust with respect to perturbation. Furthermore, while the individual components are invariant sets they need not be isolated invariant sets.

Example 3.1.4 Consider the equation
$$\dot{x} = x^2 \sin \frac{\pi}{x}.$$

Let $N = [-2, 2]$. Then, $\operatorname{Inv} N = S = [-1, 1]$ is a compact isolated invariant set. Clearly,
$$\mathcal{R}(S) = \{\pm \frac{1}{n} \mid n \in \mathbf{Z}\} \cup \{0\},$$
and hence, has an infinite number of components. In addition, observe that 0 is a component of $\mathcal{R}(S)$ but is not an isolated invariant set. Now consider the parameterized system
$$\dot{x} = x^2 \sin \frac{\pi}{x} + \lambda(x^2 - 1) \quad \lambda \in [-1, 1]$$

with corresponding flow φ^λ. Again, choose $N = [-2, 2]$ and let $S^\lambda := \operatorname{Inv}(\varphi^\lambda)$. Then, for $\lambda \neq 0$, $\mathcal{R}(S^\lambda)$ has at only finitely many components.

3.2 Attractor–Repeller Pairs

The coarsest decomposition of an invariant set is that of an attractor–repeller pair. Our discussion begins with some standard definitions from the theory of dynamical systems. Let $Y \subset X$. The *omega limit set* of Y is

$$\omega(Y) := \bigcap_{t>0} \operatorname{cl}\left(\varphi([t, \infty), Y)\right)$$

while the *alpha limit set* of Y is

$$\alpha(Y) := \bigcap_{t>0} \operatorname{cl}\left(\varphi([-\infty, -t), Y)\right).$$

Definition 3.2.1 Let S be a compact invariant set. $A \subset S$ is an *attractor* in S if there exists a neighborhood U of A such that

$$\omega(U \cap S) = A.$$

The *dual repeller* of A in S is

$$A^* := \{x \in S \mid \omega(x) \cap A = \emptyset\}.$$

The pair (A, A^*) is called an *attractor–repeller pair decomposition* of S. The set of *connecting orbits* from A^* to A in S is

$$C(A^*, A; S) := \{x \in S \mid \omega(x) \subset A, \ \alpha(x) \subset A^*\}.$$

Figure 3.1 provides a simple example of an attractor–repeller pair.

The following properties of attractors and repellers are easy to check and left to the reader.

Proposition 3.2.2 *Let S be an isolated invariant set. Let (A, R) be an attractor–repeller pair decomposition for S. Then*

1. *A and R are isolated invariant sets;*

2. *if A' is an attractor in A, then A' is an attractor in S.*

The following result, through trivial, is important enough to be designated a theorem.

Theorem 3.2.3 *Let (A, R) be an attractor–repeller pair decomposition of S. Then*
$$S = A \cup R \cup C(R, A; S).$$

The most common use of this theorem is to prove the existence of heteroclinic orbits.

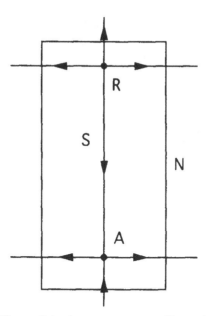

Figure 3.1: An attractor–repeller pair

Example 3.2.4 Assume that S is an isolated invariant set with an attractor–repeller pair decomposition (A, R) consisting of two fixed points. Furthermore, assume that

$$CH^n(R) \approx \begin{cases} \mathbf{Z} & \text{if } n = k+1 \\ 0 & \text{otherwise,} \end{cases} \qquad CH^n(A) \approx \begin{cases} \mathbf{Z} & \text{if } n = k \\ 0 & \text{otherwise,} \end{cases}$$

and

$$CH^*(S) \approx 0.$$

Then, by Theorem 2.2.17, $S \neq A \cup R$. Thus, by Theorem 3.2.3, $C(R, A; S) \neq \emptyset$, and hence, there exists a heteroclinic orbit from R to A.

The following result is at the heart of the proof of Theorem 3.1.3

Theorem 3.2.5 *Let S be a compact invariant set with attractor repeller pair (A, A^*). Then there exists a continuous function*

$$V : S \to [0, 1]$$

such that

(i) $V^{-1}(1) = A^*$

(ii) $V^{-1}(0) = A$

(iii) for $x \in C(A^*, A)$ and $t > 0$, $V(x) > V(\varphi(t, x))$.

Proof.   Observe that if A or A^* equals the empty set then the result is trivially true. So assume that $A \neq \emptyset \neq A^*$. The proof is now broken down into three steps. The first is to define a function $f : Y \to [0,1]$ by

$$f(x) = \frac{d(x,A)}{d(x,A) + d(x,A^*)}$$

Clearly, (since $A \cap A^* = \emptyset$) f is continuous, $f^{-1}(0) = A$, and $f^{-1}(1) = A^*$. Second, define $g : Y \to [0,1]$ by

$$g(x) = \sup_{t \geq 0} \{f(\varphi(t,x))\}.$$

Again, it is obvious that $g^{-1}(0) = A$, and $g^{-1}(1) = A^*$ and $g(\varphi(t,x)) \leq g(x)$ for all $t \geq 0$. It is left to the reader to show that g is continuous.

The third and final step is to define

$$V(x) = \int_0^\infty e^{-\xi} g(\varphi(\xi,x)) d\xi$$

Clearly (i) and (ii) are satisfied. If $x \in C(A^*, A)$, then

$$
\begin{aligned}
V(x) - V(x,t) &= \int_0^\infty e^{-\xi} g(\varphi(\xi,x)) d\xi - \int_0^\infty e^{-\xi} g(\varphi(\xi+t,x)) d\xi \\
&= \int_0^\infty e^{-\xi} [\varphi(\xi,x)) - g(\varphi(\xi+t,x))] d\xi \\
&\geq 0.
\end{aligned}
$$

Now $V(x) - V(\varphi(t,x)) = 0$ if and only if $g(\varphi(\xi,x)) - g(\varphi(\xi+t,x)) = 0$ for all $\xi \geq 0$, i.e., $g(\varphi(\xi,x)) = \bar{g}$ a constant for all $\xi \geq 0$. This implies that $\omega(x) \cap (A \cup A^*) = \emptyset$, a contradiction. Thus $V(x) - V(\varphi(t,x)) > 0$, and condition (iii) is satisfies.   ∎

The relationship between attractor–repeller pairs and the chain recurrent set is given in the following theorem.

Theorem 3.2.6 (Conley [4]) $\mathcal{R}(S) = \bigcap \{A \cup A^* \mid A \text{ is an attractor in } S\}$.

This theorem makes the following construction plausible. Let $\{A_i \mid i = 1, 2, \ldots\}$ denote the set of attractors of S. Let V_i denote the Lyapunov function corresponding to the pair (A_i, A_i^*). Now define

$$V(x) = \sum_{i=1}^\infty 3^{-i} V_i(x).$$

This V satisfies the conditions of Theorem 3.1.3.

3.3 A simple semi–conjugacy

Consider Example 3.2.4. This is a pure existence result in the sense that the conclusion is that there exists an orbit $\varphi(\mathbf{R}, x) \subset S$ such that

$$\lim_{t \to \infty} \varphi(t, x) = A \quad \lim_{t \to -\infty} \varphi(t, x) = R.$$

What it does not do is describe the dynamics on S as a whole. Now clearly, in this simple example it is obvious that the only type of orbits which can occur are the fixed points A and R and the heteroclinic orbits from R to A. However, for the sake of clarity it is worthwhile trying to recast this example in the context of semi–conjugacies. More complicated examples will be presented in later chapters, but there the technical difficulties can overwhelm the simple idea. With this in mind, therefore, we shall prove the following theorem.

Theorem 3.3.1 *Let φ be a flow for which S is an isolated invariant set with an attractor–repeller pair decomposition (A, R). Assume that*

$$CH^*(S) \not\simeq CH^*(A) \oplus CH^*(R).$$

Let $\psi : \mathbf{R} \times [-1, 1] \to [-1, 1]$ be the flow on the unit interval generated by

$$\dot{x} = (x^2 - 1).$$

Then, there exists a continuous surjective function

$$\rho : S \to [-1, 1]$$

such that

$$A = \rho^{-1}(-1) \quad R = \rho^{-1}(1)$$

and the following diagram commutes

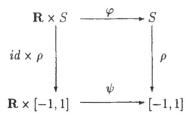

Proof.   Let $V : S \to [0, 1]$ be a Lyapunov function for φ as defined by Theorem 3.2.5. Because V is strictly decreasing off of $A \cup R$, given $x \in C(R, A; S)$ there exists a unique $t_x \in \mathbf{R}$ such that

$$V(\varphi(t_x, x)) = \frac{1}{2}.$$

Observe that the function $\tau : C(R, A; S) \to \mathbf{R}$ given by $\tau(x) = t_x$ is continuous. Now, define $\rho : S \to [0, 1]$ by

$$\rho(x) = \begin{cases} -1 & \text{if } x \in A \\ 1 & \text{if } x \in R \\ -\frac{2}{\pi} \tan^{-1} t_x & \text{if } x \in C(R, A; S) \end{cases}.$$

Since $CH^*(S) \not\approx CH^*(A) \oplus CH^*(R)$, $C(R, A; S) \neq \emptyset$, and hence, ρ is surjective.
∎

There are several points that should be remarked concerning this theorem. The first is that one uses the decomposition and the Lyapunov function to define the semi-conjugacy. The second is that it is the Conley index information that allows one to conclude that ρ is a surjection.

3.4 Morse Decompositions

The generalization of an attractor repeller pair decomposition is a Morse decomposition. This is, also, a generalization of the setting of classical Morse theory wherein one has a finite number of hyperbolic critical points for a gradient flow on a compact manifold. A starting point for much of Morse theory is a self ordering Morse function. Since the Conley index of a Morse set may contain nontrivial cohomology on several levels such a function may not exist in the context of Conley theory. Furthermore, since one of the objects of this theory is to understand invariant set throughout a parameterized family of flows, orderings that are valid at one parameter value may not be valid at another. Thus, it is perhaps appropriate that we begin with a discussion of partial orders. It may be worth commenting at this point that the insistence on using partial orders rather than total orders leads to a variety of difficulties. In particular, it is one of the causes of the non-uniqueness of connection matrices (see Chapter 4). However, it is essential for two reasons. The first is that a priori knowledge of a natural total ordering is often impossible, and second, for parameterized families of flows there may be no total ordering which is consistent for all parameter values.

A *partial order*[1] on a set $\mathcal{P}$ is a relation $>$ which satisfies:

(i) $p > p$ never holds for $p \in \mathcal{P}$.

(ii) If $p > q$ and $q > r$ then $p > r$.

If, in addition, the partial order satisfies

(iii) for all $p, q \in \mathcal{P}$, either $p > q$ or $q > p$.

Then $>$ is called a *total order*. From now on $(\mathcal{P}, >)$ will be used to denote a finite indexing set $\mathcal{P}$ with a partial order $>$.

---

[1] Actually we are defining a *strict partial order*, however, to save the use of an unnecessary adjective we shall use the above definition.

A subset $I \subset \mathcal{P}$ is called an *interval* if $p, q \in I$ and $p > r > q$ implies that $r \in I$. The set of intervals on $(\mathcal{P}, >)$ will be denoted by $\mathcal{I}(\mathcal{P}, >)$.

An *attracting interval* I has the additional property that given $p \in I$ and $p > q$ then $q \in I$. The set of attracting intervals is written as $\mathcal{A}(\mathcal{P}, >)$.

Proposition 3.4.1 *(i)* $\mathcal{A}(\mathcal{P}, >) \subset \mathcal{I}(\mathcal{P}, >)$

(ii) $\emptyset, \mathcal{P} \in \mathcal{A}(\mathcal{P}, >)$

(iii) If $I, J \in \mathcal{I}(\mathcal{P}, >)$, then $I \cap J \in \mathcal{I}(\mathcal{P}, >)$

(iv) If $I, J \in \mathcal{A}(\mathcal{P}, >)$, then $I \cap J \in \mathcal{A}(\mathcal{P}, >)$ and $I \cup J \in \mathcal{A}(\mathcal{P}, >)$.

An *adjacent n-tuple of intervals* in $(\mathcal{P}, >)$ is an ordered collection

$$(I_1, I_2, \ldots, I_n)$$

of mutually disjoint intervals satisfying:

(i) $\bigcup_{i=1}^{n} I_i \in \mathcal{I}(\mathcal{P}, >)$,

(ii) $p \in I_j, q \in I_k, j > k$ imply $q \not> p$.

The collection of adjacent n-tuples is denoted by $\mathcal{I}_n(\mathcal{P}, >)$. Notice that $\mathcal{I}(\mathcal{P}, >) = \mathcal{I}_1(\mathcal{P}, >)$. If (I, J) is an *adjacent pair* (i.e., a 2-tuple) of intervals, then we set $IJ = I \cup J$.

Finally an order $>'$ on $\mathcal{P}$ is an *extension* of $>$ if $p > q$ implies $p >' q$.

Definition 3.4.2 A finite collection

$$\mathcal{M}(S) = \{M(p) \mid p \in \mathcal{P}\}$$

of disjoint compact invariant subsets of S is a *Morse decomposition* if there exists a strict partial order $>$ on the indexing set $\mathcal{P}$ such that for every

$$x \in S \setminus \bigcup_{p \in \mathcal{P}} M(p)$$

there exists $p, q \in \mathcal{P}$ such that $p > q$ and

$$\omega(x) \subset M(q) \quad \text{and} \quad \alpha(x) \subset M(p).$$

The sets $M(p)$ are called *Morse sets*. Notice that in the definition it is not assumed that there is unique order on $\mathcal{P}$. In fact, any ordering on $\mathcal{P}$ with the above property is called *admissible*. Having chosen an admissible order $>$ we shall often write

$$\mathcal{M}(S) = \{M(p) \mid p \in (\mathcal{P}, >)\}.$$

Strictly speaking this is an abuse of notation since the Morse decomposition really only refers to the collection of Morse sets. However, in applications it is impossible to construct a Morse decomposition without somehow obtaining an admissible order. Furthermore, as the following result indicates, partial orders are of some dynamic interest.

Proposition 3.4.3 *If > is an admissible ordering for $\mathcal{M}(S) = \{M(p) \mid p \in \mathcal{P}\}$, a Morse decomposition of S, then $C(M(p), M(q)) \neq \emptyset$ implies that $p > q$.*

There is, however a unique minimal (in the sense of the number of order relations) admissible order which is called the *flow defined order*. It is this order that is of the greatest importance with respect to the dynamics of the system. However, it is often extremely difficult to obtain the flow defined order explicitly. When it is necessary to distinguish the flow defined order from another admissible order we shall use the notation $>_\varphi$ to represent the flow defined order.

Consider the system of differential equations given by

$$\dot{x} = x(x-1)$$
$$\dot{y} = y(y-2).$$

The phase portrait for this system is drawn in Figure 3.2. Observe that $S = \{(x,y) \mid 0 \leq x \leq 1,\ 0 \leq y \leq 2\}$ is an isolated invariant set and that

$$M(a) := (0,0)$$
$$M(b) := (1,0)$$
$$M(c) := (0,2)$$
$$M(d) := (1,2)$$

forms a Morse decomposition of S. Two obvious admissible orders are $a < b < c < d$ and $a < c < b < d$. These could be obtained by constructing various a Lyapunov functions for the system. However, the flow defined order, which clearly best reflects the set of connecting orbits, is $a < b < d$; $a < c < d$.

This example also raises the point that the admissible orders on the indexing set is related to Lyapunov functions of the flow. The same argument that is used to prove Theorem 3.1.3 can be used to make this observation precise.

Theorem 3.4.4 *Let S be an isolated invariant set. Let*

$$\mathcal{M}(S) = \{M(p) \mid p \in \mathcal{P}\}$$

be a finite collection of disjoint invariant compact subsets of S. Then $\mathcal{M}(S)$ is a Morse decomposition of S if and only if there exists a continuous function $V : S \to [0,1]$ such that if:

$$\forall\, x,y \in M(p) \quad \Rightarrow \quad V(x) = V(y)$$
$$x \in S \setminus \bigcup_{p \in \mathcal{P}} M(p) \quad \Rightarrow \quad V(x) > V(\varphi(t,x)) \quad \forall t > 0$$

For now we wish to observe that given a Morse decomposition $\mathcal{M}(S) = \{M(p) \mid p \in (\mathcal{P}, >)\}$ one can always coarsen it as follows.

Proposition 3.4.5 *Let $I \in \mathcal{I}(\mathcal{P}, >)$ and define*

$$M(I) = \left(\bigcup_{p \in I} M(p) \right) \cup \left(\bigcup_{p,q \in I} C(M(p), M(q)) \right).$$

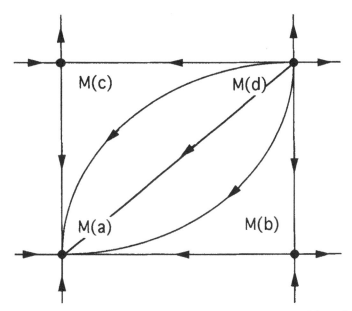

Figure 3.2: A Morse decomposition with several admissible orders.

Then $M(I)$ is an invariant set.

Proposition 3.4.6 $\mathcal{M}(S) = \{M(p) \mid p \notin \mathcal{P}\backslash I\} \cup \{M(I)\}$ *defines a Morse decomposition of S. Furthermore, an admissible partial order $>'$ on $\mathcal{P}$ is given by*

$$p >' q \quad \Leftrightarrow p > q \text{ if } p, q \in \mathcal{P}\backslash I$$
$$p >' I \quad \text{if } \exists q \in I \text{ such that } p > q$$
$$I >' p \quad \text{if } \exists q \in I \text{ such that } q > q$$

Knowing that Morse decompositions can be coarsened it is natural to ask whether given an isolated invariant set S, there exists a finest Morse decomposition. Example 3.1.4 shows that the answer is obviously no.

We began this section by stating that Morse decompositions are generalizations of with attractor–repeller pair decompositions. To see this observe that an attractor–repeller pair decomposition (A, A^*) of S defines a Morse decomposition by

$$\mathcal{M}(S) = \{M(p) \mid p = 1, 2, \quad 2 > 1\}$$

where $M(1) = A$ and $M(2) = A^*$. Conversely, we have the following lemma.

Lemma 3.4.7 *Let $I \in \mathcal{A}(\mathcal{P}, >)$ and let $I^* = \mathcal{P}\backslash I$. Then $(M(I), M(I^*))$ is an attractor repeller pair for S. More generally, if $(I, J) \in I_2(\mathcal{P}, >)$ then $M(I), M(J))$ is an attractor-repeller pair in $M(IJ)$.*

A final result concerning Morse decompositions that we shall make use of later is the following.

Proposition 3.4.8 (Conley [5, Lemma 2.2.B]) *Let ψ^μ be a continuous family of flows. Let K be an invariant set under ψ^{μ_0} and let*

$$\mathcal{M}(K) := \{M(p) \mid 1 < \ldots < p < p+1 < \ldots < P\}$$

be a Morse decomposition of K. Let U be a neighborhood of $\cup_p M(p)$.

Then for $p = 1, \ldots, P$ there are neighborhoods V_p of $M(1, \ldots, p)$ and W_p of K such that for all μ sufficiently close to μ_0 the following condition is satisfied. If $x \in V_p$, then either:

1. *$\psi^\mu((0, \infty), x) \subset U$ or*

2. *there exists a t_0 such that $\psi^\mu([0, t_0), x) \subset U$ and $\psi^\mu(t_0, x) \subset U \setminus W_p$ or*

3. *there exists a t_0 such that $\psi^\mu(t_0, x) \in V_{p-1}$.*

Furthermore, the sets can be constructed in the order W_1, V_1, W_2, $V_2, \ldots, W_P$, V_P and V_p can be constructed as small as desired without changing the previously constructed sets.

Finally, given $x \in V_n$ and μ sufficiently close to μ_0 then either there exists $t_0 > 0$ such that

$$\psi^\mu([t_0, \infty), x) \subset U$$

or there exits $t_1 > t_0 > 0$ and a p such that

$$\psi^\mu(t_0, x) \in V_p, \quad \psi^\mu([t_0, t_1], x) \subset U \quad \text{and} \quad \psi^\mu(t_2, x) \in U \setminus W_p.$$

Chapter 4

Index Filtrations and Connection Matrices

In the previous chapter it was emphasized that one of the important aspects of a Morse decomposition was that each Morse set was an isolated invariant set, and hence, had a Conley index. It is natural to ask whether there is a relationship between the Conley indices of the Morse sets and the Conley index of the total invariant set. In the context of classical Morse theory the answer is yes and can be given in terms of the Morse inequalities. In the Conley theory this relationship is expressed through a linear map known as the connection matrix. Before we can discuss the connection matrices we need to understand index filtrations. In order to motivate these two subjects we study them first in the context of attractor–repeller pair decompositions.

4.1 Index Triples

Definition 4.1.1 Let S be an isolated invariant set and let (A, R) be an attractor–repeller pair. An *index triple* for (A, R) is a collection of compact sets (N_2, N_1, N_0) where $N_0 \subset N_1 \subset N_2$ such that:

1. (N_2, N_0) is an index pair for S;

2. (N_2, N_1) is an index pair for R;

3. (N_1, N_0) is an index pair for A.

Theorem 4.1.2 *Let (A, R) be an attractor–repeller pair decomposition of an isolated invariant set S. Then, there exists an index triple (N_2, N_1, N_0).*

Proof. By Theorem 2.2.4 there exists an index pair (N_2, N_0) for S. It is left to the reader to check that there exists a closed neighborhood U of A and a constant $\epsilon > 0$ such that if $x \in U$ and $\varphi([0,t],x) \subset N_2$ then $\mathbf{d}(\varphi([0,t],x),R) > \epsilon$. Define

$$Z := \{y \in N_2 \mid \exists\, t > 0 \; x \in U \text{ such that } \varphi([0,t],x) \subset N_2 \text{ and } \varphi(t,x) = y\}.$$

Let $N_1 := N_0 \cup Z$. Then (N_2, N_1, N_0) is the desired index triple. ∎

Since $N_0 \subset N_1 \subset N_2$ there exists on the cochain level a short exact sequence

$$0 \to C^*(N_2, N_1) \to C^*(N_2, N_0) \to C^*(N_1, N_0) \to 0$$

which gives rise to the long exact sequence

$$\ldots \to H^n(N_2, N_1) \to H^n(N_2, N_0) \to H^n(N_1, N_0) \xrightarrow{\delta^*} H^{n+1}(N_2, N_1) \to \ldots \tag{4.1}$$

But this is equivalent to

$$\ldots \to CH^n(R) \to CH^n(S) \to CH^n(A) \xrightarrow{\delta^*} CH^{n+1}(R) \to \ldots \tag{4.2}$$

The following theorem relates this exact sequence to the underlying dynamics.

Theorem 4.1.3 *Let (A, R) be an attractor–repeller pair decomposition of an isolated invariant set S. If $S = A \cup R$, then $\delta^* = 0$.*

Proof. Let U and V be disjoint neighborhoods of A and R, respectively. Since $S = A \cup R$ there exists an index triple of the form

$$(N_1^A \cup N_2^R, N_1^A \cup N_0^R, N_0^A \cup N_0^R)$$

where $N_i^A \subset U$ and $N_i^R \subset V$. Equation (4.1) now takes the form

$$\begin{aligned}
\ldots \quad &\to \quad H^n(N_1^A \cup N_2^R, N_1^A \cup N_0^R) \to H^n(N_1^A \cup N_2^R, N_0^A \cup N_0^R) \\
&\to \quad H^n(N_1^A \cup N_0^R, N_0^A \cup N_0^R) \xrightarrow{\delta^*} H^{n+1}(N_1^A \cup N_2^R, N_1^A \cup N_0^R) \to \ldots
\end{aligned}$$

which by excision is equivalent to

$$\begin{aligned}
\ldots \quad &\to \quad H^n(N_2^R, N_0^R) \to H^n(N_2^R, N_0^R) \oplus H^n(N_1^A, N_0^A) \\
&\to \quad H^n(N_1^A, N_0^A) \xrightarrow{\delta^*} H^{n+1}(N_2^R, N_0^R) \to \ldots
\end{aligned}$$

Since the sequence splits, $\delta^* = 0$. ∎

As usual it is the converse of this theorem which is of use in applications.

Corollary 4.1.4 *If $\delta^* \neq 0$, then $S \neq A \cup R$, i.e $C(R, A; S) \neq \emptyset$.*

Remark 4.1.5 Example 2.2.16 shows that the converse of this corollary is not true.

In preparation for the rather complicated algebra which will be introduced in Section 4.3, we will now try to present the information that is carried by δ^* in the form of a matrix. To simplify the presentation we shall now only consider cohomology with *field* coefficients. The information which we have consists of three parts.

1. A collection of graded vector spaces arising from the indices of the attractor and repeller,

$$CH^*(A) \oplus CH^*(R).$$

2. The index of the total invariant set S,

$$CH^*(S).$$

3. The *connecting homomorphism* δ^* which is a degree 1 operator, i.e. it sends n-level cohomology to $n + 1$-level cohomology. We shall, from now on, denote this homomorphism by

$$\delta^*(A, R) : CH^*(A) \to CH^*(R).$$

We now pose the following question. Let us view $CH^*(A) \oplus CH^*(R)$ as a cochain complex with a coboundary operator Δ. Can we choose Δ in such a way that the resulting cohomology $H\Delta^*$ is isomorphic to $CH^*(S)$? To be more precise, let

$$\Delta^n : CH^n(A) \oplus CH^n(R) \to CH^{n+1}(A) \oplus CH^{n+1}(R).$$

denote the obvious restriction of Δ to each cohomology level of the indices. Then, by definition

$$H\Delta^n := \frac{ker(\Delta^n)}{image(\Delta^{n-1})}.$$

Thus we are asking whether it is possible to choose Δ in such a way that

$$H\Delta^n \approx CH^n(S) \quad n = 0, 1, 2, \ldots$$

Observe that if $S = A \cup R$, then $CH^*(S) \approx CH^*(A) \oplus CH^*(R)$. In this case we can choose $\Delta = 0$. This is a special case of the following proposition.

Proposition 4.1.6 *If we define*

$$\Delta = \begin{bmatrix} 0 & 0 \\ \delta^*(A, R) & 0 \end{bmatrix} : CH^*(A) \oplus CH^*(R) \to CH^*(A) \oplus CH^*(R)$$

then

$$H\Delta \approx CH^*(S).$$

Proof. We begin by checking that Δ is a coboundary operator, i.e. that Δ is degree $+1$ and $\Delta \circ \Delta = 0$. The first is obvious since δ^* is of degree $+1$. The second condition is equally obvious.

We now need to show that

$$H\Delta^n \approx CH^n(S) \quad n = 0, 1, 2, \ldots$$

Observe that

$$ker(\Delta^n) = ker\left(\delta^n(A, R)\right) \oplus CH^n(R)$$

and
$$image(\Delta^{n-1}) = image(\delta^{n-1}(A, R)).$$

Thus, we need to show that

$$
\begin{aligned}
CH^n(S) &\approx \frac{ker(\delta^n(A, R)) \oplus CH^n(R)}{image(\delta^{n-1}(A, R))} \\
&\approx ker(\delta^n(A, R)) \oplus \frac{CH^n(R)}{image(\delta^{n-1}(A, R))}.
\end{aligned}
$$

Now consider the sequence

$$\ldots \to CH^n(R) \overset{i^n}{\to} CH^n(S) \overset{j^n}{\to} CH^n(A) \overset{\delta^*}{\to} CH^{n+1}(R) \to \ldots$$

which since we are working with field coefficients can be written as

$$\ldots \to CH^n(R) \to L^n \oplus ker(j^n) \to CH^n(A) \overset{\delta^*}{\to} CH^{n+1}(R) \to \ldots$$

By exactness
$$L^n \approx ker(\delta^n(A, R))$$

and

$$ker(j^n) \approx \frac{CH^n(R)}{image(\delta^{n-1}(A, R))}.$$

∎

4.2 Index Filtrations

In the previous section we obtained the connecting homomorphism $\delta^*(A, R)$ from the long exact sequence of an index triple. To obtain similar and compatible homomorphisms for a Morse decomposition we need to generalize index triples to index filtrations.

Definition 4.2.1 An *index filtration* for the Morse decomposition

$$\mathcal{M}(S) = \{M(p) \mid p \in (\mathcal{P}, >)\}$$

is a collection of compact sets

$$\mathcal{N} = \mathcal{N}(\mathcal{M}(S)) = \{N(I) \mid I \in \mathcal{A}(\mathcal{P}, >)\}$$

satisfying:

1. for every $I \in \mathcal{A}(\mathcal{P}, >)$, $(N(I), N(\emptyset))$ is an index pair for $M(I)$;

2. for all $I, J \in \mathcal{A}(\mathcal{P}, >)$,

$$
\begin{aligned}
N(I) \cap N(J) &= N(I \cap J) \\
N(I) \cup N(J) &= N(I \cup J).
\end{aligned}
$$

Remark 4.2.2 Let $K \in \mathcal{I}(\mathcal{P}, >)$, then there exist $I, J \in \mathcal{A}(\mathcal{P}, >)$ such that $(J, K) \in \mathcal{J}_2(\mathcal{P}, >)$, i.e., (J, K) is an adjacent pair and $J \cup K = I$. We shall refer to (J, K) as an $A - R$ decomposition of I.

Proposition 4.2.3 Given $\mathcal{N} = \{N(I) \mid I \in \mathcal{A}(\mathcal{P}, >)\}$ an index filtration and $K \in \mathcal{I}(\mathcal{P}, >)$, there exist $I, J \in \mathcal{A}(\mathcal{P}, >)$ such that $(N(I), N(J))$ is an index pair for $M(I)$.

Proof. Choose $I, J \in \mathcal{A}(\mathcal{P}, >)$ such that (J, K) is an $A - R$ decomposition of I. Now notice that $(M(J), M(K))$ is an attractor repeller decomposition of $M(I)$. Furthermore $(N(I), N(\phi))$ is an index pair for $M(I)$ and $(N(J), N(\phi))$ is an index pair for $M(J)$. Now it is a fairly straightforward argument to check that this implies that $(N(I), N(J))$ is an index pair for $M(K)$. ∎

What needs to be shown is that index filtrations exist. We begin by choosing (N_1, N_0) an index pair for S. Let $I \in \mathcal{A}(\mathcal{P}, >)$. Then $(M(I), M(P \backslash I))$ is an attractor repeller pair decomposition of S. Therefore by Theorem 4.1.2, there exists N_I such that (N_1, N_I, N_0) is an index triple for $(M(I), M(P \backslash I))$. Observe that if $p \in I$ then $M(p) \subset \text{int}(N_I \backslash N_0)$ and if $p \notin I$ then $M(p) \subset \text{int}(N_1 \backslash N_I)$. For every $p \in P$ define D_p to be an open set such that

$$M(p) \subset D_p = \bigcap_{p \in I \in \mathcal{A}(\mathcal{P}, >)} \text{int}(N_I \backslash N_0) \cap \bigcap_{p \notin I \in \mathcal{A}(\mathcal{P}, >)} \text{int}(N_1 \backslash N_I)$$

(Note the intersection is taken over the attracting intervals I.) Given D_p define

$$E_p = \{x \in N_1 \mid \exists \, t > 0 \ni \varphi(t, x) \in D_p \text{ and } \varphi([0, t], x) \subset N_1\}.$$

Finally, for every $I \in \mathcal{A}(\mathcal{P}, >)$ set

$$N(I) = N_1 \Big\backslash \bigcup_{p \in P \backslash I} E_p.$$

In [17] the following theorem is proven.

Theorem 4.2.4 $\mathcal{N} = \{N(I) | I \in \mathcal{A}(\mathcal{P}, >)\}$ *is an index filtration for the Morse decompositions* $\mathcal{M}(S) = \{M(p) | p \in (\mathcal{P}, >)\}$.

4.3 Connection Matrices

To define the connection matrix for a Morse decomposition

$$\mathcal{M}(S) = \{\mathcal{M}(p) | p \in (\mathcal{P}, >)\}$$

we fix an index filtration

$$\mathcal{N} = \{N(I) | I \in \mathcal{A}(\mathcal{P}, >)\}.$$

Given $I \in \mathcal{I}(\mathcal{P}, >)$ define

$$Q_I = \{q \in \mathcal{P} \mid q \not> p \text{ for all } p \in I\}.$$

Observe that $Q_I \in \mathcal{A}(\mathcal{P}, >)$ as is $Q_I I := Q_I \cup I$. Also, since $\mathcal{N}$ is an index filtration

$$CH^*(M(I)) \approx H^*(N(Q_I I), N(Q_I)).$$

Now let $(I, J) \in \mathcal{I}_2(\mathcal{P}, >)$. Then $(M(I), M(J))$ is an attractor–repeller decomposition of $M(IJ)$, and hence,

$$(N(Q_{IJ}IJ), N(Q_{IJ}I), N(Q_{IJ}))$$

is an index triple for $(M(I), M(J))$. Thus as in Section 4.1 we have a sequence of inclusion maps

$$(N(Q_{IJ}I), N(Q_{IJ})) \xrightarrow{\iota} (N(Q_{IJ}IJ), N(Q_{IJ})) \xrightarrow{\rho} (N(Q_{IJ}IJ), N(Q_{IJ}I))$$

which give rise to the short exact sequence on the cochain level

$$0 \to C^*\left(N(Q_{IJ}IJ), N(Q_{IJ}I)\right) \xrightarrow{\rho^{\#}} C^*\left(N(Q_{IJ}IJ), N(Q_{IJ})\right) \xrightarrow{\iota^{\#}}$$
$$C^*\left(N(Q_{IJ}I), N(Q_{IJ})\right) \to 0.$$

Passing to cohomology results in the long exact sequence

$$\dots \to CH^n(M(J)) \xrightarrow{\rho^*} CH^n(M(IJ)) \xrightarrow{\iota^*} CH^n(M(I)) \xrightarrow{\delta^*} CH^{n+1}(M(J)) \to \dots.$$

Remark 4.3.1 If (I, J) and (J, I) are simultaneously elements of $\mathcal{I}_2(\mathcal{P}, >)$, then we have two short exact sequences

$$0 \to C^*\left(N(Q_{IJ}IJ), N(Q_{IJ}I)\right) \xrightarrow{\rho(IJ,J)} C^*\left(N(Q_{IJ}IJ), N(Q_{IJ})\right) \xrightarrow{\iota(I,IJ)}$$
$$C^*\left(N(Q_{IJ}I), N(Q_{IJ})\right) \to 0.$$

and

$$0 \to C^*\left(N(Q_{IJ}IJ), N(Q_{IJ}J)\right) \xrightarrow{\rho(IJ,I)} C^*\left(N(Q_{IJ}IJ), N(Q_{IJ})\right) \xrightarrow{\iota(J,IJ)}$$
$$C^*\left(N(Q_{IJ}J), N(Q_{IJ})\right) \to 0.$$

Furthermore, it can be checked that

$$\iota(J, IJ)\rho(IJ, J) = \text{id}_{C(N(Q_{IJ}IJ), N(Q_{IJ}I))}$$

Finally, let us consider $(I, J, K) \in \mathcal{I}_3(\mathcal{P}, >)$ and let $Q = Q_{IJK}$. Then we have the commutative diagram of short exact sequences displayed in Figure 4.1.

We now capture all these ideas in the following definition.

Definition 4.3.2 A *cochain complex braid* over a partially order set $(\mathcal{P}, >)$ is a collection $\mathcal{C}(\mathcal{P}, >)$ consisting of cochain complexes and chain maps satisfying:

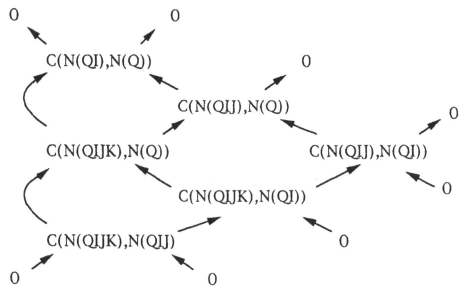

Figure 4.1: Short exact sequences in a chain complex braid.

1. for all $I \in \mathcal{I}(\mathcal{P}, >)$, there exists a cochain complex $C(I)$;

2. for all $(I, J) \in \mathcal{I}_2(\mathcal{P}, >)$ there exist chain maps

$$\iota(I, IJ) : C(IJ) \rightarrow C(I)$$
$$\rho(IJ, J) : C(J) \rightarrow C(IJ)$$

such that

(a)

$$0 \rightarrow C(J) \xrightarrow{\rho} C(IJ) \xrightarrow{\iota} C(I) \rightarrow 0$$

is exact, and

(b) if, simultaneously, $(J, I) \in \mathcal{I}_2(\mathcal{P}, >)$, then

$$\iota(I, IJ)\rho(IJ, I) = \mathrm{id}_{C(I)}$$

3. for all $(I, J, K) \in \mathcal{I}_3(\mathcal{P}, >)$ the diagram of Figure 4.1 commutes.

Let $H(I)$ denote the homology of the chain complex $C(I)$. Passing to cohomology, the diagram of Figure 4.1 of short exact sequences becomes a diagram, Figure 4.2, of long exact sequences called the *braid diagram*

Again we encapsulate this information into a definition.

Definition 4.3.3 A *graded module braid* $\mathcal{H}(\mathcal{P}, >)$ over the partially ordered set $(\mathcal{P}, >)$ is a collection of graded modules and linear maps between the graded modules such that:

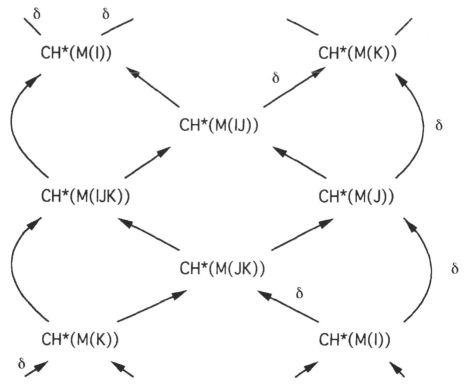

Figure 4.2: Braid diagram of a graded module braid.

1. for all $I \in \mathcal{I}(\mathcal{P}, >)$, there exists a graded module $H^*(I)$;

2. for all $(I, J) \in \mathcal{I}_2(\mathcal{P}, >)$ there exist linear maps

$$
\begin{array}{llll}
\iota^*(I, IJ) : H^*(IJ) & \to & H^*(I) & \text{(degree 0)} \\
\rho^*(IJ, J) : H^*(J) & \to & H^*(IJ) & \text{(degree 0)} \\
\delta^*(J, I) : H^*(I) & \to & H^*(J) & \text{(degree } +1)
\end{array}
$$

such that

(a)

$$
\ldots \to H^n(J) \xrightarrow{\rho^*} H^n(IJ) \xrightarrow{\iota^*} H^n(I) \xrightarrow{\delta^*} H^{n+1}(J) \to \ldots
$$

is exact, and

(b) if, simultaneously, $(J, I) \in \mathcal{I}_2(\mathcal{P}, >)$, then

$$
\iota(I, IJ)\rho(IJ, I) = \mathrm{id}_{C(I)}
$$

3. for all $(I, J, K) \in \mathcal{I}_3(\mathcal{P}, >)$ the braid diagram of Figure 4.2 commutes.

The proof of the following proposition is immediate from the definitions.

Proposition 4.3.4 *Let $\mathcal{C}(\mathcal{P}, >)$ be a cochain complex braid. Passing to cohomology gives rise to a graded module braid which will be denoted by $\mathcal{HC}(\mathcal{P}, >)$.*

We will need to be able to compare graded module braids which leads the next definition.

Definition 4.3.5 Let $\mathcal{H}(\mathcal{P}, >)$ and $\mathcal{H}'(\mathcal{P}, >)$ be graded module braids. A *graded module braid homomorphism*

$$
\Psi : \mathcal{H} \to \mathcal{H}'
$$

is a collection of graded module homomorphisms

$$
\psi(I) : H(I) \to H'(I)
$$

defined for all $I \in \mathcal{I}(\mathcal{P}, >)$ such that for every pair of intervals $(I, J) \in \mathcal{I}_2(\mathcal{P}, >)$

$$
\begin{array}{ccccccccc}
\to & H(J) & \xrightarrow{\rho^*} & H(IJ) & \xrightarrow{\iota^*} & H(I) & \xrightarrow{\delta^*} & H(J) & \to \\
& \downarrow{\psi(J)} & & \downarrow{\psi(IJ)} & & \downarrow{\psi(I)} & & \downarrow{\psi(J)} & \\
\to & H'(J) & \xrightarrow{\rho^*} & H'(IJ) & \xrightarrow{\iota^*} & H'(I) & \xrightarrow{\delta^*} & H'(J) & \to
\end{array}
$$

commutes.

If $\psi(I)$ is an isomorphism for every interval I, then Ψ is called a graded module braid isomorphism

We have at this point imposed a rather monumental edifice of definitions, but we are finally in the position to start defining the connection matrix itself. Let

$$\Delta : \bigoplus_{p \in \mathcal{P}} CH^*(M(p)) \to \bigoplus_{p \in \mathcal{P}} CH^*(M(p))$$

be a linear map which we shall write as a matrix

$$\Delta = [\Delta(p,q)]$$

where $\Delta(p,q) : CH^*(M(q)) \to CH^*(M(p))$. We shall say that Δ is *lower triangular* with respect to the partial order $>$ if $p \not> q$ implies $\Delta(p,q) = 0$. Δ is a coboundary operator if it is a degree $+1$ map and $\Delta \circ \Delta = 0$. Finally, for each interval I define

$$\Delta(I) = [\Delta(p,q)]_{p,q \in I}$$

Lemma 4.3.6 *If Δ is a coboundary operator which is lower triangular with respect to $>$, then so is $\Delta(I)$ for every interval I.*

Let

$$\iota^*(I, IJ) : \bigoplus_{p \in IJ} CH^*(M(p)) \quad \to \quad \bigoplus_{p \in I} CH^*(M(p))$$

$$\rho^*(IJ, J) : \bigoplus_{p \in J} CH^*(M(p)) \quad \to \quad \bigoplus_{p \in IJ} CH^*(M(p))$$

denote the canonical inclusion and projection maps. Then one can prove the following proposition.

Proposition 4.3.7 $\mathcal{C}\Delta(\mathcal{P}, >) := \{C\Delta(I), \iota^*(I, IJ), \rho^*(IJ, J)\}$ *is a chain complex braid.*

The graded module braid obtained from $\mathcal{C}\Delta(\mathcal{P}, >)$ will be denoted by $\mathcal{H}\Delta(\mathcal{P}, >)$.

Definition 4.3.8 Δ is a *connection matrix* if the graded module braid $\mathcal{H}\Delta(\mathcal{P}, >)$ is isomorphic to the graded module braid obtained from the index filtration.

We shall let $\mathcal{CM}(S, \mathcal{P}, >)$ denote the set of connection matrices for the Morse decomposition $\mathcal{M}(S)$. The following remarkable theorem due to R. Franzosa states that connection matrices always exist.

Theorem 4.3.9 *For any isolated invariant set S and any Morse decomposition $\mathcal{M}(S) = \{M(p) \mid p \in (\mathcal{P}, >)\}$,*

$$\mathcal{CM}(S, \mathcal{P}, >) \neq \emptyset.$$

Note that this theorem does not state that the connection matrix is unique. In fact, in many instances there is no unique connection matrix and it remains an important open problem to understand what the non-uniqueness represents. The reader is referred to [14, 15, 16, 17, 18, 30, 40, ?] for more details.

Chapter 5

Periodic Orbits

In the previous sections the various elements of the Conley index theory were introduced. For the remainder of these notes we consider the problem using the Conley index to understand the dynamics of isolated invariant sets. The simplest results in this vein are existence results. In this section it will be shown that the Conley index can be used to prove the existence of periodic orbits. As Example 2.2.16 demonstrates it is not enough to merely know that the index is that of a periodic orbit. In fact, in that example the invariant set was not even recurrent. What is surprising, however, is that this is the only obstruction to the existence of a periodic orbit.

5.1 Poincaré Sections and Suspensions

To insure recurrence we shall make use of a Poincaré section. Since we are adopting the point of view that it is the isolating neighborhood which is the observable object rather than the invariant set, we want to formulate all of our hypotheses in terms of N, rather than S. Therefore, we employ the following definition.

Definition 5.1.1 $\Xi \subset X$ is a *Poincaré section* for N under φ if Ξ is a local section, $\Xi_N := \Xi \cap N$ is closed, and for every $x \in N$ there exists $t_x > 0$ such that

$$\varphi(t_x, x) \in \Xi. \tag{5.1}$$

Observe that it is not necessary to know S in order to find a Poincaré section. Also, Ξ is not required to be a subset of N. Indeed, if N has an exit set, no subset of N can be a Poincaré section, as there will be points in N who's orbits exit N before they cross the section.

Given a Poincaré section Ξ for N, there exists a subset Ξ_0 of Ξ, open in Ξ, such that $S \cap \Xi \subset \Xi_0$ and such that, for every $x \in \Xi_0$, there exists a unique minimal strictly positive time $\tau_\Xi(x)$ with $\varphi(x, [0, \tau_\Xi(x)]) \subset N$ and $\varphi(x, tau_\Xi(x)) \in \Xi$.

The *Poincaré map*

$$\pi : \Xi_0 \to \Xi$$

associated with the Poincaré section Ξ is then defined by $x \mapsto \varphi(x, \tau_\Xi(x))$. This defines a discrete dynamical system on an open subset of Ξ. If $K = S \cap \Xi$, then one can prove the following result [32].

Proposition 5.1.2 π *is continuous,* $K \subset \Xi_0$, *and* K *is an isolated invariant set under* π.

Observe that there is a one to one correspondence between periodic points of π and periodic orbits in φ. Thus, the problem of proving the existence of a periodic orbit is equivalent to prove the existence of a periodic point of π. This, in turn, can be reduced to proving the existence of a fixed point of π^n for some n.

If $\chi^*(K, \pi^n)$, the index automorphism for the nth iterate of π, were known, and if its Lefschetz number were nonzero, then by [43] π^n would have a fixed point. However, requiring knowledge of $\mathrm{Con}^*(K, \pi)$ is too strong of an assumption. Constructing the map π that describes the dynamics on the section requires detailed knowledge of the flow φ; more detail than one can expect in applications. In practice, little about the Poincaré section Ξ and even less concerning the Poincaré map π can be assumed.

With this in mind it should be remarked that our definition of a Poincaré section is rather weak. Consider the example of the unstable periodic orbit exhibited in Figure 5.1. For the isolating neighborhood N both sets

$$\Xi_1 \ := \ \{(x, 0) \mid \frac{1}{2} \le x \le \frac{5}{2}\} \tag{5.2}$$

$$\Xi_2 \ := \ \{(x, 0) \mid \frac{1}{2} \le |x| \le \frac{5}{2}\} \tag{5.3}$$

are Poincaré sections. As will be clear in the application in Section 7.3 it is essential for the ease of applications that we adopt such a lax notion of Poincaré section.

However, it should come as no surprise that there is a relation between the Conley index of the flow and the Conley index of the Poincaré map. An exact statement of this will be discussed in the first subsection. This will then allow us in Section 5.3 to prove the following existence result.

Theorem 5.1.3 *If* N *is an isolating neighborhood for* φ *which admits a Poincaré section* Ξ *and for all* $n \in \mathbf{Z}$ *either*

$$\dim CH^{2n}(N, \varphi) = \dim CH^{2n+1}(N, \varphi) \tag{5.4}$$

or

$$\dim CH^{2n}(N, \varphi) = \dim CH^{2n-1}(N, \varphi) \tag{5.5}$$

where not all the above dimensions are zero, then φ *has a periodic trajectory in* N.

Corollary 5.1.4 *Under the above hypothesis, if* N *has the Conley index of a hyperbolic periodic orbit, then* $\mathrm{Inv}\, N$ *contains a periodic orbit.*

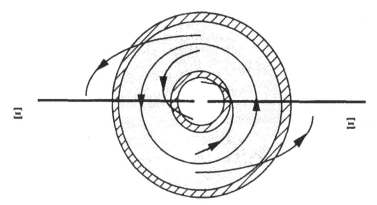

Figure 5.1: A Poincaré section which is not connected

5.2 The Index Suspension Sequence

In the previous section we discussed, in general terms, the relationship between the dynamics of π and φ. In this section we will prove and discuss the consequences of the following theorem which relates the Conley indices of K and S.

Theorem 5.2.1 *Assume N is an isolating neighborhood for the flow φ and assume that N admits a Poincaré section Ξ. Let Π denote the corresponding Poincaré map, $S = \text{Inv}(N, \varphi)$, and $K = \Xi \cap S$. Then, there is the following exact sequence of cohomology Conley indices:*

$$\cdots \to CH^n(S, \varphi) \to CH^n(K, \Pi) \xrightarrow{id - \chi^n(K, \Pi)} CH^n(K, \Pi) \to CH^{n+1}(S, \varphi) \to \cdots \tag{5.6}$$

As will be shown this theorem is essentially the Mayer–Vietoris sequence obtained by decomposing N into two sets, which are obtained by translating the Poincaré section under the flow. An analogy of this sequence can be found in [20].

In order to simplify the notation of the proof we shall assume that

$$\pi(x) = \varphi(1, x). \tag{5.7}$$

This is not a serious assumption in the setting of flows on locally compact spaces. If φ is an arbitrary flow and N is an isolating neighborhood with a Poincaré section, then there exists a simple reparameterization of time such that (5.7) holds [31, Section 5].

To simplify the notation let $\varphi^t : X \to X$ be the homeomorphism given by

$$\varphi^t(x) := \varphi(t, x).$$

Proof of Theorem 5.2.1. The proof begins with the introduction of necessary notation. Let $\tilde{\Xi}_0 \subset \Xi_0$ be an isolating neighborhood for K under π. Define

$$\tilde{N} := \varphi([0, 1], \tilde{\Xi}_0)$$

and observe that $\tilde{N}$ is an isolating neighborhood for $\text{Inv}(N, \varphi)$.

Define

$$
\begin{aligned}
A_1 &:= \varphi([0, \tfrac{1}{2}], \Xi_0) \\
A_2 &:= \varphi([\tfrac{1}{2}, 1], \Xi_0) \\
B_1 &:= \Xi_0 \\
B_1 &:= \varphi^{\frac{1}{2}}(\Xi_0)
\end{aligned}
$$

On these sets there are the obvious discrete dynamical systems with open domains given by

$$
\begin{aligned}
a_i &: A_i \multimap A_i, & a_i(x) &= \varphi^1(x) \\
b_i &: B_i \multimap B_i, & b_i(x) &= \varphi^1(x).
\end{aligned}
$$

Set $S_i = S \cap A_i$, $K_i = S \cap B_i$. Observe that $K_1 = K$ and $b_1 = \pi$. Let

$$\mu_{i,j} : B_j \to A_i$$

denote the inclusion maps.

Define

$$
\begin{aligned}
\theta_{2,1} &: A_1 \to B_2 \\
\theta_{1,2} &: A_2 \to B_1
\end{aligned}
$$

as follows. Let $x \in A_1$. Then there exists a unique $\xi_x \in \Xi_0$ and $t_x \in [0, \tfrac{1}{2}]$ such that $x = \varphi(t_x, \xi_x)$. Set

$$
\begin{aligned}
\theta_{2,1} &= \varphi(\tfrac{1}{2} - t_x, x) \\
&= \varphi(\tfrac{1}{2} - t_x, \varphi(t_x, \xi_x)) \\
&= \varphi(\tfrac{1}{2}, \xi_x) \in B_2.
\end{aligned}
$$

Similarly, let

$$\theta_{1,2}(x) = \varphi(1 - t_x, \xi_x)$$

where $t_x \in [\frac{1}{2}, 1]$ and $\xi_x \in \Xi$ are defined accordingly. It is easily seen that $\theta_{i,j}$ represents "pushing" points in A_j forward to B_i. Furthermore, the maps $\theta_{i,j} : B_j \rightarrow B_i$ are equivalent to $\varphi^{\frac{1}{2}} : B_j \rightarrow B_i$.

Finally, choose an index pair (P, Q) in $\hat{N}$ under φ such that (P_i, Q_i) and (P'_i, Q'_i) are index pairs for S_i and K_i, respectively, where

$$P_i := P \cap A_i \qquad Q_i := Q \cap A_i$$
$$P'_i := P \cap B_i \qquad Q'_i := Q \cap B_i$$

Using the Mayer–Vietoris sequence one obtains the following commutative diagram

$$
\begin{array}{ccccccc}
\rightarrow & H^n(P, Q) & \overset{\iota}{\rightarrow} & H^n(P_1, Q_1) \oplus H^n(P_2, Q_2) & \overset{\nu}{\rightarrow} & H^n(P'_1, Q'_1) \oplus H^n(P'_2, Q'_2) & \rightarrow \\
& \downarrow{\scriptstyle F_{\varphi^1}} & & \downarrow{\scriptstyle F_{a_1} \oplus F_{a_2}} & & \downarrow{\scriptstyle F_{b_1} \oplus F_{b_2}} & \\
\rightarrow & H^n(P, Q) & \overset{\iota}{\rightarrow} & H^n(P_1, Q_1) \oplus H^n(P_2, Q_2) & \overset{\nu}{\rightarrow} & H^n(P'_1, Q'_1) \oplus H^n(P'_2, Q'_2) & \rightarrow
\end{array}
$$

where ι and ν are the inclusion induced maps and in particular

$$\nu = \begin{bmatrix} \mu^*_{1,1} & -\mu^*_{2,1} \\ \mu^*_{1,2} & -\mu^*_{2,2} \end{bmatrix}.$$

Applying the direct limit functor to this diagram gives the long exact sequence

$$\cdots \rightarrow CH^n(\tilde{N}, \varphi) \rightarrow CH^n(A_1, a_1) \oplus CH^n(A_2, a_2) \overset{\nu}{\rightarrow} \qquad (5.8)$$
$$\rightarrow CH^n(B_1, b_1) \oplus CH^n(B_2, b_2) \rightarrow CH^{n+1}(\tilde{N}, \varphi) \rightarrow \cdots$$

It remains to be shown that this sequence is equivalent to the suspension sequence. As a first step the terms $CH^n(A_1, a_1) \oplus CH^n(A_2, a_2)$ and $CH^n(B_1, b_1) \oplus CH^n(B_2, b_2)$ need to be replaced by $CH^n(B_1, b_1) \oplus CH^n(B_1, b_1)$. Define

$$\begin{bmatrix} (\mu^*_{1,1})^{-1} & 0 \\ 0 & \theta^*_{1,2} \end{bmatrix} \begin{bmatrix} id & 0 \\ id & id \end{bmatrix} : H^n(P'_1, Q'_1) \quad \oplus \quad H^n(P'_1, Q'_1)$$
$$\rightarrow \quad H^n(P_1, Q_1) \oplus H^n(P_2, Q_2)$$

and

$$\begin{bmatrix} id & -id \\ id & 0 \end{bmatrix} \begin{bmatrix} id & 0 \\ 0 & (\varphi^{\frac{1}{2}*}_2)^{-1} \end{bmatrix} : H^n(P'_1, Q'_1) \quad \oplus \quad H^n(P'_2, Q'_2)$$
$$\rightarrow \quad H^n(P'_1, Q'_1) \oplus H^n(P'_1, Q'_1)$$

where $\varphi^{\frac{1}{2}}_2 : B_2 \rightarrow B_1$ and $\varphi^{\frac{1}{2}}_1 : B_1 \rightarrow B_2$ are the restrictions of $\varphi^{\frac{1}{2}}$. These maps define change of variables if $\mu^*_{1,1}$, $\theta^*_{1,2}$, and $\varphi^{\frac{1}{2}*}_2$ are isomorphisms. Since φ is a flow $\varphi^{\frac{1}{2}}$ is a homeomorphism, and hence, $\varphi^{\frac{1}{2}*}$ is an isomorphism. Similarly, the fact that φ is a flow shows that $\theta_{1,2}$ and $\mu_{1,1}$ are homotopic to the identity, and hence, $\theta^*_{1,2}$ and $\mu^*_{1,1}$ are isomorphisms.

Now passing to the direct limit one obtains

$$
\begin{bmatrix} \mathcal{L}\left((\mu_{1,1}^*)^{-1}\right) & 0 \\ 0 & \mathcal{L}\left(\theta_{1,2}^*\right) \end{bmatrix} \begin{bmatrix} \mathrm{id} & 0 \\ \mathrm{id} & \mathrm{id} \end{bmatrix}
$$

and

$$
\begin{bmatrix} \mathrm{id} & -\mathrm{id} \\ \mathrm{id} & 0 \end{bmatrix} \begin{bmatrix} \mathrm{id} & 0 \\ 0 & \mathcal{L}\left((\varphi_2^{\frac{1}{2}*})^{-1}\right) \end{bmatrix}
$$

Again, it needs to be checked that $\mathcal{L}\left((\mu_{1,1}^*)^{-1}\right)$, $\mathcal{L}\left(\theta_{1,2}^*\right)$, and $\mathcal{L}\left((\varphi_2^{\frac{1}{2}*})^{-1}\right)$ are isomorphisms between the corresponding Conley indices. This follows from the commutativity property of the Conley index. For example, to see that

$$
\mathcal{L}\left((\varphi_2^{\frac{1}{2}*})^{-1}\right) : CH^*(B_2, b_2) \to CH^*(B_1, b, 1)
$$

is an isomorphism, observe that $b_1 = \varphi_2^{\frac{1}{2}} \circ \varphi_1^{\frac{1}{2}}$ and $b_2 = \varphi_1^{\frac{1}{2}} \circ \varphi_2^{\frac{1}{2}}$.

Substituting these maps into the sequence (5.8) gives rise to the long exact sequence

$$
\cdots \to CH^n(\widetilde{N}, \varphi) \to CH^n(K, \pi) \oplus CH^n(K, \pi) \xrightarrow{\nu'} \tag{5.9}
$$
$$
CH^n(K, \pi) \oplus CH^n(K, \pi) \to CH^{n+1}(\widetilde{N}, \varphi) \to \cdots
$$

where

$$
\nu' = \begin{bmatrix} \mathrm{id} & -\mathrm{id} \\ \mathrm{id} & 0 \end{bmatrix} \begin{bmatrix} \mathrm{id} & 0 \\ 0 & \mathcal{L}\left((\varphi_2^{\frac{1}{2}*})^{-1}\right) \end{bmatrix} \begin{bmatrix} \mathcal{L}(\mu_{1,1}^*) & -\mathcal{L}(\mu_{2,1}^*) \\ \mathcal{L}(\mu_{1,2}^*) & -\mathcal{L}(\mu_{2,2}^*) \end{bmatrix}
$$
$$
\begin{bmatrix} \mathcal{L}\left((\mu_{1,1}^*)^{-1}\right) & 0 \\ 0 & \mathcal{L}\left(\theta_{1,2}^*\right) \end{bmatrix} \begin{bmatrix} \mathrm{id} & 0 \\ \mathrm{id} & \mathrm{id} \end{bmatrix}
$$

$$
= \begin{bmatrix} \mathrm{id} & -\mathrm{id} \\ \mathrm{id} & 0 \end{bmatrix}
$$
$$
\begin{bmatrix} \mathrm{id} & -\mathcal{L}(\mu_{2,1}^*)\mathcal{L}\left(\theta_{1,2}^*\right) \\ \mathcal{L}\left((\varphi_2^{\frac{1}{2}*})^{-1}\right)\mathcal{L}(\mu_{1,2}^*)\mathcal{L}\left((\mu_{1,1}^*)^{-1}\right) & -\mathcal{L}\left((\varphi_2^{\frac{1}{2}*})^{-1}\right)\mathcal{L}(\mu_{2,2}^*)\mathcal{L}\left(\theta_{1,2}^*\right) \end{bmatrix}
$$
$$
\begin{bmatrix} \mathrm{id} & 0 \\ \mathrm{id} & \mathrm{id} \end{bmatrix}
$$

$$
= \begin{bmatrix} \mathrm{id} & -\mathrm{id} \\ \mathrm{id} & 0 \end{bmatrix} \begin{bmatrix} \mathrm{id} & -\mathrm{id} \\ \chi^*(K, \pi) & -\mathrm{id} \end{bmatrix} \begin{bmatrix} \mathrm{id} & 0 \\ \mathrm{id} & \mathrm{id} \end{bmatrix}
$$
$$
= \begin{bmatrix} \mathrm{id} - \chi^*(K, \pi) & 0 \\ 0 & -\mathrm{id} \end{bmatrix}
$$

At this point we can eliminate one factor of

$$
CH^*(B_1, b_1) = CH^*(K, \pi)
$$

and obtain the desired sequence. ∎

It is important to observe that $\mathrm{Con}^*(K, \pi)$ contains more information than $\mathrm{Con}^*(S, \varphi)$.

Corollary 5.2.2 *In the setting of Theorem 5.2.1, the Conley index for S under the semiflow is determined from* $\mathrm{Con}^*(K, \pi)$ *by the exact sequence*

$$0 \to \mathrm{coker}(id - \chi^{n-1}(K, \pi)) \to CH^n(S, \varphi) \to \ker(id - \chi^n(K, \pi)) \to 0.$$

If field coefficients are used for the cohomology, $\mathrm{coker}(id - \chi^n(K, \pi)) \cong \ker(id - \chi^n(K, \pi))$ *and* $CH^n(S, \varphi) \cong \ker(id - \chi^{n-1}(K, \pi)) \oplus \ker(id - \chi^n(K, \pi))$.

However, the index of K under the Poincaré map cannot be recovered from $\mathrm{Con}^*(S, \varphi)$.

Example 5.2.3 Recall from 2.4.9, that the index of the "G-horseshoe" with rational coefficients is

$$\mathrm{Con}^n(K, \pi) = \left\{ \begin{array}{ll} (\mathbf{Q}, 2id) & n = 1 \\ 0 & n \neq 1 \end{array} \right.$$

In particular, $\ker(id - \chi^n(K, \pi)) = 0$ for all n. Now take the suspension flow φ of π. The suspension of K is an isolated invariant set under φ, but by Corollary 5.2.2 its rational cohomology index is trivial.

This example shows that, if we had started with the knowledge that $CH^*(S, \varphi) = 0$, we would not have been able to deduce that $\mathrm{Con}^*(K, \pi)$ was nonzero. More generally, it is possible for sections with different discrete indices to yield semiflows with the same continuous index. In fact, as is indicated by the Poincaré section defined by (5.2) and (5.3) it is possible to have sections for the same isolating neighborhood with different Conley indices.

5.3   The Existence Theorem

By assumption K is an isolated invariant set for the local map $\pi : \Xi \multimap \Xi$, and hence, the set of periodic points of π does not lie on $\partial N \cap \Xi$. This implies that the fixed point index $i(K, \pi^n)$ is defined for all positive integers n [10]. Recall that the cohomological zeta function of π on K is defined as a formal power series

$$\zeta_{K,f}(t) := \exp\left(\sum_{n=1}^\infty \frac{i(K, f^n)}{n} t^n \right).$$

Thus by definition if $\zeta_{K,\pi} \neq 1$, then π has a periodic point in K. The key to the proof of Theorem 5.1.3 is showing that $\zeta_{K,\pi} \neq 1$ from knowledge of $CH^*(S, \varphi)$.

Mrozek [43] showed that the Conley index and zeta functions are related as follows.

Theorem 5.3.1 *Assume $f : X \to X$ is a continuous map and K is an isolated invariant set for f. Then*

$$\zeta_{K,f} = \prod_{n=0}^\infty \det(id - t\chi^n(K, f))^{(-1)^{n+1}}$$

Finally, we shall make use of the following lemma, the proof of which is left to the reader.

Lemma 5.3.2 *Assume A is an endomorphism of a finite dimensional vector space. If 1 is an eigenvalue for A then 1 is an eigenvalue of A^n for all n. If 1 is not an eigenvalue of A then for infinitely $n \in \mathbf{Z}^+$, 1 is not an eigenvalue for A^n.*

Proof of Theorem 5.1.3 Given the space we are in $\mathrm{Con}^*(K, \pi)$ is finite. So let

$$
\begin{aligned}
s_n &:= \dim CH^n(S, \varphi) \\
k_n &:= \dim \ker(\mathrm{id} - \chi^n(K, \pi)).
\end{aligned}
$$

Corollary 5.2.2 implies that

$$
s_n = k_n + k_{n-1} \qquad \text{for } n \in \mathbf{Z}.
$$

Solving these equations leads to

$$
k_n = s_n - s_{n-1} + s_{n-2} - \cdots \pm s_0 \qquad \text{for } n \in \mathbf{Z}.
$$

Under the first assumption (5.4) (in the other case the proof is analogous) this implies that $k_{2n+1} = 0$, $k_{2n} = s_{2n}$ for $n \in \mathbf{N}$. This shows that 1 is not an eigenvalue of

$$
\chi^{\mathrm{odd}}(K, \pi) := \bigoplus_{n=0}^{\infty} \chi^{2n+1}(K, \pi)
$$

but it is an eigenvalue of

$$
\chi^{\mathrm{even}}(K, \pi) := \bigoplus_{n=0}^{\infty} \chi^{2n}(K, \pi),
$$

because not all s_{2n} are zero. By Theorem 5.3.1, $\zeta_{K, \bar{\pi}^m} \neq 1$. Thus $\bar{\pi}^m$ has a periodic point in K, which is obviously also a periodic point of π and φ. ∎

Chapter 6

A Global Semi–Conjugacy

As was remarked in the introduction one of our goals is to show how the Conley index can be used to describe the set of connecting orbits between Morse sets. In this chapter we shall discuss such a result. As presented here this result may appear to be quite artificial. It is purposely stated in a very concrete and simple manner to explain the ideas. However the general theorem (which will be stated at the very end of the chapter) was motivated by and is applicable to several different partial differential equations which are similar only to the extent that they exhibit bistable gradient like dynamics. The reader is referred to [35] for further information concerning the applicability of the result and for the details of the proof.

6.1   The Semi–Conjugacy

Let
$$D^2 := \{z = (z_0, z_1, z_2) \mid \|z\| \le 1\} \subset \mathbf{R}^2$$
be the closed unit ball in $\mathbf{R}^2$ and $S^1 = \partial D^2$ be the unit sphere. Let $\mathbf{e}_p^{\pm} = (0, \ldots, \pm 1, \ldots, 0)$ denote the unit vectors in the pth direction. Consider $\psi : \mathbf{R} \times D^2 \to D^2$ the flow generated by the following system of ordinary differential equations

$$
\begin{aligned}
\dot{\zeta} &= Q\zeta - \langle Q\zeta, \zeta \rangle \zeta \quad \zeta \in S^1 & (6.1) \\
\dot{r} &= r(1 - r) \qquad r \in [0, 1] & (6.2)
\end{aligned}
$$

where
$$Q = \begin{bmatrix} 1 & 0 \\ 0 & \frac{1}{2} \end{bmatrix}$$

The dynamics of ψ which is indicated in Figure 6.1 is best understood by realizing that (6.1) is obtained by projecting the linear system $\dot{z} = Qz$ onto the unit sphere.

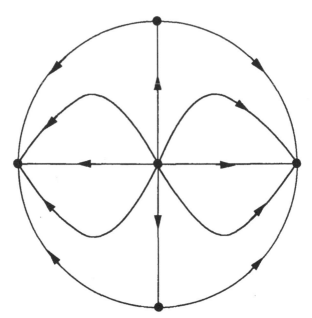

Figure 6.1: The dynamics of ψ.

Using the language we have been developing we can describe the dynamics of ψ as follows.

Proposition 6.1.1 *Let* $\Pi(p^\pm) = e_p^\pm$ *and* $\Pi(2) = 0$, *then*

$$\mathcal{M}(D^2) = \{\Pi(p^\pm) \mid p = 0, 1\} \cup \{\Pi(2)\}$$

is a Morse decomposition of D^2 *under* ϕ *with an admissible order given by* $0^\pm < 1^\pm < 2$.

Furthermore, the Conley index of each Morse set is given by

$$CH^n(\Pi(p^\pm)) \approx \begin{cases} \mathbf{Z} & \text{if } n = p \\ 0 & \text{otherwise} \end{cases}$$

and

$$CH^n(\Pi(2))) \approx \begin{cases} \mathbf{Z} & \text{if } n = 2 \\ 0 & \text{otherwise} \end{cases}$$

A brute force computation provides us with the following result.

Proposition 6.1.2 *There is a unique connection matrix for* $\mathcal{M}(D^P)$ *and acting on the space*

$$CH^0(\Pi(0^-)) \oplus CH^0(\Pi(0^+)) \oplus CH^1(\Pi(1^-)) \oplus CH^1(\Pi(1^+)) \oplus CH^2(\Pi(2))$$

it takes the form

$$\Delta = \begin{bmatrix} 0 & 0 & 0 & 0 & 0 \\ 0 & 0 & 0 & 0 & 0 \\ 1 & 1 & 0 & 0 & 0 \\ -1 & -1 & 0 & 0 & 0 \\ 0 & 0 & 1 & -1 & 0 \end{bmatrix}$$

We can now ask to what extent do Propositions 6.1.1 and 6.1.2 characterize the dynamics of an invariant set? In other words let us rewrite the propositions as hypothesis thereby making the following assumptions.

A1 A *is the global attractor of a flow φ on $\mathbf{R}^n$.*

A2

$$\mathcal{M}(A) = \{M(p^{\pm}) \mid p = 0, 1\} \cup \{M(2)\}$$

with ordering $0^{\pm} < 1^{\pm} < 2$ is a Morse decomposition of A.

A3 *The cohomology Conley indices of the Morse sets are*

$$CH^n(M(2))) \approx \begin{cases} \mathbf{Z} & \text{if } n = 2 \\ 0 & \text{otherwise} \end{cases}$$

and for $p = 0, 1$

$$CH^n(M(p^{\pm})) \approx \begin{cases} \mathbf{Z} & \text{if } n = p \\ 0 & \text{otherwise} \end{cases}$$

A4 *The connection matrix for $\mathcal{M}(A)$ is the same as in Proposition 6.1.2.*

The following theorem states that these assumptions on the Morse decomposition and the connection matrix are enough to regain the dynamics of ψ up to a semi-conjugacy. Observe that no explicit assumptions are being made on the geometry or topology of the invariant set A or the Morse sets. Of course by Theorem 2.2.11 we know that they contain fixed points, but they may, also, contain considerably more dynamics.

Theorem 6.1.3 *Given assumptions* **A1** - **A4** *there exists a continuous surjective map $\rho : A \to D^2$ and $\tilde{\varphi}$ a flow obtained by an order preserving time re-parameterization of φ such that the following diagram commutes*

$$
\begin{array}{ccc}
\mathbf{R} \times A & \xrightarrow{\;id \times \rho\;} & \mathbf{R} \times D^2 \\
{\scriptstyle \tilde{\varphi}} \downarrow & & \downarrow {\scriptstyle \psi} \\
A & \xrightarrow{\;\;\rho\;\;} & D^2
\end{array}
$$

where $M(p^{\pm}) = \rho^{-1}(e_p^{\pm})$ for $p = 0, 1$, and $M(2) = \rho^{-1}(0)$.

Some of the basic ideas used in the proof of this theorem already appeared in the almost trivial semi-conjugacy constructed in Section 3.3. In this case there are three steps:

1. Define a function $f : \mathcal{A} \to X$ (which is not necessarily even continuous) from which the semi–conjugacy ρ will be derived. To construct this map we need to impose a "coordinate" system on $\mathcal{A}$. As in Section 3.3, we can use the existence of a Lyapunov function to obtain "global" coordinates. "Local" coordinates, which describe how trajectories behave in the vicinity of the Morse sets also have to be derived. The product of these coordinates then define the space X.

2. The next step is to define an equivalence relation $\sim$ on X in such a way that the induced quotient space $X/\sim$ is homeomorphic to D^2, the induced map $\rho : \mathcal{A} \to D^2$ is continuous, and such that the induced flow on D^2 is conjugate to the flow ψ.

3. Finally, it needs to be shown that ρ is surjective.

6.2 Constructing ρ

To better distinguish between the points of the different spaces we shall always assume that u denotes an element of $\mathcal{A}$.

As was indicated in the introduction, our theorem can be stated more clearly if we use a time reparameterized flow $\tilde{\varphi}$ rather that the actual flow φ. Though obviously important if one wishes to construct general semi–conjugacies, the details of performing the reparameterization are rather tedious and thus the reader is referred to [31, Section 5]. For our purposes we shall be satisfied by merely asserting the existence of the following result.

Proposition 6.2.1 *Given a flow $\varphi : \mathbf{R} \times \mathcal{A} \to \mathcal{A}$ satisfying A2 there exists:*

(i) $\tilde{\varphi} : \mathbf{R} \times \mathcal{A} \to \mathcal{A}$ *a flow obtained via an order preserving reparameterization of time;*

(ii) *sets N_p, L_p^+, L_p^- for $p = 0, 1, 2$ (let $2 = 2^+$) such that:*

    (a) $N_{p\pm}$ *are isolating neighborhoods of $M(p^\pm)$, respectively, and $N_{p+} \cap N_{p-} = \emptyset$;*

    (b) $\partial N_p = L_p^+ \cup L_p^-$;

    (c) $L_p^\pm$ *are local sections of $\tilde{\varphi}$;*

    (d) $(N_{p\pm}, L_{p\pm}^-)$ *is a regular index pair for $M(p^\pm)$ under $\tilde{\varphi}$;*

    (e) $(N_{p\pm}, L_{p\pm}^+)$ *is a regular index pair for $M(p^\pm)$ under $\tilde{\varphi}'$, where $\tilde{\varphi}'(t, u) = \tilde{\varphi}(-t, u)$;*

    (f)

$$\tilde{\varphi}(\mathbf{R}, u) \cap N_{p\pm} = \tilde{\varphi}(I_{p\pm}(u), u) \cap N_{p\pm}$$

    where $I_{p\pm}$ is a closed interval;

(iii) a Lyapunov function

$$\tilde{V} : \mathcal{A} \to [0, 2]$$

such that

(a) if $u \in M(p^{\pm}) \cup (L_p^+ \cap L_p^-)$, then $\tilde{V}(u) = p$;

(b) if

$$\tilde{\varphi}([0, t], x) \cap \bigcup_{p=0}^{2} N_{p\pm} = \emptyset$$

then

$$\tilde{V}(\tilde{\varphi}(t, x)) = \tilde{V}(x) - t.$$

Since this proposition guarantees that $I_p(x)$ is a closed interval, we shall write

$$I_p(x) = [a_p(x), b_p(x)]$$

with the understanding that if $I_p(x) = \emptyset$ then $a_p(x)$ and $b_p(x)$ are not defined, and if $I_p(x)$ is unbounded then $a_p(x) = -\infty$ and/or $b_p(x) = \infty$. Let

$$\Theta_p := \{x \in \mathcal{A} \mid I_p(x) \neq \emptyset\},$$

then the fact that $(N_p, L_p^{\pm})$ are regular index pairs gives rise to the following lemma.

Lemma 6.2.2 *The functions $a_p, b_p : \Theta_p \to [-\infty, \infty]$ are continuous.*

We now begin by considering the local coordinates. For each Morse set $M(p^*)$ we want to measure how "close" the orbit of $u \in \mathcal{A}$ passes by $M(p^*)$. Thus we shall define a function

$$\tau_p : \mathcal{A} \to [-1, 1], \quad p = 1, \dots, P-1$$

with the property that $\tau_p(u) = 1$ implies that $\omega(u)$ or $\alpha(u) \subset M(p^+)$, $\tau_p(u) = -1$ implies that $\omega(u)$ or $\alpha(u) \subset M(p^-)$, while $\tau_p(u) = 0$ implies that the orbit of u does not intersect $\text{int}(N_p)$, the isolating neighborhood of $M(p)$. Guided by the idea that the longer the trajectory stays in the isolating neighborhood the closer it is to the Morse set, define

$$\lambda_{p\pm}(u) = \begin{cases} \infty & \text{if } b_{p\pm}(u) = \infty \text{ or } a_{p\pm}(u) = -\infty \\ 0 & \text{if } I_{p\pm}(u) = \emptyset \\ b_{p\pm}(u) - a_{p\pm}(u) & \text{otherwise} \end{cases} \tag{6.3}$$

Set

$$\tau_p(u) = \frac{2}{\pi} \tan^{-1}(\lambda_{p+}(u) - \lambda_{p-}(u)) \tag{6.4}$$

where $\tan^{-1}(\pm\infty) = \pm\frac{\pi}{2}$.

Lemma 6.2.3 *For* $p = 0, 1$,

$$\tau_p : \Theta_p \to [-1, 1]$$

is a continuous function. Furthermore,

(i) $\tau_p(u) = 1$ *if and only if* $\omega(u)$ *or* $\alpha(u)$ *is contained in* $M(p^+)$.

(ii) $\tau_p(u) = -1$ *if and only if* $\omega(u)$ *or* $\alpha(u)$ *is contained in* $M(p^-)$.

(iii) $\tau_p(u) = 0$ *if and only if* $\tilde{\varphi}(\mathbf{R}, u) \cap N_{p\pm} = \emptyset$ *or* $\tilde{\varphi}(\mathbf{R}, u) \cap N_{p\bullet} \subset L_{p\bullet} \cap L_{p\bullet}^*$.

Using the Lyapunov function $\tilde{V}$ we can construct a more suitable Lyapunov function V.

Lemma 6.2.4 *The Lyapunov function V is continuous. Furthermore,*

(i) *if* $\tilde{\varphi}([0, t], u) \cap \left(\bigcup_{p=0}^{2} N_p \right) = \emptyset$, *then*

$$V(\tilde{\varphi}(t, u)) = V(u) - t.$$

(ii) *if* $\tilde{\varphi}([0, t], u) \subset N_p$, $\tilde{\varphi}([0, t], u') \subset N_p$, $\tau_p(u) = \tau_p(u')$, *and* $V(u) = V(u')$, *then*

$$V(\tilde{\varphi}(t, x)) = V(\tilde{\varphi}(t, y)).$$

We are finally ready to define

$$f(u) = (V(u), \tau_0(u), \tau_1(u)).$$

Lemma 6.2.4 guarantees that f is well defined. While it is obvious that $f : \mathcal{A} \to [0, 2] \times [-1, 1] \times [-1, 1]$ we wish to get a better approximation of the range of f. Since we have a Morse decomposition of $\mathcal{A}$ if $u \in \mathcal{A}$, then both $\omega(u)$ and $\alpha(u)$ belong to some Morse set. If $u = M(2)$ then

$$f(u) = (2, 0, 0, \ldots, 0) \in \partial([0, 2] \times [-1, 1] \times [-1, 1]).$$

If $u \notin M(2)$, then $\omega(u) \in M(0^\pm) \cup M(1^\pm)$ which implies that $\tau_0(u) = \pm 1$ or $\tau_1(u) = \pm 1$. In either case

$$f(u) \in \partial([0, 2] \times [-1, 1] \times [-1, 1]).$$

Thus we can write

$$f : \mathcal{A} \to \partial \left([0, P] \times [-1, 1]^P \right).$$

In fact, we can still further refine our knowledge of the image of f as follows. If $V(u) = 2$, then $u \in M(2)$ hence $f(u) = (2, 0, 0)$. If $1 < V(u) < 2$, then $\alpha(u) \subset M(2)$. If in addition $\omega(u) \subset M(1^\pm)$, then $\tau_1(u) =$ and $\tau_0(u) = 0$ (since $V(\varphi(t, u)) > 1$ for all $t \in \mathbf{R}$ and hence $\varphi(\mathbf{R}, u) \cap N_{0\pm} = \emptyset$). Thus, $f(u) = (V(u), 0, 1^\pm)$. Continuing this type of analysis we see that the image of

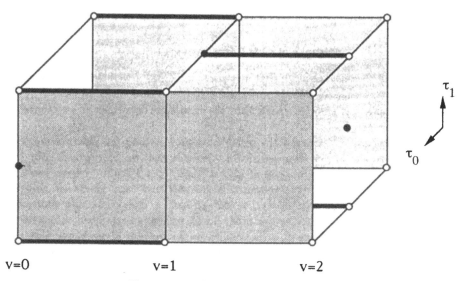

v=0   v=1   v=2

Figure 6.2: The image of $\mathcal{A}$ under f.

$\mathcal{A}$ under f is a subset of the shaded region in Figure 6.2. We will refer to this
shaded region as X.

Now observe that in Figure 6.2 the dark dots represent the images of the
Morse sets and the dark lines the images of the connecting orbits between adjacent Morse sets. Also, observe that the lines are disconnected from the dots.
This represents the discontinuities of the map f. How did these discontinuities
arise? V is a continuous function defined on all of $\mathcal{A}$, hence the point at which
the discontinuities must be due to τ_0 and τ_1. By Lemma 6.2.3 these discontinuities occur in $\mathcal{A} \setminus \Theta_p$ where it is not too hard to check that they arise from the
following type of phenomenon. Let $\{u_n\}$ be a sequence of points in $\mathcal{A}$ and $\{t_n\}$ a
sequence of positive times, such that $u_n \to u$, $u_n(t_n) \to M(1^{\pm})$, $\omega(u_n) \subset M(0^{\pm})$,
and $\alpha(u_n) \subset M(2)$. Clearly, $\tau_0(u_n) = \pm 1$. But since $\omega(u) \subset M(p^{\pm})$, $\tau_0(u) = 0$.

With this in mind we define the following equivalence class on the shaded
region.

$$
\begin{aligned}
(0,1,0) &\sim (0,1,\pm 1) \\
(0,-1,0) &\sim (0,-1,\pm 1) \\
(1,0,1) &\sim (1,\pm 1,1) \\
(1,0,-1) &\sim (1,\pm 1,-1) \\
(2,0,0) &\sim (2,0,\pm 1)
\end{aligned}
$$

Let $Q : X \to X/\sim$ be the resulting quotient map. We leave it to the reader to
check that $X/\sim$ is homeomorphic to D^2. If the reader believed our discussion

concerning the discontinuities of f then it is easy to see that

$$\rho := Q \circ f : \mathcal{A} \to D^2$$

is continuous. Finally, observe that any point in X has a well defined trajectory in $\mathcal{A}$ using the local and global coordinate systems. These trajectories carry over to the D^2 and it is easy to check that the resulting dynamics is conjugate to that of ψ.

Thus, we have constructed the semi–conjugacy ρ using the information provided to us in the Morse decomposition. Observe,that we have potentially collapsed out a tremendous amount of information. In particular, we are sending each Morse set to point. Our assumption **A3** was that each Morse set had the index of a fixed point, but, as was mentioned before, does not imply that each Morse set is a fixed point. On the other hand, unless we are given information concerning each Morse set it is impossible to do better. If we know that the Morse sets have the index of periodic orbits and that their corresponding isolating neighborhoods have Poincaré sections then we can do better [30, 36, 22].

6.3 ρ is surjective

As in the example of Section 3.3, the proof of the surjectivity of ρ depends on the algebra of the Conley index. In this case we need to use slightly more sophisticated techniques. We begin therefore with the simple observation that if (N, L) is an index pair for an isolated invariant set S, the cup product defines a pairing

$$H^p(N) \otimes H^q(N, L) \to H^{p+q}(N, L).$$

Since the collection

$$\{N_\alpha \mid (N_\alpha, L_\alpha) \text{ is an index pair}\}$$

is cofinal under inclusion with the set of neighborhoods of S, this pairing defines another pairing

$$H^p(S) \otimes CH^q(S) \to CH^{p+q}(S).$$

This pairing exists for any invariant set in any flow. If $T_I \in CH^n(M(I))$ is a generator, there is a map

$$\iota_I^n : H^p(M(I)) \to CH^{p+n}(M(I))$$

defined by

$$\iota_I^n(z) = z \cup T_I.$$

Obviously the behavior of the index under semi-conjugacies is of considerable importance to us. The essence of the matter is that is that the index theory is natural with respect to semi-conjugacies, as long as one works with pre-images, rather than images (cf. [28, 29]). To be more precise, if $f : X \to Y$ is a semi-conjugacy and S an isolated invariant set in Y with index pair (N, L), then

$T = f^{-1}(S)$ is an isolated invariant set in X with index pair $(f^{-1}(N), f^{-1}(L))$. The naturality implies that there are maps $f_* : CH_*(T) \to CH_*(S)$ and $f^* : CH^*(S) \to CH^*(T)$ and that the pairing defined above commutes with this map. In other words, there is a commutative diagram

$$
\begin{array}{ccc}
H^p(S) \otimes CH^q(S) & \overset{\iota^q}{\to} & CH^{p+q}(S) \\
\downarrow f^* \otimes f^* & & \downarrow f^* \\
H^p(T) \otimes CH^q(T) & \overset{\iota^q}{\to} & CH^{p+q}(T)
\end{array}
$$

Similarly, if $\{M(p)\}$ is a Morse decomposition of S, then $\{T(p) = f^{-1}(M(p))\}$ is a Morse decomposition of T, and any admissible ordering on S gives an admissible ordering on T. Thus we can use the same ordering for both decompositions, and if I is an interval in that ordering, there is a map $CH^*(M(I)) \to CH^*(T(I))$. Moreover, the attractor-repeller sequence is natural: if I and J are adjacent intervals with $I < J$, there is a commutative diagram

$$
\begin{array}{ccccccc}
\overset{\delta}{\to} & CH^p(M(J)) & \to & CH^p(M(IJ)) & \to & CH^p(M(J)) & \to \\
& \downarrow f^* & & \downarrow f^* & & \downarrow f^* & \\
\overset{\delta}{\to} & CH^p(T(J)) & \to & CH^p(T(IJ)) & \to & CH^p(T(J)) & \to
\end{array}
$$

Now recall the following elementary fact from algebraic topology.

Lemma 6.3.1 *If $f : X \to S^n$ is continuous and $f^* : H^n(S^n) \to H^n(X)$ is non-zero, then f is surjective.*

The strategy is to reduce the cohomology index information to an application of this lemma.

Recall that $\{\Pi(p^\pm) \mid p = 0, 1\} \cup \{\Pi(12)\}$ with the flow defined partial order $2 > 1^\pm > 0^\pm$ is a Morse decomposition for D^2 under ψ. Thus the indexing set and flow defined order for $\mathcal{M}(D^2)$ and $\mathcal{M}(\mathcal{A})$ are identical. This implies that I is an interval in $\mathcal{M}(D^2)$ if and only if it is an interval for $\mathcal{M}(\mathcal{A})$. For I an interval define

$$ \rho_I := \rho \mid_{M(I)} : M(I) \to D^2. $$

This of course implies that $\rho_I(M(I)) \subset \Pi(I)$.

Since both $\mathcal{A}$ and D^P are attractors, [31, Proposition 7.4] implies that for any interval I

$$ \rho^* : CH^*(\Pi(I)) \to CH^*(M(I)) $$

is an isomorphism. From now on let

$$ I = \{0^\pm, 1^\pm\}. $$

This implies that I is an attracting interval, and hence, $M(I)$ and $\Pi(I)$ are attractors. Therefore, the maps $\iota_I^0 : H^k(\Pi(I)) \to CH^k(\Pi(I))$ and $\iota_I^{\prime 0} : H^k(M(I)) \to CH^k(M(I))$ are isomorphisms.

By [31, Proposition 7.4] there exists a commutative diagram

$$
\begin{array}{ccc}
H^k(\Pi(I)) & \xrightarrow{\iota_I^0} & CH^k(\Pi(I)) \\
\downarrow \rho_I^* & & \downarrow \rho^* \\
H^k(M(I)) & \xrightarrow{\iota_I'^0} & CH^k(M(I))
\end{array}
\tag{6.5}
$$

This now forces ρ_I^* to be an isomorphism. Since $\Pi(I) = S^1$, Lemma 6.3.1 implies that $\rho : M(I) \to \Pi(I)$ is onto.

Since $\rho(M(P)) = \Pi(P)$, it only remains to be shown that $C(M(P), M(I))$ maps onto $C(\Pi(P), \Pi(P))$. To do this we use the fact that

$$
\mathcal{W} := \{ u \in \mathcal{A} \mid V(u) = P - \frac{1}{2} \}
$$

is a local section for $\widetilde{\varphi}$ and

$$
W := Q\left(\{ (\frac{3}{2}, \tau_0, \tau_1) \} \subset X \right)
$$

is a local section for ψ. Since $\widetilde{\varphi}(\mathbf{R}, \mathcal{W}) = C(M(P), M(I))$ and $\psi(\mathbf{R}, W) = C(\Pi(P), \Pi(I))$, it is sufficient to show that $\rho : \mathcal{W} \to W$ is onto. Observe that $Q\left(\{ (\frac{3}{2}, \tau_0, \tau_1) \} \subset X \right)$, and hence W, is homeomorphic to S^1.

Define

$$
N := Q\left(\bigcup_{v=\frac{3}{2}}^{2} \{ (v, \tau_0, \tau_1) \} \subset X \right).
$$

Then, (N, W) is an index pair for $\Pi(P)$. Furthermore $(\mathcal{N}, \mathcal{W}) := (\rho^{-1}(N), \mathcal{W})$ is an index pair for $M(P)$. We can relate this information via the following commutative diagram.

$$
\begin{array}{ccccccccc}
\to & H^{P-1}(N) & \to & H^{P-1}(W) & \xrightarrow{\delta} & H^P(N, W) & \to & H^P(N) & \to \\
& \downarrow \rho^* & & \downarrow \rho^* & & \downarrow \rho^* & & \downarrow \rho^* & \\
\to & H^{P-1}(\mathcal{N}) & \to & H^{P-1}(\mathcal{W}) & \xrightarrow{\delta'} & H^P(\mathcal{N}, \mathcal{W}) & \to & H^P(\mathcal{N}) & \to
\end{array}
$$

Using the fact that N is homeomorphic to D^2, W is homeomorphic to S^1, and the cohomology index information, this reduces to

$$
\begin{array}{ccccccccc}
\to & 0 & \to & \mathbf{Z} & \xrightarrow{\delta} & \mathbf{Z} & \to & 0 & \to \\
& \downarrow & & \downarrow \rho^* & & \downarrow \approx & & \downarrow & \\
\to & H^1(\mathcal{N}) & \to & H^1(\mathcal{W}) & \xrightarrow{\delta'} & H^2(\mathcal{N}, \mathcal{W}) & \to & H^2(\mathcal{N}) & \to
\end{array}
$$

Therefore, $\rho^* : H^1(W) \to H^1(\mathcal{W})$ is injective. By Lemma 6.3.1 this implies that $\rho : \mathcal{W} \to W$ is onto. ∎

6.4 The General Theorem

We conclude this chapter with the statement of the general result.

We begin by defining the model flow ψ. Let

$$D^P := \{z = (z_0, \ldots, z_{P-1}) \mid \|z\| \le 1\} \subset \mathbf{R}^P$$

be the closed unit ball in $\mathbf{R}^P$ and $S^{P-1} = \partial D^P$ be the unit sphere. Let $\psi^P :$ $\mathbf{R} \times D^P \to D^P$ denote the flow generated by the following system of ordinary differential equations

$$\begin{aligned} \dot{\zeta} &= Q\zeta - \langle Q\zeta, \zeta \rangle \zeta \quad \zeta \in S^{P-1} \\ \dot{r} &= r(1-r) \qquad r \in [0,1] \end{aligned}$$

where

$$Q = \begin{bmatrix} 1 & 0 & \cdots & 0 \\ 0 & \frac{1}{2} & & \\ \vdots & & \ddots & \\ 0 & & & \frac{1}{P} \end{bmatrix}$$

Let $e_p^{\pm} = (0, \ldots, \pm 1, \ldots, 0)$ be the unit vectors in the pth direction.

The general assumptions are as follows.

A1 $\mathcal{A}$ *is a global compact attractor for a semi-flow Φ on a Banach space. Furthermore, if φ denotes the restriction of Φ to $\mathcal{A}$ then φ defines a flow on $\mathcal{A}$.*

A2 *Under the flow $\varphi : \mathbf{R} \times \mathcal{A} \to \mathcal{A}$*

$$\mathcal{M}(\mathcal{A}) = \{M(p^{\pm}) \mid p = 0, \ldots, P-1\} \bigcup \{M(P)\}$$

with ordering $P > P - 1^{\pm} > \ldots > 1^{\pm} > 0^{\pm}$ is a Morse decomposition of $\mathcal{A}$.

A3 *The cohomology Conley indices of the Morse sets are*

$$CH^k(M(P))) \approx \begin{cases} \mathbf{Z} & \text{if } k = P \\ 0 & \text{otherwise} \end{cases}$$

and for $p = 0, \ldots, P-1$

$$CH^k(M(p^{\pm})) \approx \begin{cases} \mathbf{Z} & \text{if } k = p \\ 0 & \text{otherwise} \end{cases}$$

A4 *The connection matrix for $\mathcal{M}(\mathcal{A})$ is given by*

$$\Delta = \begin{bmatrix} 0 & 0 & 0 & \cdots & & 0 \\ D_0 & 0 & 0 & & & \vdots \\ & D_1 & & \ddots & & 0 \\ & & & & 0 & 0 \\ \vdots & & & & & \\ 0 & & \cdots & & D_{P-1} & 0 \end{bmatrix}$$

where, up to a choice of orientation, for $p = 0, \ldots, P - 2$

$$D_p : CH^p(M(p^-)) \oplus CH^p(M(p^+)) \rightarrow CH^{p+1}(M(p+1^-)) \oplus CH^{p+1}(M(p+1^+))$$

is given by

$$D_p = \begin{bmatrix} 1 & -1 \\ 1 & -1 \end{bmatrix}$$

and

$$D_{P-1} : CH^{P-1}(M(P-1^-))) \oplus CH^{P-1}(M(P-1^+)) \rightarrow CH^P(M(P))$$

is given by

$$D_P = [1, -1]$$

Furthermore, rank $D_p = 1$ for all p.

Theorem 6.4.1 *Given assumptions* **A1** *-* **A4** *there exists a continuous surjective map $\rho : \mathcal{A} \rightarrow D^P$ and $\tilde{\varphi}$ a flow obtained by an order preserving time reparameterization of φ such that the following diagram commutes*

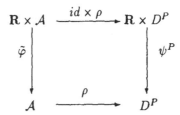

where $M(p^\pm)) = \rho^{-1}(e_p^\pm)$ for $0 \leq p \leq P - 1$, and $M(P) = \rho^{-1}(0)$.

Chapter 7

A Local Semi-Conjugacy

In the previous chapter we discussed how to show the existence of a semi-conjugacy which captured the global dynamics of a Morse decomposition, i.e. the structure of the connecting orbits. In this chapter we turn our attention to local dynamics, that is dynamics within Morse sets. Inorder for this type of result to be interesting the dynamics have to be complicated recurrent motion. The most common examples of such dynamics are horseshoes. Following the lead of the last chapter we shall recast the example of the G-horseshoe into the language of the Conley index theory and then ask whether the index captures enough of the dynamics inorder to obtain a semi–conjugacy back onto the horseshoe.

7.1 The G–Horseshoe

We begin by examining the G-horseshoe map f further. Refering to Figure 7.1, the the two components of $f^{-1}(Q)$ are denoted by N_0 and N_1, and set $N :=
N_0 \cup N_1$. Observe that $\operatorname{Inv} N = \operatorname{Inv} Q$. Thus given $x \in \operatorname{Inv} Q$ its trajectory lies in $N = N_0 \cup N_1$, and hence, there is a well defined sequence of the symbols 0 and 1 which indicate how x travels through Q. This of course defines a map $\rho : \operatorname{Inv} Q \to \Sigma^2$, the space of bi–infinite sequences consisting of 0 and 1. For this example it is easy to show that ρ is a homeomorphism, and that if $\sigma : \Sigma^2 \to \Sigma^2$ denotes the shift map then

$$\rho \circ f = \sigma \circ \rho$$

i.e. ρ is a conjugacy.

Now how do we relate this to the Conley index theory? Let us define

$$N_{ij} = N_i \cap f(N_j).$$

Now observe that since N is an isolating neighborhood and N_1 and N_0 are disjoint compact sets whos union is N, that N_1 and N_0 are also isolating neighborhoods. Continuing in this manner one can also show that N_{ij} and any union of these sets is an isolating neighborhood. Furthermore,

$$\operatorname{Con}^*(\operatorname{Inv} N_{00}) \approx \operatorname{Con}^*(\operatorname{Inv} N_0) \approx (\mathbf{Z}, \operatorname{id})$$

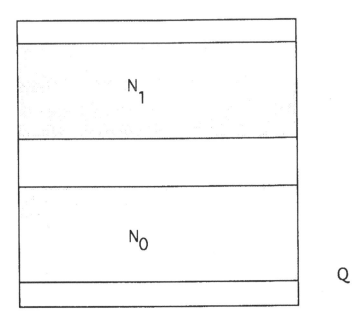

Figure 7.1: An isolating neighborhood for Smale's horseshoe.

and

$$\mathrm{Con}^*(\mathrm{Inv}\, N_{11}) \approx \mathrm{Con}^*(\mathrm{Inv}\, N_1) \approx (\mathbf{Z}, \mathrm{id})$$

Now define

$$
\begin{aligned}
N(01) &:= N_{00} \cup N_{01} \cup N_{11} \\
N(10) &:= N_{00} \cup N_{10} \cup N_{11}.
\end{aligned}
$$

The $\mathrm{Inv}\, N(01)$ contains a connecting orbit from the fixed point in N_1 to the fixed point in N_0. Likewise, $\mathrm{Inv}\, N(01)$ contains the connecting orbit from N_0 to N_1. These heteroclinic connections play a fundamental role in the dynamics of the horseshoe since they serve to "carry" the orbits from one region to the other. Thus we should ask whether we can detect these orbits via the Conley index. This leads to the following theorem.

Theorem 7.1.1 *1. $(\mathrm{Inv}\, N_k, \mathrm{Inv}\, N_l)$ is an attractor-repeller pair in $\mathrm{Inv}\, N(kl)$.*

2. The Conley indices of $\mathrm{Inv}\, N_k, \mathrm{Inv}\, N_l$ and $\mathrm{Inv}\, N(kl)$ are related by the following commutative diagram with exact rows

$$
\begin{array}{ccccccccc}
0 & \to & CH^*(\mathrm{Inv}\, N_l) & \to & CH^*(\mathrm{Inv}\, N(kl)) & \to & CH^*\,\mathrm{Inv}\, N_k) & \to & 0 \\
& & \downarrow{\scriptstyle \chi(\mathrm{Inv}\, N_l)} & & \downarrow{\scriptstyle \chi(\mathrm{Inv}\, N(kl))} & & \downarrow{\scriptstyle \chi(\mathrm{Inv}\, N_k)} & & \\
0 & \to & CH^*(\mathrm{Inv}\, N_l) & \to & CH^*(\mathrm{Inv}\, N(kl)) & \to & CH^*\,\mathrm{Inv}\, N_k) & \to & 0
\end{array}
$$

$$(7.1)$$

in which vertical arrows denote the corresponding index maps.

3.
$$CH^*(\text{Inv } N(kl)) = CH^*(\text{Inv } N_k) \oplus CH^*(\text{Inv } N_l). \tag{7.2}$$

4. If

$$\chi(\text{Inv } N(kl)) \text{ and } \chi(\text{Inv } N_k) \oplus \chi(\text{Inv } N_l) \text{ are not conjugate}$$

then there exists a connecting orbit from $\text{Inv } N_l$ *to* $\text{Inv } N_k$ *in* $\text{Inv } N(kl)$, *i.e. an element* $x \in \text{Inv } N(kl)$ *such that* $\alpha(x) \subset \text{Inv } N_k$ *and* $\omega(x) \subset \text{Inv } N_l$.

For the horseshoe, observe that this implies that in matrix form

$$\chi^1(\text{Inv } N(01)) = \begin{bmatrix} 1 & \mu \\ 0 & 1 \end{bmatrix}$$

in which μ can depend on the choice of the complement to $CH^1(\text{Inv } N_k)$ in the decomposition (7.2) given by (7.1). However, since all such matrices which represent the same map are conjugate, the number

$$\alpha_{kl} := \begin{cases} 0 & \text{if } \mu = 0 \\ 1 & \text{if } \mu \neq 0 \end{cases}$$

is an invariant of the attractor-repeller pair $(\text{Inv } N_k, \text{Inv } N_l)$ in $\text{Inv } N(kl)$.

We shall call α_{kl} a *connection number*, because $\alpha_{kl} \neq 0$ implies $\chi^m(\text{Inv } N(kl))$ and $\chi^m(\text{Inv } N_k) \oplus \chi^m(\text{Inv } N_l)$ are not conjugate. Hence, by Theorem 7.1.1, if $\alpha_{kl} = 1$, then there exists a connection from $\text{Inv } N_l$ to $\text{Inv } N_k$

For the horseshoe it can be checked that $\alpha_{10} = 1$ and $\alpha_{01} = 1$.

We can now state the desired theorem.

Theorem 7.1.2 *Let N be an isolating neighborhood. Assume that $N = N_0 \cup N_1$ where N_0 and N_1 are disjoint compact sets, and that for some fixed m and $k = 0, 1$*

$$\text{Con}^m(\text{Inv}(N_k)) \approx (G, \text{id}).$$

Let α_{kl} denote the connection number obtained from $N(kl)$ and assume $\alpha_{10} = \alpha_{01} = 1$. Then there exists $d \in \mathbf{N}$ and a continuous surjection

$$\rho : \text{Inv } N \to \Sigma_2$$

such that the following diagram commutes

$$
\begin{array}{ccc}
\text{Inv}(N) & \overset{f^d}{\to} & \text{Inv}(N) \\
\downarrow{\scriptstyle \rho} & & \downarrow{\scriptstyle \rho} \\
\Sigma_2 & \overset{\sigma}{\to} & \Sigma_2
\end{array}
$$

7.2 The Proof

This section would be better titled sketch of proof. Complete proofs can be found in [37]. Her we will not provide all the details, but rather only those which we feel are essential to understanding the basic ideas. What is being left out, for the most part, are arguments on the level of the Alexander–Spanier cochains.

We begin with a description of a portion of the proof of Theorem 7.1.1. Without loss of generality we can assume that N is an *isolating block*, i.e.

$$f^{-1}(N) \cap N \cap f(N) \subset \text{int } N.$$

Let

$$M = \bigcup_{k,l=0,1} N_{kl}.$$

This is, of course, a more refined isolating neighborhood for Inv N. Set

$$E := M \setminus f^{-1}(M).$$

Observe that since N is an isolating block E acts as an exit set for M. In the construction of the connection matrix the first step was to find an index filtration which provided all the indices of each of the Morse sets and, also, provided a means for determining the relations between these indices. The proof of Theorem 7.1.1 and in particular the existence of the commutative diagram (7.1) requires a something similiar. In this case the spaces and the cohomology indices of interest are the following:

$$
\begin{aligned}
V &= H^*(M, E) \\
V_0 &= H^*(M, E \cup N_{11} \cup N_{10}) \\
V_l &= H^*(M, E \cup N_{00} \cup N_{01}) \\
V_{00} &= H^*(M, E \cup N_{11} \cup N_{10} \cup N_{01}) \\
V_{11} &= H^*(M, E \cup N_{00} \cup N_{10} \cup N_{01}) \\
V_{01} &= H^*(M, E \cup N_{11} \cup N_{10} \cup N_{00}) \\
V_{10} &= H^*(M, E \cup N_{11} \cup N_{00} \cup N_{01})
\end{aligned}
$$

One now checks (this is not trivial) that

$$V = V_0 \oplus V_l.$$

Similarly, for $i = 0, 1$,

$$V_i = V_{i0} \oplus V_{i1}.$$

$f_M^* : V \to V$ can be decomposed as

$$f_M^* = \begin{bmatrix} F_{00} & F_{01} \\ F_{10} & F_{11} \end{bmatrix} \qquad \text{where } F_{ij} : V_j \to V_i \tag{7.3}$$

and each F_{ij} can be further decomposed as

$$F_{ij} = \begin{bmatrix} B_{i0j0} & B_{i0j1} \\ B_{i1j0} & B_{i1j1} \end{bmatrix} \qquad \text{where } B_{imjn} : V_{jn} \rightarrow V_{im}. \tag{7.4}$$

Since $f(N_{ij}) \subset N_{ki} \cup N_{li}$, one can once again argue that

$$f_M^* = \begin{bmatrix} B_{0000} & B_{0001} & 0 & 0 \\ 0 & 0 & B_{0110} & B_{0111} \\ B_{1000} & B_{1001} & 0 & 0 \\ 0 & 0 & B_{1110} & B_{1111} \end{bmatrix}. \tag{7.5}$$

Using (7.3) one can show that

$$\text{Con}(N_k) = \mathcal{L}\left(V, \begin{bmatrix} F_{00} & F_{01} \\ 0 & 0 \end{bmatrix}\right) = \mathcal{L}(V_k, F_{00}) \tag{7.6}$$

and

$$\text{Con}(N_1) = \mathcal{L}\left(V, \begin{bmatrix} 0 & 0 \\ F_{10} & F_{11} \end{bmatrix}\right) = \mathcal{L}(V_1, F_{11}) \tag{7.7}$$

where $\mathcal{L}$ denotes the direct limit functor. In a similar manner, using (7.4) one can show that

$$\text{Con}(\text{Inv } N(kl)) = \mathcal{L}(V, \bar{F}_{01}) \quad \text{with} \quad \bar{F}_{01} = \begin{bmatrix} F_{00} & F_{01} \\ 0 & F_{11} \end{bmatrix}. \tag{7.8}$$

Let $\iota : V_k \rightarrow V$ and $\pi : V \rightarrow V_l$ denote the inclusion and projection corresponding to the direct sum $V = V_k \oplus V_l$. Then, we have the commutative diagram with exact rows

$$\begin{array}{ccccccccc} 0 & \rightarrow & V_k & \xrightarrow{\iota} & V & \xrightarrow{\pi} & V_l & \rightarrow & 0 \\ & & \downarrow{\scriptstyle F_{00}} & & \downarrow{\scriptstyle \bar{F}_{01}} & & \downarrow{\scriptstyle F_{11}} & & \\ 0 & \rightarrow & V_k & \xrightarrow{\iota} & V & \xrightarrow{\pi} & V_l & \rightarrow & 0 \end{array} \tag{7.9}$$

Applying the functor $\mathcal{L}$ and using the (7.6), (7.7), and (7.8) gives the commutative diagram (7.1). Since, the direct limit preservees exact sequences, the rows in (7.1) are exact.

Turning now to Theorem 7.1.2 let m be the level at which

$$\text{Con}^m(\text{Inv } N_i) \approx (\mathbf{G}, \text{id}).$$

From now on all maps and spaces are assumed to be with respect to the mth level. The basic idea of the proof is that we can use the index automorphisms $\chi(\text{Inv } N_i)$ and $\chi(\text{Inv } N(ij))$ to find trajectories which follow any prescibe sequence on the symbols 0 and 1. The difficulty is how does one relate these automorphisms to actual trajectories.

We begin by defining the semi–conjugacy. Define the map $\beta : N \to \{0,1\}$ by

$$\beta(x) := \begin{cases} 0 & \text{if } x \in N_0 \\ 1 & \text{if } x \in N_1 \end{cases}$$

and the map $\rho : \text{Inv } N \to \Sigma_2$ by

$$\rho(x) := (\beta(f^{-nd}(x)))_{n \in \mathbf{Z}}.$$

In this case the continuity of β and ρ are obvious, as is the fact that

$$\rho \circ f^d = \sigma \circ \rho.$$

It remains to show that ρ is surjective and again this is where the algebraic information must be used. We begin our analysis on the with the claim that it can be shown that F_{ii} has a unique (up to a constant) 1–eigenvector v_i. Observe that this is not immediate. The assumptions are all on the level of the index which is obtained by applying the direct limit functor to maps F_{ii}. What needs to be shown is that if the automorphism obtained via the direct limit is the identity, then the map which defines the directed system has an eigenvalue 1 with a one dimensional eigenspace.

Since we want to show that we can find trajectories from N_i to N_j and then back again we need to show that $F_{10}v_0 \notin \text{gker } F_{11}$. If $F_{10}v_0 \in \text{gker } F_{11}$ then $F_{11}^p F_{10}v_0 = 0$ for some natural p. Put

$$w_0 = \left(v_0, \sum_{j=0}^{p-1} F_{11}^j F_{10}v_0 \right), \qquad w_1 = (0, v_1)$$

It is straightforward to verify that w_0 and w_1 are linearly independent 1–eigenvectors of $\bar{F}_{10}$. Also, an easy computation shows that $D_{10} = L(\bar{F}_{10})$ has two linearly independent 1–eigenvectors. This implies that $\alpha_{10} = 0$, a contradiction. Hence $F_{10}v_0$ is not in the generalized kernel of F_{11}.

At this point one can argue that there exists a positive integer d_1 such that $F_{11}^{d_1} F_{10}v_0$ is a 1–eigenvector of F_{11}. Similarly, there exists d_2 such that $F_{00}^{d_2} F_{01}v_1$ is a 1–eigenvector of F_{00}.

Take $d := 1 + \max(d_1, d_2)$. Then

$$F_{ii}^{d-1} F_{ij}v_j \text{ is a 1–eigenvector of } F_{ii} \text{ for any } i, j \in \{0,1\}, i \neq j. \tag{7.10}$$

We now show that being able to map eigenvectors to eigenvectors allows us to obtain any finite sequence of symbols from the dynamics. Let $\alpha : \mathbf{Z} \to \{0,1\}$ and fix a positive integer n. Let $\phi_i \in v_i$ be an Alexander–Spanier cochain representative of v_i. Let $|\phi_i|$ denote the support of ϕ_i. Define recursively a sequence

$$\psi_{-n} := \phi_{\alpha_{-n}}$$
$$\psi_i := [f_M^\#(\alpha_i, \alpha_i)^{d-1} \circ f_M^\#(\alpha_i, \alpha_{i-1})](\psi_{i-1}) \quad i = -n+1, -n+2, \ldots \tag{7.11}$$

Then ψ_i is a cochain for the pair (N, E_{α_i}). In particular $|\psi_i| \subset \mathrm{cl}(N \backslash E_{\alpha_i}) \subset N_{\alpha_i}$. This in turn implies that

$$f^j(|\psi_i|) \subset N, \qquad j = 0, 1, \ldots d$$

and

$$f^d(|\psi_i|) \subset |\psi_{i-1}| \qquad i = -n+1, -n+2, \ldots, n,$$

from which we can conclude that

$$f^j(|\psi|) \subset N, \qquad j = 0, 1, \ldots 2nd \tag{7.12}$$

and

$$f^{md}(|\psi_n|) \subset N_{\alpha_{n-m}} \qquad m = 0, 1, \ldots, 2n. \tag{7.13}$$

By (7.10) and (7.11) the cohomology class of ψ_n is a 1-eigenvector of $F_{\alpha_n \alpha_n} = f^k_M(\alpha_n, \alpha_n)$. But by hypothesis on the index this class is different from zero and therefore $|\psi_n| \neq \emptyset$. Choose $y_n \in |\psi_n|$ and set $x_n := f^{nd}(y_n)$. Then, by (7.12)

$$f^i(x_n) \in N, \qquad i = -nd, nd+1, \ldots nd \tag{7.14}$$

and by (7.13)

$$f^{-id}(x_n) = f^{(n-i)d}(y_n) \in f^{(n-i)d}(|\psi_n|) \subset N_{\alpha_i} \qquad i = -n, -n+1, \ldots, n. \tag{7.15}$$

In particular $x_n \in N_{\alpha_0}$. Since N_{α_0} is compact, one can assume that $x_n \to x_* \in N_{\alpha_0}$. From (7.14) one obtains $x_* \in \mathrm{Inv}\, N$ and from (7.15)

$$f^{-id}(x_*) \in N_{\alpha_i} \qquad \text{for } i \in \mathbf{Z},$$

i.e. $\beta(f^{-id}(x_*)) = \alpha_i$ for $i \in \mathbf{Z}$. This shows that $\rho(x_*) = \alpha$ and proves that ρ is surjective.

Chapter 8

Singular Perturbations

While we may believe that typical compact sets are isolating neighborhoods, showing that a particular set N isolates is a problem in global nonlinear analysis, and hence, in general difficult. If one is lucky then the system of interest can be viewed as a perturbation of a simpler or lower dimensional system. However, an added complication may occur in that N need not be an isolating neighborhood for the simpler system. Consider the following classic example

$$
\begin{aligned}
\dot{x} &= y + x(x^2 - 1) \\
\dot{y} &= -\epsilon x.
\end{aligned} \tag{8.1}
$$

Let N be the compact set shown in Figure 8.1. Observe that for $\epsilon = 0$ the curve $S := \{(x, y) \mid y = -x(x^2 - 1)\}$ is an invariant set consisting of fixed points. Since $S \cap \partial N \neq \emptyset$, N is not an isolating neighborhood for flow when $\epsilon = 0$. However, as the following argument shows, N isolates for all small $\epsilon > 0$. If $z \in \partial N \setminus S$, then by definition there exists a time (in this case positive) such that under the $\epsilon = 0$ flow the point is carried out of N. On the other hand, if $z \in S \cap \partial N$ then one can explicitly check that z leaves N under all flows generated by $\epsilon > 0$ but small. There are of course two problems with this argument. The first is that the amount of time required for z to leave N depends both on ϵ and z and the second is for $z \in S \cap \partial N$ the explicit calculation can be difficult. A practical solution to this problem has been obtained by Conley [5] and is the focus of the first section.

However, Conley's ideas, though over 15 years old by now, do not seem to have been used. Perhaps one of the reasons for this, and again this touches upon the theme of these lectures, is that isolating neighborhoods alone provide little information concerning the dynamics of the invariant set.

Thus, it is natural to ask whether it is possible to compute the index of N for $\epsilon > 0$ by studying the flow at $\epsilon = 0$. The second section discusses a result which, under certain conditions, provides an affirmative answer to this question.

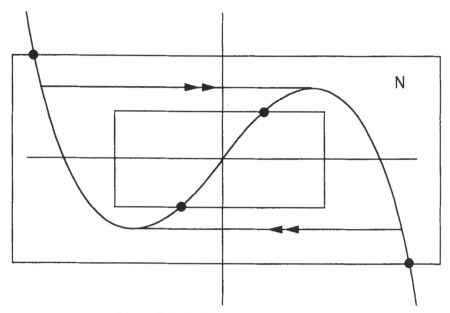

Figure 8.1: A simple fast–slow system

8.1 Singular Isolating Neighborhoods

Consider the family of differential equations on $\mathbf{R}^n$ given by

$$\dot{x} = f(x, \epsilon, \lambda) = f_0(x, \lambda) + \sum_{i=1}^{m} \epsilon^i f_i(x, \lambda) + o(\epsilon^m). \qquad (8.2)$$

where $\epsilon \in [0, \infty)$ and $\lambda \in \Lambda$. The corresponding flow is denoted by $\varphi^{\epsilon, \lambda}$ and the parameterized flow by Φ. ϵ is viewed as the "singular" parameter, i.e it is assumed that the family of flows $\varphi^{0, \lambda}$ is understood and the goal is to understand the dynamics of $\varphi^{\epsilon, \lambda}$ for $\lambda \in \Lambda$ and $\epsilon > 0$ but small.

If N is an isolating neighborhood for φ^{0, λ_0} then the standard perturbation and continuation results apply. Therefore, we begin with the additional assumption that $N \subset \mathbf{R}^n$ is not an isolating neighborhood for φ^{0, λ_0}.

Definition 8.1.1 A compact set $N \subset \mathbf{R}^n$ is called a *singular isolating neighborhood* for φ^{0, λ_0} if N is not an isolating neighborhood for φ^{0, λ_0}, but there is an $\bar{\epsilon} > 0$ and a neighborhood Λ_0 of λ_0 such that for all $(\epsilon, \lambda) \in (0, \bar{\epsilon}] \times \Lambda_0$, N is an isolating neighborhood for $\varphi^{\epsilon, \lambda}$.

Observe that a compact set N is not an isolating neighborhood under the flow φ^{0, λ_0} if and only if there exists a point $x \in \mathrm{Inv}(N, \varphi^{0, \lambda_0}) \cap \partial N$. However, if N is a singular isolating neighborhood, then $x \notin \mathrm{Inv}(N, \varphi^{\epsilon, \lambda})$ for all $\epsilon > 0$ sufficiently small and λ close to λ_0. Thus under all nearby flows associated with positive ϵ, x must leave N either in forward or in backward time. This leads to the following definition.

Definition 8.1.2 Let N be a compact set and let $x \in \mathrm{Inv}(N, \varphi^{0,\lambda_0})$. x is a *slow exit (entrance) point* if there exist neighborhoods $U \subset \mathbf{R}^n$ of x and $\Lambda_0 \subset \Lambda$ of λ_0, an $\bar{\epsilon} > 0$, and a time $T(\epsilon, U, \Lambda_0) > 0$ $(T(\epsilon, U, \Lambda_0) < 0)$, such that for all $\epsilon \in (0, \bar{\epsilon}]$ and $\lambda \in \Lambda_0$,

$$\varphi^{\epsilon, \lambda}(T(\epsilon, U, \Lambda_0), U) \cap N = \emptyset.$$

This definition makes the following theorem a triviality.

Theorem 8.1.3 *Let N be a compact set. If $\mathrm{Inv}(N, \varphi^{0,\lambda_0}) \cap \partial N$ consists of slow exit and slow entrance points, then N is a singular isolating neighborhood.*

Proof. It needs to be shown that there exists $\bar{\epsilon} > 0$ and a neighborhood Λ_0 of λ_0 such that if $x \in \partial N$ and $(\epsilon, \lambda) \in (0, \bar{\epsilon}] \times \Lambda_0$, then there is a $t_x(\epsilon, \lambda) \in \mathbf{R}$ such that $\varphi^{\epsilon, \lambda}(t_x(\epsilon, \lambda), x) \notin N$.

Let $x \in \mathrm{Inv}(N, \varphi^{0,\lambda_0}) \cap \partial N$. Then there exists U_x a neighborhood of $x, \epsilon_x > 0$, a neighborhood $\Lambda_x \subset \Lambda$ of λ_0 and a time $T(\epsilon, U_x, \Lambda_x)$ such that

$$\varphi^\epsilon(T(\epsilon, U_x, \Lambda_x), U_x) \cap N = \emptyset$$

for all $\epsilon \in (0, \epsilon_x]$. Since $\mathrm{Inv}(N, \varphi^{0,\lambda_0}) \cap \partial N$ is compact we can cover it with a finite collection of sets $U_{x_1}, \ldots, U_{x_k}$. Let

$$U = \cup_{i=1}^k U_{x_i}, \quad T(\epsilon) = \max_{i=1,\ldots k} |T(\epsilon, U_x, \Lambda_x)|, \quad \bar{\epsilon}_1 = \min_{i=1,\ldots k} \epsilon_{x_i}, \quad \Lambda_0 = \cap_{i=1}^k \Lambda_i.$$

Then for every $x \in U$ and $(\epsilon, \lambda) \in (0, \bar{\epsilon}] \times \Lambda_0$, $\varphi^{\epsilon, \lambda}([-T(\epsilon), T(\epsilon)], x) \not\subset N$.

On the other hand, $\partial N \setminus U$ is compact and for all $x \in \partial N \setminus U$, there exists a $t_x \in \mathbf{R}$ such that $\varphi^{0,\lambda}(t_x, x) \notin N$. Therefore, by uniform continuity there exists $\bar{\epsilon}_2 > 0$ such that for all $\epsilon \in (0, \bar{\epsilon}_2]$, and for all $x \in \partial N \setminus U$ there exists $t_x(\epsilon)$ such that $\varphi^{\epsilon, \lambda}(t_x(\epsilon), x) \notin N$. Let $\bar{\epsilon} = \min\{\bar{\epsilon}_1, \bar{\epsilon}_2\}$. ∎

The proof of Theorem 8.1.3 has been included because it clearly shows the essential difficulty which arises in demonstrating the results of this section. If $x \notin \mathrm{Inv}(N, \varphi^0)$, then for all ϵ sufficiently small $x \notin \mathrm{Inv}(N, \varphi^\epsilon)$. Therefore, points of this type pose no problem and all the effort needs to be expended on understanding the behavior of the points in $\mathrm{Inv}(N, \varphi^0)$ under perturbation. In Theorem 8.1.3 this difficulty was defined away through the notion of slow exit and slow entrance point. Thus, what is needed are theorems which identify slow exit and slow entrance points. This is, in fact, the fundamental contribution of Conley alluded to earlier.

The following notation will be used throughout this section. $N \subset \mathbf{R}^n$ is a compact set and $S^\lambda := \mathrm{Inv}(N, \varphi^{0,\lambda})$. $S^{\lambda-}$ $(S^{\lambda+})$ denote the set of slow exit (entrance) points. $S_\partial^\lambda = S^\lambda \cap \partial N$ and $S_\partial^{\lambda\pm} = S_\partial^\lambda \cap S^{\lambda\pm}$.

Definition 8.1.4 Let K be an invariant set and let φ be a flow defined on K. The *average of g on K*, $\mathrm{Ave}(g, K)$, is the limit as $T \to \infty$ of the set of numbers $\{\frac{1}{T} \int_0^T g(\varphi(s, x)) \, ds \mid x \in K\}$. If $\mathrm{cl}(\mathrm{Ave}(g, K)) \subset (0, \infty)$, then we say g has *strictly positive averages* on K.

An important property of averages is the following [5].

Proposition 8.1.5 *Let φ^μ be a continuous family of flows and let K be an invariant set of φ^{μ_0}. Given a neighborhood J of $\mathrm{Ave}(g, K)$, there is a neighborhood U of K and a $T > 0$ such that for μ sufficiently close to μ_0 if $t \geq T$ and $\varphi^\mu([0, t], x) \subset U$, then*

$$\frac{1}{t} \int_0^t g(\varphi^\mu(s, x)) \, ds \in J. \tag{8.3}$$

Remark 8.1.6 Observe that if g has strictly positive averages on K, then U and the set of parameter values μ can be chosen such that there is a constant $k > 0$, for which $\varphi^\mu([0, t], x) \subset U$ implies that

$$\int_0^t g(\varphi^\mu(s, x)) \, ds \geq -k.$$

The following theorem is stated in more general terms than Conley's result [5], however, it follows almost verbatim from Conley's proof.

Theorem 8.1.7 *$x \in S^{\lambda_0}$ is a slow exit point if there exists a compact set $K_x^{\lambda_0} \subset S^{\lambda_0}$, a neighborhood U_x of $\mathcal{R}(K_x^{\lambda_0})$ and a function $\ell : \mathrm{cl}(U_x) \times [0, \bar{\epsilon}] \times \Lambda_0 \to \mathbf{R}$ for some $\bar{\epsilon} > 0$ and $\Lambda_0 \subset \Lambda$ for which the the following conditions can be satisfied.*

1. *$K_x^{\lambda_0}$ is invariant under the flow φ^{0,λ_0} and $\omega(x, \varphi^{0,\lambda_0}) \subset K_x^{\lambda_0}$.*

2. *ℓ is of the form*

$$\ell(z, \epsilon, \lambda)) = \ell_0(z, \lambda) + \epsilon\ell_1(z, \lambda) + \ldots + \epsilon^m \ell_m(z).$$

3. *If $L_0^\lambda = \{ z \mid \ell_0(z, \lambda) = 0 \}$ then*

$$K_x^\lambda \cap \mathrm{cl}(U_x) = S^\lambda \cap L_0^\lambda \cap \mathrm{cl}(U_x)$$

and furthermore, $\ell_0|_{K_x \cap \mathrm{cl}(U_x)} = 0$, and $\ell_0|_{S \cap \mathrm{cl}(U_x)} \leq 0$.

4. *Let*

$$g_j(z, \lambda) = \nabla_z \ell_0(z, \lambda) \cdot f_j(z, \lambda) + \nabla_z \ell_1(z, \lambda) \cdot f_{j-1}(z, \lambda) + \ldots + \nabla_z \ell_j(z, \lambda) \cdot f_0(z, \lambda).$$

Then, for some m, $g_j \equiv 0$ if $j < m$, and g_m has strictly positive averages on $\mathcal{R}(K_x)$.

Remark 8.1.8 The functions ℓ and g_j may seem unnatural until one observes that

$$\nabla_z \ell(z, \lambda) \cdot f(z, \lambda) = \sum_{j=0}^m \epsilon^j g_j + \epsilon^{m+1} r(z, \lambda).$$

Thus, ℓ acts a local Lyapunov function and the average of g determines whether orbits increase or decrease in value with respect to this Lyapunov function.

Definition 8.1.9 A slow exit point which satisfies the conditions of Theorem 8.1.7 is called a *C-slow exit point*.

Remark 8.1.10 It follows immediately from the definition that if x is a slow exit point, then under the transformation $t \to -t$ (time reversal) x becomes a slow entrance point. Thus, Theorem 8.1.7 can, also, be used to test for slow entrance points. Slow entrance points of this form will be called *C-slow entrance points*.

Remark 8.1.11 If x is a C-slow exit point of S^{λ_0}, then $K_x^{\lambda_0}$, $U_x^{\lambda_0}$, and ℓ will denote sets which satisfy the conditions of Theorem 8.1.7. If it is clear from context what the parameter λ_0 is we shall exclude it from the notation.

Corollary 8.1.12 *Let $x \in S$ be a C-slow exit point and let $y \in \mathcal{R}(K_x)$. Then y is a C-slow exit point.*

Proof.   Since $y \in \mathcal{R}(K_x) \subset S$, $y \in S$. Furthermore, $\omega(y, \varphi^0) \subset K_x$. Thus we can choose $K_y = K_x$, $U_y = U_x$ and the same function ℓ.   ∎
The main step in the proof of Theorem 8.1.7 is the following proposition.

Proposition 8.1.13 *Let x be a C-slow exit point for S^{λ_0}. Then, there exist neighborhoods U_x of $\mathcal{R}(K_x)$, W of S^{λ_0}, Λ_0 of λ_0, and an $\bar{\epsilon}$ with the following property. For each $y \in U_x$ there exists $0 < t_0(y, \epsilon) \leq t_1(y, \epsilon)$ such that if $(\epsilon, \lambda) \in (0, \bar{\epsilon}] \times \Lambda_0$ then*

$$\varphi^{\epsilon, \lambda}([0, t_0(y, \epsilon)], y) \subset U_x$$
$$\varphi^{\epsilon, \lambda}([t_0(y, \epsilon), t_1(y, \epsilon)], y) \cap W = \emptyset$$
$$\varphi^{\epsilon, \lambda}(t_1(y, \epsilon), y) \notin N.$$

Proof.   Since g_m (as defined by 8.1.7.5) is assumed to have positive averages on U_x, by Remark 8.1.6 we can without loss of generality assume that U_x, $\bar{\epsilon}_1 > 0$, and $\Lambda_1 \subset \Lambda$, a neighborhood of λ_0, have been chosen such that if $(\epsilon, \lambda) \in (0, \bar{\epsilon}_1] \times \Lambda_1$, $t > T$, and $\varphi^{\epsilon, \lambda}([0, t], x) \subset U_x$, then

$$\int_0^t g(\varphi^{\epsilon, \lambda}(s, x)) \, ds \geq -k \tag{8.4}$$

for some positive constant k.
By definition K_x is an invariant set for φ^{0, λ_0}. Since U_x is a neighborhood of $\mathcal{R}(K_x)$ there exists a Morse decomposition

$$\mathcal{M}(K_x) = \{M(p) \mid 1 < \ldots < p < p + 1 < \ldots < P\}$$

of K_x such that

$$\bigcup_{p=1}^{P} M(p) \subset U_x.$$

We can now outline the idea of the proof. As was demonstrated in the proof of Theorem 8.1.3 the only trajectories we need to worry about are those that pass close to both the invariant set and the boundary of N. Such an orbit must of course pass through U_x for some $x \in S_\partial$. On U_x and for $\epsilon > 0$ the function ℓ acts as a Lyapunov function "pushing" the orbit away from $\mathcal{R}(K_x)$ since the average of g_m is positive. Away from U_x such an orbit is guided by the connecting trajectories of the Morse decomposition.

The behavior of orbits passing near the Morse decomposition is understood by means of Proposition 3.4.8. In particular, we can assume that $\bar{\epsilon}_2 > 0$ and $\Lambda_2 \subset \Lambda_1$, a neighborhood of λ_0, have been chosen such that the proposition applies. We shall now construct the open sets V_p and W_p of Proposition 3.4.8 inductively. So assume the sets $V_1, W_1, \ldots, W_k$ have been chosen.

By definition (Proposition 3.4.8) $\mathcal{R}(K_x) \subset W_k$ and (Theorem 8.1.7.4)

$$K_x \cap \mathrm{cl}(U_x) = S \cap L_0 \cap \mathrm{cl}(U_x).$$

Therefore,

$$\mathrm{cl}\,(U_x \cap (S \setminus W_k)) \cap L_0 = \emptyset.$$

Furthermore, by Theorem 8.1.7.4 we can assume that

$$\ell_0|_{\mathrm{cl}(U \cap (S \setminus W_k))} < -5\delta_k \tag{8.5}$$

where $\delta_k > 0$.

Again using Theorem 8.1.7.4 we can choose V_k, a sufficiently small neighborhood of $M(1, \ldots, k)$, such that

$$\ell_0|_{V_k \cap U} > -\delta_k. \tag{8.6}$$

Using this procedure we construct all the sets V_p and W_p, $p = 1, \ldots, P$. Let

$$\delta = \min\{\delta_k\}.$$

By Proposition 3.4.8 given $y \in \mathcal{R}(K_x)$ there exists $\bar{\epsilon}_3 > 0$ and $\Lambda_3 \subset \Lambda_2$, a neighborhood of λ_0, such that either

$$\varphi^{\epsilon,\lambda}([0,\infty), y) \subset U_x \tag{8.7}$$

or there exits $t_1 > 0$ and a p such that

$$\varphi^{\epsilon,\lambda}(0, y) \in V_p \quad \varphi^{\epsilon,\lambda}([0,t_1], x) \subset U_x \quad \text{and} \quad \varphi^{\epsilon,\lambda}(t_2, x) \in U_x \setminus W_p. \tag{8.8}$$

We shall now use the Lyapunov function ℓ to observe the behavior on U_x. Observe that

$$\begin{aligned}
\nabla_y \ell(y, \lambda) \cdot f(y, \lambda) &= \left(\sum_{i=1}^m \epsilon^i \varphi^{\epsilon,\lambda} \ell_i(y, \lambda) \right) \cdot \left(\sum_{i=1}^m \epsilon^i f_i(y, \lambda) + h.o.t. \right) \\
&= \sum_{j=0}^m \epsilon^j g_j + \epsilon^{m+1} r(y, \lambda)
\end{aligned}$$

Now let

$$K := \sup_{y \in \mathrm{cl}(U_x), \, \lambda \in \Lambda_3} \{|r(y, \lambda)|, |\ell_j(y, \lambda)|, j = 0, \ldots, m\}.$$

Then on U_x

$$
\begin{aligned}
|\ell_0(y, \lambda) - \ell(y, \lambda)| &= \left| \sum_{i=1}^{m} \epsilon^i \ell_i(y, \lambda) \right| \\
&\leq (\epsilon + \ldots + \epsilon^{m+1}) K.
\end{aligned}
$$

Now choose $\bar{\epsilon}_4 > 0$ and $\Lambda_4 \subset \Lambda_3$, a neighborhood of λ_0, such that

$$|\ell_0(y, \lambda) - \ell(y, \lambda)| < \delta$$

on U_x. Then for $(\epsilon, \lambda) \in (0, \bar{\epsilon}_4] \times \Lambda_4$ and $\varphi^{\epsilon, \lambda}([0, t], y) \subset U_x$

$$\ell_0\left(\varphi^{\epsilon, \lambda}(t, y)\right) - \ell_0(y) \geq \ell\left(\varphi^{\epsilon, \lambda}(t, y)\right) - \ell(y) - 2\delta.$$

But

$$
\begin{aligned}
\ell_0\left(\varphi^{\epsilon, \lambda}(t, y)\right) - \ell_0(y) &= \int_0^t \varphi^{\epsilon, \lambda} \ell \cdot f \\
&= \sum_{j=0}^{m} \epsilon^j \int_0^t g_j + \epsilon^m \int_0^t g_m + \epsilon^{m+1} \int_0^t r \\
&= \epsilon^m \int_0^t g_m + \epsilon^{m+1} \int_0^t r
\end{aligned}
$$

where the last line follows from Theorem 8.1.7.5.

To proceed we need estimates on $\int_0^t g_m$. The first comes from equation (8.4) and is used for the case that $t \leq T$ where T is defined in Proposition 8.1.5;

$$\ell_0\left(\varphi^{\epsilon, \lambda}(t, y)\right) - \ell_0(y) \geq -\epsilon^m k - \epsilon^{m+1} KT.$$

For the remaining case (when $t \geq T$) observe that since $\mathrm{Ave}(g_m, S_0^\lambda)$ is strictly positive,

$$\mathrm{Ave}(g_m, S_0^\lambda) \subset J = [J_-, J_+]$$

where $J_- > 0$. Thus by Proposition 8.1.5

$$
\begin{aligned}
\ell_0\left(\varphi^{\epsilon, \lambda}(t, y)\right) - \ell_0(y) &\geq \epsilon^m t J_- - \epsilon^{m+1} t K \\
&= \epsilon^m (J_- - \epsilon K) t \\
&> 0
\end{aligned}
$$

where the last line is valid for ϵ sufficiently small. However, for fixed ϵ this shows that $\ell_0\left(\varphi^{\epsilon, \lambda}(t, y)\right) - \ell_0(y) \to \infty$ as $t \to \infty$. But, ℓ is bounded on U_x and hence this second case cannot occur. In particular, we only need to worry about the situation described by (8.8).

Now choose $\bar{\epsilon}_5 \in (0, \bar{\epsilon}_4)$ such that $-\epsilon^m k - \epsilon^{m+1} KT > -\delta$. Observe that

$$
\begin{aligned}
\ell_0\left(\varphi^{\epsilon,\lambda}(t_1, y)\right) - \ell_0(y) &\geq \ell_0\left(\varphi^{\epsilon,\lambda}(t_1, y)\right) - \ell_0(y) - 2\delta \\
&\geq -3\delta \\
&\geq -3\delta_k.
\end{aligned} \tag{8.9}
$$

With these estimates we can now start to construct the set W of S^{λ_0}. Let W_k' be a neighborhood of $S \setminus W_k$ such that

$$
\ell_0|_{W_k' \cap U} < -4\delta_k.
$$

Observe that

$$
S \subset \bar{W}_k := W_k \cup W_k'.
$$

Finally define

$$
W := \bigcap_{p=0}^{P} \bar{W}_p.
$$

Now observe that by (8.8), $\varphi^{\epsilon,\lambda}(t_1, y) \notin W_p$. Also, by (8.9) $\varphi^{\epsilon,\lambda}(t_1, y) \notin W_p'$ and therefore

$$
\varphi^{\epsilon,\lambda}(t_1, y) \notin W.
$$

∎

Proof of Theorem 8.1.7. Let $x \in S$ be a C-slow exit point. Let U_x be as in Proposition 8.1.13. Observe that by Proposition 8.1.13 there exists a neighborhoods $U_y \subset U_x$ of y and Λ_0 of λ_0, $\bar{\epsilon} > 0$, and a time $t_y(\epsilon, \lambda)$ such that for $(\epsilon, \lambda) \in (0, \bar{\epsilon}] \times \Lambda_0$, $\varphi^{\epsilon,\lambda}(t_y(\epsilon, \lambda), U_y) \cap N = \emptyset$.

Observe that $\omega(x, \varphi^{0,\lambda_0}) \subset \mathcal{R}(K_x) \subset U_x$. Thus, there exists $T_x > 0$ such that $\varphi^{0,\lambda_0}(T_x, x) \in U_x$. Since by the argument in the previous paragraph $\varphi^{0,\lambda_0}(T_x, x)$ is a slow exit point, x is a slow exit point. ∎

8.2 Singular Index Pairs

With the knowledge of how to identify singular isolating neighborhoods we now turn to the question of how to capture the index from the singular flow. We begin with two observations. The first is that an index pair (N, L) for a given flow is not in general an index pair for nearby flows. The second is that the quantity $H^*(N, L)$ does represent the index for all nearby flows. This leads to the following definition.

Definition 8.2.1 A pair of compact sets (N, L) with $L \subset N$ is a *singular index pair* if $\mathrm{cl}(N \setminus L)$ is a singular isolating neighborhood and there exists $\bar{\epsilon} > 0$ such that for all $\epsilon \in (0, \bar{\epsilon}]$

$$
H^*(N, L) \approx CH^*((\mathrm{Inv}(\mathrm{cl}(N \setminus L), \varphi^{\epsilon})).
$$

Remark 8.2.2 Observe that in the definition of a singular index pair it is *not* assumed that (N, L) is an index pair for any $\epsilon > 0$. It remains an open question whether it is even possible in general to fine a pair (N, L) that would act as an index pair for all sufficiently small positive ϵ in the case in which N is a singular isolating neighborhood.

This definition forces us to consider the following question. Given a set N which is a singular isolating neighborhood, can one find a set L such that (N, L) is a singular index pair for Inv N? Recall, however, Remark 2.2.5 where an example was given of an isolating neighborhood which could not be used to construct an index pair. To avoid this problem we are required to introduce the following restriction.

Definition 8.2.3 A slow entrance point x is a *strict slow entrance point* if there exists a neighborhood Θ_x of x and an $\bar{\epsilon} > 0$ such that if $y \in \Theta_x \cap N$ and $\epsilon \in (0, \bar{\epsilon}]$, then there exists $t_y(\epsilon) > 0$ for which

$$\varphi^\epsilon([0, t_y], y) \subset N.$$

We will let S_∂^{++} denote the strict slow entrance points.

Theorem 8.2.4 stated below gives a recipe for finding such an L under certain conditions on N. The idea behind the construction of L is quite natural; try to mimic the three conditions which must be satisfied by an exit set. In particular, L will need to contain the *immediate exit* set for N which is defined by

$$N^- := \{x \in \partial N \mid \varphi^0((0, t), x) \not\subset N \text{ for all } t > 0\}.$$

Positive invariance leads to the following construction. Given $Y \subset N$ its *push forward set* in N under the flow φ^0 is defined to be

$$\rho(Y, N, \varphi^0) := \{x \in N \mid \exists z \in Y, t \geq 0 \text{ such that } \varphi^0([0, t], z) \subset N, \varphi^0(t, z) = x\}.$$

Observe that $\rho(Y, N, \varphi^0)$ consists of the set of points which can be reached from Y by a positive trajectory in N and that $Y \subset \rho(Y, N, \varphi^0)$. The *unstable set* of an invariant set $Y \subset N$ under φ^0 is

$$W_N^u(Y) := \{x \in N \mid \varphi^0((-\infty, 0), x) \subset N \text{ and } \alpha(x, \varphi^0) \subset Y\}.$$

Notice that $Y \subset W_N^u(Y)$. Since we will only be interested in the unstable set under the singular flow φ^0 we do not include it in the notation.

Theorem 8.2.4 *Let N be a singular isolating neighborhood. Assume*

1. *S_∂ consists of strict slow entrance and C-slow exit points.*

2. *$S_\partial \setminus S_\partial^+ \cap \text{cl}(N^-) = \emptyset$.*

For each $x \in S_\partial^-$, let K_x denote the compact invariant set in Theorem 8.1.7. Define

$$L := \rho(cl(N^-), N, \varphi^0) \bigcup W_N^u \left(\bigcup_{x \in S_\partial^-} \mathcal{R}(K_x) \right).$$

If L is closed, then (N, L) is a singular index pair for the family of flows φ^ϵ.

A complete proof of this theorem can be found in [39]. Here we shall only present a sketch of the proof in order to demonstrate the difficulties and how they can be overcome.

As was mentioned earlier it is not known how to construct a pair (N, L) which serves as an index pair for all small $\epsilon > 0$. However, since the definition of a singular index pair only requires agreement on the cohomological level one expects to be able to take advantage of the weak continuity property of the Alexander–Spanier cohomology. In particular, one would like to find a sequence of index pairs (N, L^{ϵ_i}), where $\epsilon_i \to 0$ as $i \to \infty$, with the properties that $L \subset L^{\epsilon_{i+1}} \subset L^{\epsilon_i}$ for all i and $L = \cap_i L^{\epsilon_i}$. Using the inclusion induced homomorphisms this give rise to a direct limit of the cohomology groups $H^*(N, L^{\epsilon_i})$ for which

$$H^*(N, L) \approx \lim_{\to} H^*(N, L^{\epsilon_i}).$$

However, as is easily imagined it is difficult to control the behavior of the dynamics of φ^ϵ over all of N. This problem can be overcome by introducing the following bump functions. Given $x \in \mathbf{R}^n$ and $\eta > 0$, let $B_\eta(x) := \{y \in \mathbf{R}^n \mid ||x - y|| < \eta\}$ and given $Y \subset \mathbf{R}^n$, let $B_\eta(Y) := \cup_{y \in Y} B_\eta(y)$. Define $Q_\eta^- = B_\eta(\cup_{x \in S_\partial^-} \mathcal{R}(K_x))$, $Q_\eta^+ = B_\eta(S_\partial^{++})$, and $Q_\eta = Q_\eta^- \cup Q_\eta^+$. Our family of bump functions will consist of smooth functions $\mu_\eta : \mathbf{R}^n \to [0, 1]$ satisfying

- supp $\mu_\eta \subset Q_\eta$.

- $B_{\eta/2}(\cup_{x \in S_\partial} \mathcal{R}(K_x)) \subset \mu^{-1}(1)$.

Now consider the two parameter singular perturbation problem given by the equation

$$\dot{x} = f(x, \epsilon) = f_0(x) + \mu_\eta(x)[\sum_{i=1}^{m} \epsilon^i f_i(x) + o(\epsilon^m)] \tag{8.10}$$

and let $\psi^{\epsilon, \eta}$ denote its flow. Notice that for η sufficiently large, $\psi^{\epsilon, \eta} = \varphi^\epsilon$. However, for η small, $\psi^{\epsilon, \eta}$ and φ^0 differ only on a small neighborhood of the $\mathcal{R}(K_x)$, exactly the region on which one has control due to the local Lyapunov functions ℓ.

One can now construct index pairs $(N, L^{\epsilon_i, \eta_i})$ for the flow $\psi^{\epsilon_i, \eta_i}$ where $\epsilon_i, \eta_i \to 0$ as $i \to \infty$ with the above mentioned properties. Thus,

$$H^*(N, L) \approx \lim_{\to} H^*(N, L^{\epsilon_i, \eta_i}).$$

However, one can now check that the inclusion induced homomorphisms are isomorphisms. In fact, one one shows is the the inclusion maps are homotopic to the maps used to prove the continuation property of the Conley index. Therefore, one has the stronger result that

$$H^*(N, L) \approx H^*(N, L^{\epsilon_i, \eta_i}).$$

Finally, one uses Theorem 8.1.7 to conclude that there exits $\epsilon_0 > 0$ for which N is an isolating neighborhood for all $\eta > 0$. Using the standard continuation theorem for the Conley index it then follows that

$$H^*(N, L) \approx CH^*(\mathrm{Inv}(N \setminus L, \varphi^\epsilon))$$

for $\epsilon \in (0, \bar{\epsilon}]$.

8.3　An Example

To demonstrate the potential applicability of these singular techniques we shall now prove that the flow φ^ϵ generated by (8.1) possesses a periodic orbit for $\epsilon > 0$ but small. Obviously, this result can be obtained via much more elementary techniques, but it serves to demonstrate the constructions of the last section. Let N be the compact set shown in Figure 8.1.

Theorem 8.3.1 *For $\epsilon > 0$, but sufficiently small, $\mathrm{Inv}(N, \varphi^\epsilon)$ contains a periodic orbit.*

Proof.　By Lemma 8.3.2

$$CH^n(N, \varphi^\epsilon) \approx \begin{cases} \mathbf{Z} & \text{if } n = 1, 2 \\ 0 & \text{otherwise.} \end{cases}$$

The curve Ξ indicated in Figure 8.2 is a Poincaré section for N if $\epsilon > 0$. Therefore, by Theorem 5.1.3 $\mathrm{Inv}(N, \varphi^\epsilon)$ contains a periodic orbit.　∎

Thus, the problem has been reduced to computing the index of N for small positive ϵ.

Lemma 8.3.2

$$CH^n(N, \varphi^\epsilon) \approx \begin{cases} \mathbf{Z} & \textit{if } n = 1, 2 \\ 0 & \textit{otherwise.} \end{cases}$$

Proof.　Following the arguments of the last two sections. The proof breaks up into three steps.

Step 1. We need to show that $S_\partial = \partial N \cap \{(x, y) \mid y = -x(x^2 - 1)\}$ consists of C–slow exit and strict slow entrance points. Since each element of S_∂ is a fixed point we can choose $\mathcal{R}(K_z) = z$ for each $z \in S_\partial$. Also, we can choose a Lyapunov function of the form

$$\ell(z) = \ell(x, y) = c \pm x$$

where c is a constant chosen such that $\ell(z) = 0$ and the choice of a plus or minus sign is determined by the conventions of Theorem 8.1.7. Observe that all points of S_∂ are C-slow exit points.

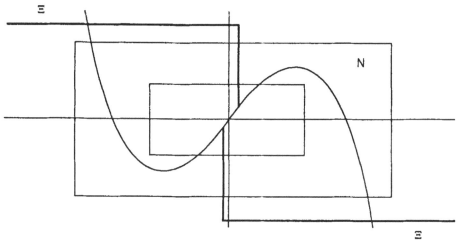

Figure 8.2: A Poincaré section for small $\epsilon > 0$.

Step 2. We now need to construct L. It is easy to see that N^- consists of precisely the vertical sides of N. The outer vertical sides are closed, but the inner sides are not. Following the flow on the inner horizontal boundaries of N we see that $\rho(\text{cl}(N^-), N, \varphi^0)$ consists of the vertical boundaries and the inner horizontal boundaries excluding the two elements of S_∂. If $z \in S_\partial$ and z is on one of the outer horizontal boundaries of N, then

$$W_N^u\left(\mathcal{R}(K_z)\right) = W_N^u\left(z\right)$$

consists of the entire horizontal boundary segment. On the inner horizontal segments,

$$W_N^u\left(\mathcal{R}(K_z)\right) = z.$$

Thus,

$$L = \partial N.$$

Step 3. Clearly, L is closed, which by Theorem 8.2.4 implies that for $\epsilon > 0$

$$CH^*(N, \varphi^\epsilon) \approx H^*(N, L).$$

∎

Bibliography

[1] V. Benci, A new approach to the Morse–Conley theory and some applications, *Ann. Mat. Pura. Appl.* **4** (1991), 231–305.

[2] V. Benci and M. Degiovanni, *Morse–Conley theory* (in preparation).

[3] R. Bott and L. W. Tu, *Differential Forms in Algebraic Topology*, Springer–Verlag 1982.

[4] C. Conley, *Isolated Invariant Sets and the Morse Index.* CBMS Lecture Notes **38** A.M.S. Providence, R.I. 1978.

[5] C. Conley, A qualitative singular perturbation theorem, *Global Theory of Dynamical Systems*, (eds. Z. Nitecki and C. Robinson) Lecture Notes in Math. **819** Springer–Verlag 1980, 65–89.

[6] C. Conley and R. Easton, Isolated invariant sets and isolating blocks, *Trans. AMS* **158** (1971) 35–61.

[7] C. C. Conley and P. Fife, *Critical manifolds, travelling waves and an example from population genetics*, J. Math. Bio. **14** (1982) 159-176.

[8] C. C. Conley and R. Gardner, *An application of the generalized Morse index to travelling wave solutions of a competitive reaction diffusion model*, Ind. Univ. Math. Jour. **33**, (1989) pp. 319-343.

[9] M. Degiovanni and M. Mrozek, The Conley index for maps in the absence of compactness, *Proc. Roy. Soc. Edin.* **123A** (1993) 75–944.

[10] A. Dold, *Lectures on Algebraic Topology*, Springer–Verlag, New York, 1972.

[11] R. Easton, Isolating blocks and symbolic dynamics, *JDE* **17** (1975) 96–118.

[12] R. Easton, Isolating Blocks and Epsilon Chains for Maps, *Physica D* **39** (1989) 95–110.

[13] M. Eidenschink and K. Mischaikow, A numerical algorithm for finding isolating neighborhoods, in progress.

[14] R. Franzosa, Index filtrations and the homology index braid for partially ordered Morse decompositions, *Trans. AMS* **298**, (1986) pp. 193–213.

[15] R. Franzosa, The connection matrix theory for Morse decompositions, *Trans. AMS* **311**, (1989) pp. 561–592.

[16] R. Franzosa, The continuation theory for Morse decompositions and connection matrices, *Trans. AMS* **310**, (1988) pp. 781–803.

[17] R. Franzosa and K. Mischaikow, The connection matrix theory for semiflows on (not necessarily locally compact) metric spaces, *Journal of Differential Equations* **71**, (1988) 270-287.

[18] R. Franzosa and K. Mischaikow, Algebraic Transition Matrices, preprint.

[19] A. Floer, A refinement of the Conley index and an application to the stability of hyperbolic invariant sets, *Erg. Thy. Dyn. Sys.* **7** (1987) 93–103.

[20] A. Floer, A topological persistence theorem for normally hyperbolic manifolds via the Conley index, *Trans. AMS* **321**, (1990) pp. 645–657.

[21] R. Gardner, *Existence of travelling wave solution of predator-prey systems via the Conley index*, SIAM J. Appl. Math. **44**, (1984), pp. 56-76.

[22] T. Gedeon and K. Mischaikow, Global Dynamics of Cyclic Feedback Systems, *J. Dif. Eq. Dyn.* (to appear).

[23] J. K. Hale, L. T. Magalhães, and W. M. Oliva, *An Introduction to Infinite Dimensional Dynamical Systems - Geometric Theory*, Appl. Math. Sci. **47**, Springer-Verlag 1984.

[24] M. Hirsch, *Differential Topology*, Springer-Verlag 1976.

[25] B. R. Hunt, T. Sauer, and J. A. Yorke, Prevalence: a trnslation–invariant "almost everywhere" on infinite dimensional spaces, preprint 1991.

[26] T. Kaczynski and M. Mrozek, Conley Index for Discrete Multivalued Dynamical Systems, preprint.

[27] W. S. Massey, *Homology and Cohomology Theory*, Marcel Dekker, New York, 1978.

[28] C. McCord, Mappings and homological properties in the homology Conley index. *Erg. Th. & Dyn. Sys.* **8*** (1988) 175 - 198.

[29] C. McCord, Mappings and Morse decompositions in the homology Conley index. *Indiana Univ. Math. J.* **40** (1991) 1061 - 1082.

[30] C. McCord and K. Mischaikow, Connected simple systems, transition matrices, and heteroclinic bifurcations, *Trans. AMS* **333**, (1992) pp. 379–422.

[31] C. McCord and K. Mischaikow, On the global dynamics of attractors for scalar delay equations, preprint CDSNS92-89.

[32] C. McCord, K. Mischaikow, and M. Mrozek. Zeta functions, periodic trajectories, and the Conley index, *JDE* (to appear).

[33] J. Milnor, *Morse Theory*, Annals Study 51, Princeton University Press, Princeton, 1963.

[34] J. Milnor and J. D. Stasheff, *Characteristic Classes*, Princeton University Press, Princeton, NJ, 1974.

[35] K. Mischaikow, Global asymptotic dynamics of gradient-like bistable equations, *SIAM Math. Anal.* (to appear).

[36] K. Mischaikow and Y. Morita, Dynamics on the Global Attractor of a Gradient Flow arising in the Ginzburg-Landau Equation, *JJIAM* (to appear).

[37] K. Mischaikow and M. Mrozek, Isolating Neighborhoods and Chaos, *JJIAM* (to appear).

[38] K. Mischaikow and M. Mrozek, Chaos in the Lorenz Equations: a Computer Assisted Proof, *Bull. AMS* (to appear).

[39] K. Mischaikow, M. Mrozek, and J. Reineck, Singular Index Pairs, preprint.

[40] R. Moeckel, Morse decompositions and connection matrices, *Erg. Th. & Dyn. Sys.* **8*** (1988) 227-550.

[41] M. Mrozek, Leray functor and cohomological index for discrete dynamical systems. *Trans. A. M. S.* **318** (1990) 149-178.

[42] M. Mrozek, A cohomological index of Conley type for multivalued admissible flows, *JDE* **84** (1990) 15-51.

[43] M. Mrozek, Open index pairs, the fixed point index and rationality of zeta functions, *Ergod. Th. & Dyn. Sys.*, (1990) **10**, 555-564.

[44] M. Mrozek, Shape index and other indices of Conley type for local maps on locally compact Hausdorff spaces, CDSNS92-106.

[45] M. Mrozek, Topological invariants, multivalued maps and computer assisted proofs in dynamics, in preparation.

[46] M. Mrozek and K. P. Rybakowski, A cohomological Conley index for discrete dynamical systems, *J. D. E.* **90** (1991), 143-171.

[47] J. Reineck, Connecting orbits in one-parameter families of flows, *Ergodic Theory and Dynamical Systems* 8*(1988), 359-374.

[48] J.W. Robbin and D. Salamon, Dynamical systems, shape theory and the Conley index, *Ergodic Theory and Dynamical Systems* 8*(1988), 375-393.

[49] K. P. Rybakowski, *The Homotopy Index and Partial Differential Equations*, Universitext, Springer-Verlag 1987.

[50] D. Salamon, Connected simple systems and the Conley index of isolated invariant sets. *Trans. A. M. S.* **291** (1985) 1 - 41.

[51] S. Smale, Differentiable Dynamical Systems, *Bull. AMS* **73**, 1967.

[52] J. Smoller, *Shock Waves and Reaction Diffusion Equations*, Springer Verlag, New York, 1980.

[53] E. H. Spanier, *Algebraic Topology*, McGraw–Hill, 1966, Springer–Verlag, 1982.

[54] A. Szymczak, The Conley Index for Discrete Semidynamical Systems. preprint 1994.

[55] Todd Young, Personal communication.

Dynamics of partial differential equations on thin domains

Geneviève Raugel

Introduction

The purpose of this course is to study the dynamics of PDE on thin domains. The main motivation of this kind of problems comes from two sources.

The first source is physics; thin domains problems are encountered in solid mechanics (thin rods, plates, shells), fluid dynamics (lubrication, meteorology problems, ocean dynamics, etc ...). For instance in ocean dynamics, one is dealing with large scale fluid problems, in which the fluid regions are thin compared to the horizontal length scales (see the recent work of Camassa and Holm (1992), for example). Some meteorology problems involve fluids circulating between two concentric spheres of nearly equal radii, which models the atmosphere. Problems in mechanics and especially in elasticity concerning thin rods, plates and shells have already been studied for over a century. The study of the equilibrium states of a thin plate $\Omega \times (-\epsilon, +\epsilon)$ under external forces where Ω is a smooth domain in $\mathbb{R}^2$ and ϵ is a small parameter, leads to a two-dimensional modelling of plates (see Ciarlet and Destuynder (1979), Destuynder (1981), for instance). In lubrication problems, the Reynolds equation can be justified mathematically from the Stokes equation on thin domains (Bayada and Chambat (1986)). In reaction-diffusion equations we can also consider some models for which all the diffusion coefficients are large; after rescaling, we obtain a PDE on a thin domain around a point. The modelling of multi-structures involves more complicated thin domains. Roughly speaking, a multi-structure problem is described by a PDE given on a connected domain in $\mathbb{R}^n$ consisting of several parts, some of which can be relatively thin in one or several directions. Typical examples are encountered in elasticity, in heat diffusion, etc ... and lead to problems with junctions (see Ciarlet, Le Dret and Nzengwa (1987, 1989), Ciarlet (1990), Le Dret (1989a, 1989b, 1991)). For sake of simplicity and of lowering the costs of computation, one tries to replace the n-dimensional multi-structure by a multi-structure, the dimension of which is smaller in the thin parts (see the just mentioned papers). Thin domains problems have been considered by many authors from different points of view, including modelling, control, homogenization questions, ... and using different techniques like asymptotic expansions, singular perturbations methods, ... ; see, for instance, Lions (1973), Lagnese and Lions (1988), Ciarlet (1988, 1990), Le Dret (1991) and the references therein. Let me also refer to the following papers (Aganovic and Tutek (1986), Caillerie (1981), Cimetière, Geymonat, Le Dret, Raoult and Tutek (1988), Kohn and Vogelius (1984, 1985, 1986), Raoult (1988)). For the study of Navier–Stokes equations in thin domains, we also refer to Dridi (1982), Besson, Laydi and Touzani (1990), Besson and Laydi (1992).

Let us suppose that we are given a partial differential evolutionary equation (P) with homogeneous Neumann boundary conditions on the $(n + k)$-dimensional product domain $\Omega \times (-\epsilon, +\epsilon)^k$, where Ω is a smooth n-dimensional domain. In such a situation it is natural to ignore the thin directions and to expect to obtain most of the important qualitative and quantitative informations by considering the same evolutionary equation (P) on the lower dimensional domain Ω. This type of reduction in the dimension of the domain is still valid for more general thin domains and often leads to a simpler problem for which we can give a more complete description of the dynamics. In the discussion of a PDE on a thin domain Q_ϵ, it is convenient to perform a change of variables, which transforms Q_ϵ to a canonical fixed domain Q, and the equation (P) to an equation (P_ϵ). This permits to do the analysis in function spaces on the fixed domain Q and injects the parameters describing the domain into the differential operators of the problem. The first step in the analysis then consists in defining the formal limit equation (P_0) of the equation (P_ϵ). Since (P_0) is given on a lower dimensional domain, it will sometimes be called the reduced equation. One aim of these lectures is the comparison of the dynamics of the full equation (P_ϵ) on Q with the dynamics of the equation (P_0) and, especially, the comparison of their long time behaviours. In particular, in the case of dissipative equations, we try to compare the attractors of (P_ϵ) and (P_0). The spirit of these lectures is somewhat different from most of the above mentioned works. Indeed many papers dealt with asymptotic expansions of the solutions of the problem (P_ϵ) or with the obtaining of the formal limit equation (P_0). Some papers showed weak convergence (or even strong convergence) of some solutions of (P_ϵ) to some solutions of (P_0), but did not really try to deduce properties of the problem (P_ϵ) from those of the problem (P_0). Here we seek "error" estimates between the solutions of both equations. We also try to understand in which extends the dynamics of (P_0) reflect the ones of (P_ϵ) (see also Yanagida (1990)). In this perspective, the comparison of the spectra of the differential operators in (P_ϵ) and in (P_0) also plays an important role (Hale and Raugel (1992b), see also Bourquin and Ciarlet (1989), Ramm (1985)).

The second motivation of the study of problems on thin domains is mathematical. At first, the study of the dynamics of PDE on thin domains is only a particular case of the general question concerning the effects of the variation in the domain on the dynamics of the PDE. If the domain variations are regular, the dependence in the domain of the equilibria of a PDE or of the spectra of linear differential operators has been studied for a long time (Hadamard (1907), Garabedian and Schiffer (1952), Nickel (1962), Micheletti (1973), Stummel (1976), Uhlenbeck (1976), Saut and Temam (1979), Simon (1981), Henry (1985, 1987, 1988), ...). If the domain variations are not regular, the behaviour of the dependence in the domain of the PDE is not yet well understood. The thin domain problems enter into this category of questions. In this perspective, an interesting question is how the shape of the thin domain reflects in the limit equation (P_0). Another example of singular variation of domains is given by the dumbbell shaped domain, that is, a domain consisting of a very thin channel connecting to larger domains (see for example Matano (1979), Matano and Mimura (1983), Hale and Vegas (1984), Jimbo (1988,1989), Kohn and Sternberg (1989), Morita (1989), Jimbo and Morita (1992), Vegas (1983, 1989), Dancer (1988), Arrieta (1994) and the references therein). In these papers, the spectral problems as well as some dynamical questions have been studied. If the

PDE is dissipative, an interesting question is the changes in the flow on the global attractor coming from the variations in the domain.

Finally, the study of the dynamics of an evolution equation on a thin domain, which leads to a comparison of the flows of two PDE ($\mathbf{P}_\epsilon$) and ($\mathbf{P}_0$), enters into the general program of studying the flows of semigroups $T_\epsilon(t)$ depending on a parameter ϵ. If we perturb a semigroup $T_0(t)$ in some singular (but not too singular) way, how can we compare the flow of $T_0(t)$ with the one of the perturbed semigroup $T_\epsilon(t)$? For instance, if $T_\epsilon(t)$ and $T_0(t)$ admit global attractors $\mathcal{A}_\epsilon$ and $\mathcal{A}_0$ respectively, how does converge $\mathcal{A}_\epsilon$ towards $\mathcal{A}_0$? Can we compare the flows of $T_\epsilon(t)\big|_{\mathcal{A}_\epsilon}$ and $T_0(t)\big|_{\mathcal{A}_0}$? In the case of perturbations of finite dimensional dynamical systems, such questions have been successfully answered (see for instance the papers about topological equivalence, use of Conley index, ...). There have been some attempts to adapt the ideas of the finite dimensional case to the infinite dimensional case. Some general results can be obtained and we will show that some properties of PDE on thin domains are a consequence of these general results. We can also find some analogy with other singular perturbation problems. For instance, if we consider a system of reaction diffusion equations for which all the diffusion coefficients are large, then the dynamics is well described by the ones of an ODE involving only reaction terms (see Conway, Hoff and Smoller (1978), Hale (1986)). In reaction diffusion systems, if some diffusion coefficients are large relative to others, then the dynamics is described by a *shadow system* consisting of some PDE coupled with some ODE (Nishiura (1982), Hale and Sakamoto (1989)).

The content of these lectures comes for a large part from joint papers with J. Hale and G. Sell (Hale and Raugel (1989, 1991, 1992a, 1992b, 1992c, 1992d, 1992e, 1993a, 1993b, 1993c) and Raugel and Sell (1989, 1992, 1993a, 1993b)).

Acknowledgment: I would like to thank R. Johnson, R. Conti and P. Zecca for their kind invitation to give a CIME Course on Dynamics on thin domains.

Outline of the lectures.

Chapter 1 : Generalities and examples of PDE on thin domains
- Examples of thin domains (simple and non-simple, L–shaped, T–shaped, ...).
- Specific model PDE on thin domains.
- Strategy of the analysis of PDE on thin domains. Abstract setting.
- Formal limit of some differential operators on thin domains.

Chapter 2 : Abstract comparison theorems
- Upper semicontinuity of attractors.
- Lower semicontinuity of attractors in gradient systems.
- Comparison of equilibrium points.
- Comparison of eigenvalues.

Chapter 3 : Description of the results in our specific examples of PDE.
- Reaction diffusion equations (inertial manifolds, Morse–Smale property, ...).
- Ginzburg–Landau equations.

- Damped wave equations.

Chapter 4 : Convergence to singletons.

- A general convergence result.
- Convergence results for some gradient PDE on thin domains around an arc of curve.

Chapter 5 : Global existence of solutions and global stability.

- The Navier–Stokes equations : classical results and statements of global existence on thin 3D domains.
- Proofs in the case of the Navier–Stokes equations with periodic boundary conditions.
- Remarks about the Euler equations on a thin domain.

Chapter 1

Generalities and examples of PDE on thin domains.

1.1 Examples of thin domains.

1.1.1 "Simple" thin domains (bounded domains). If $\Omega \subset \mathbb{R}^{n+k}$ is a compact connected n-dimensional smooth submanifold with boundary of $\mathbb{R}^{n+k}$, there is a general canonical way to construct "simple", relatively regular thin domains Q_ϵ of dimension $n + m$ with $m \leq k$, which contain Ω (see (Hale and Raugel (1991, 1992d)). If Ω is a smooth submanifold without boundary, Q_ϵ will still be a smooth domain. If Ω has a boundary, then, in general, Q_ϵ will be a domain "with corners". The idea of the construction is very simple: we modify the domain a small amount along some directions in the normal space at each point of Ω, we also allow some distorsions along the normal directions. Since the general setting is rather long and introduces useless technicalities, we shall only give some examples and choose later a model (simple) thin domain.

• *Thin product domains.* Roughly speaking, such domains Q_ϵ can be written as $Q_\epsilon = \Omega \times (0, \epsilon)$.

Thin rectangle

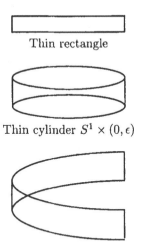

Thin cylinder $S^1 \times (0, \epsilon)$

Thin cylinder $\mathcal{C} \times (0, \epsilon)$ where $\mathcal{C} = \{(x(s), y(s)) \mid a < s < b\}$

Full thin cylinder $\Omega \times (0, \epsilon)$

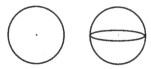

Thin domains around a point

The metric on Q_ϵ is "the direct sum of the metric on Ω with the one on $\mathbb{R}^m$".
• *Uniform tubular neighbourhoods of radius ϵ, of dimension $n + m$, containing Ω $(m \leq k)$.*

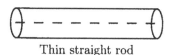

Thin straight rod

Uniform tubular neighbourhoods of radius ϵ,
of dimension 1+2 (resp. 1+1), containing $\mathcal{C}$

• *Domain between two concentric spheres of radius R_0 and $R_0 + \epsilon$.*
• *A Möbius band which is thickened with a thickness of ϵ.*
• *A thin torus (i.e. a uniform tubular neighbourhood of radius ϵ, of dimension $1 + 2$, containing S^1).*
• *A thin uniform section of a cone.*

• *Modified and distorted thin product domains.*

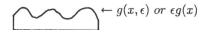

$\leftarrow g(x, \epsilon)$ or $\epsilon g(x)$

One-sided thin domain

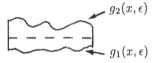

Two-sided thin domain

- *Modified thin tubular neighbourhoods.*

Thin domain around a point

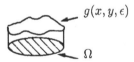

$$Q_\epsilon = \{(r, \varphi) \; : \; 0 \le \varphi \le 2\pi, \; 1 - g_2(\varphi, \epsilon) < r < 1 + g_1(\varphi, \epsilon)\}$$

- *Our model (simple) thin domain* will be the following:

(1.1) $$Q_\epsilon = \{(\tilde{x}, \tilde{y}) \in \mathbb{R}^{n+1} \; ; \; 0 < \tilde{y} < g(\tilde{x}, \epsilon), \; \tilde{x} \in \Omega\}$$

where Ω is a smooth domain or a C^1–polygonal domain in $\mathbb{R}^n$ and $g : \overline{\Omega} \times [0, \epsilon_0] \to \mathbb{R}$ is a function of class C^3 satisfying :

(1.2) $$\begin{cases} g(\tilde{x}, 0) = 0, \; g(\tilde{x}, \epsilon) > 0, \; \forall \tilde{x} \in \overline{\Omega}, \; \epsilon > 0 \\ g_0(\tilde{x}) \equiv \dfrac{\partial g}{\partial \epsilon}(x, 0) > 0, \; \forall \tilde{x} \in \overline{\Omega} \, . \end{cases}$$

1.1.2 Some examples of non–simple thin domains (bounded domains).

- *L–shaped and T–shaped thin domains.*

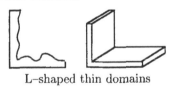

L–shaped thin domains

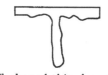

T–shaped thin domain

Comments
- This type of thin domains is encountered in mechanics, in the multi–structures problems (Ciarlet, Le Dret and Nzengwa (1987, 1989), Ciarlet (1990), Le Dret (1989a, 1989b, 1991)) and also in problems of electrical circuits.
- Such type of thin domains can be decomposed into N simple thin subdomains. The junction region is then counted N times, because it is contained in the N simple subdomains.
- This leads to formal limits consiting of a system of N equations with compatibility conditions.

• *Thin–domains with discontinuities. An example.*

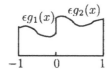

Here $g_1(0) \neq g_2(0)$ (in the more general case, we can choose $(g_1(x, \epsilon), g_2(x, \epsilon))$ with $\frac{\partial g_1}{\partial \epsilon}(0, 0) \neq \frac{\partial g_2}{\partial \epsilon}(0, 0)$). Again we decompose the domain into several simple thin domains.

• *Thin domains with varying order of thinness.*

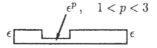

This case has been considered by Ciuperca (1994). Again the thin domain can be decomposed into several simple thin domains.

• *N–parameters thin domains* (see Hale and Raugel (1992d)). Thin domain around a point :
$$Q_{\epsilon,\mu} = \{(x, y) \ : \ y = \epsilon g(x/\epsilon, \mu), \ 0 < x < \mu\}$$
where ϵ and μ go to 0. The limit is a system of ODE.

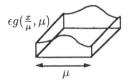

Thin domain around a line segment

$$Q_{\epsilon,\mu} = \{(x,y,z) \; : \; 0 < y < 1, \; z = \epsilon g(x/\epsilon, \mu), \; 0 < x < \mu\}$$

where ϵ and μ go to 0.

1.1.3 Unbounded thin domains.

In Kirchgässner and Raugel (1994a, b), we study the stability of travelling waves for a KPP system in a thin strip $(g(x) = 1)$.

1.2. Specific examples of PDE on thin domains.

Let Q_ϵ be a given simple thin domain. Below we give some examples of *model* equations on Q_ϵ that can be studied with our framework.

1.2.1 Reaction–diffusion equations.

$$\text{(RD)} \quad \begin{cases} u_t - \Delta u + \alpha u = -f(u) - G & \text{in } Q_\epsilon \\ \dfrac{\partial u}{\partial \nu_\epsilon} = 0 & \text{in } \partial Q_\epsilon, \end{cases}$$

where ν_ϵ is the outward normal to ∂Q_ϵ, $\alpha > 0$, $f \in \mathcal{C}^2(\mathbb{R}; \mathbb{R})$ satisfies some sign and growth conditions.

This is a *gradient system*, that is , admits a Lyapunov functional. It is a parabolic equation which also means that the associated semigroup has smoothing properties for positive time. We could also replace $-f(u) - G$ by $-f(x, u)$.

1.2.2 The complex Ginzburg–Landau equation. This is another example of parabolic equation.

$$\text{(CGL)} \quad \begin{cases} u_t - (\nu + ia)\Delta u + (\kappa + ib)|u|^2 u - du = 0 & \text{in } Q_\epsilon \\ \dfrac{\partial u}{\partial \nu_\epsilon} = 0 & \text{in } \partial Q_\epsilon, \end{cases}$$

where ν, a, κ, b, d are real numbers, and $\nu > 0$, $\kappa > 0$. In general, this equation is not a gradient system. (For some special values of the parameter, it can be reduced to a gradient system).

1.2.3 Damped wave equations. *The following equations are examples of hyperbolic gradient systems.* In these examples, the associated semigroup has no smoothing properties in finite time.

(DW)
$$\begin{cases} u_{tt} + \beta u_t - \Delta u + \alpha u = -f(u) - G & \text{in } Q_\epsilon \\ \dfrac{\partial u}{\partial \nu_\epsilon} = 0 & \text{in } \partial Q_\epsilon \end{cases}$$

where β is a constant.

We can also consider *locally damped wave equations*, i.e.,

(LDW)
$$\begin{cases} u_{tt} + \widetilde{\beta}_\epsilon(x,y)u_t - \Delta u + \alpha u = -f(u) - G & \text{in } Q_\epsilon \\ \dfrac{\partial u}{\partial \nu_\epsilon} = 0 & \text{in } \partial Q_\epsilon \end{cases}$$

where $\widetilde{\beta}_\epsilon$ is a nonnegative function, the support of which may not be all of Q_ϵ.

We could also consider wave equations with non–linear damping.

If, in the equation (DW), we put $\beta = 0$, we obtain the *conservative wave equation*

(W)
$$\begin{cases} u_{tt} - \Delta u + \alpha u = -f(u) - G & \text{in } Q_\epsilon \\ \dfrac{\partial u}{\partial \nu_\epsilon} = 0 & \text{in } \partial Q_\epsilon \ . \end{cases}$$

1.2.4 More general parabolic equations.

(PB)
$$\begin{cases} u_t - \Delta u + \alpha u = F(u, \nabla u) & \text{in } Q_\epsilon \\ \dfrac{\partial u}{\partial \nu_\epsilon} = 0 & \text{in } \partial Q_\epsilon \ . \end{cases}$$

1.2.5 Cahn–Hilliard equation. This is a gradient system containing a differential operator of order 4.

(CH)
$$\begin{cases} u_t + \Delta^2 u - \Delta f(u) = 0 & \text{in } Q_\epsilon \\ \dfrac{\partial u}{\partial \nu_\epsilon} = \dfrac{\partial \Delta u}{\partial \nu_\epsilon} = 0 & \text{in } \partial Q_\epsilon \ . \end{cases}$$

where

$$f(u) = \sum_{j=1}^{2p-1} a_j u^j \ , \qquad a_{2p-1} > 0 \ ,$$

and $p > 1$ is an integer. The Cahn–Hilliard equation (CH) on a thin domain Q_ϵ has been studied by Abounouh (1994).

1.2.6 Navier–Stokes equations. We shall consider the Navier–Stokes equations on the thin product domain $Q_\epsilon = (0, \ell_1) \times (0, \ell_2) \times (0, \epsilon)$ or $\Omega \times (0, \epsilon)$. The boundary conditions will be periodic or mixed ones (see Chapter 5).

(NS)
$$\begin{cases} U_t - \nu \Delta U + (U \cdot \nabla)U + \nabla P = F \\ \nabla \cdot U = 0 \end{cases}$$

where $\nu > 0$, $F(x, y, t)$ is a forcing term.

1.2.7 Euler equations. When ν is taken equal to 0, we obtain the (conservative) Euler equations. We have studied them on the thin domain Q_ϵ between two concentric spheres, with a thin gap

(Eu)
$$\begin{cases} U_t + (U \cdot \nabla)U + \nabla P = F \\ \nabla \cdot U = 0 \\ U \cdot \nu_\epsilon = 0 \ . \end{cases}$$

If we consider the Euler equations with periodic boundary conditions, the limit equations are slightly different and simpler (see Chapter 5).

1.2.8 The Kuramoto–Sivashinsky equation in thin 2D domains (see Sell and Taboada (1992)).

(KS)
$$\begin{cases} u_t + \nu \Delta^2 u + \Delta u + \frac{1}{2}|\nabla u|^2 = 0 \quad \text{in } Q_\epsilon \\ \text{periodic boundary conditions} , \end{cases}$$

where $Q_\epsilon = (0, 2\pi) \times (0, 2\pi)$. Sell and Taboada have shown the existence of strong solutions for big initial data in $D(A^{1/4})$, where $A = \Delta^2$.
If we set $U_1 = D_x u$, $U_2 = D_y u$, $U = (U_1, U_2)$, we obtain the system of equations:

$(\widetilde{\text{KS}})$
$$\begin{cases} U_t + \nu \Delta^2 U + \Delta U + (U \cdot \nabla)U = 0 \quad \text{in } Q_\epsilon \\ \operatorname{curl} U = 0 , \end{cases}$$

with periodic boundary conditions.

1.2.9 Some comments and generalizations.
1. The above problems are model problems. Of course, we could also consider systems of p such equations with coupling.
2. In the analysis below, we could always replace the operator $-\Delta$ by the more general selfadjoint operator

$$Au = - \sum_{i,j=1}^{N} \frac{\partial}{\partial x_i}\left(a_{ij}(x)\frac{\partial u}{\partial x_j}\right) + \alpha(x)u ,$$

where $x = (x_1, \ldots, x_N)$, with the corresponding homogeneous Neumann boundary conditions

$$\sum_{i=1}^{N} \left(\sum_{j=1}^{N} \frac{\partial}{\partial x_i} (a_{ij}(x) \frac{\partial u}{\partial x_j}) \right) \cdot \nu_{\epsilon,i} = 0,$$

where $\nu_\epsilon = (\nu_{\epsilon,1}, \ldots, \nu_{\epsilon,N})$, $a_{ij}(x) = a_{ji}(x)$, $\alpha(x) \geq \alpha_0 > 0$, $\sum_{i,j=1}^{N} a_{ij}(x)\xi_i\xi_j \geq \alpha_1(\sum_{i=1}^{N} \xi_i^2)$ with $\alpha_1 > 0$.

Most of the time, we can replace $-\Delta$ by the operator

$$A = -\sum_{i,j=1}^{N} \frac{\partial}{\partial x_i}(a_{ij}(x)\frac{\partial u}{\partial x_j}) + \sum_{i=1}^{N} b_i(x)\frac{\partial u}{\partial x_i} + \alpha(x)u .$$

3. *Comments about the important role of the boundary conditions.* The choice of the boundary conditions in the thin direction plays an important role. Let us give an illustration with simple examples.

Homogeneous boundary conditions. We again consider the thin model domain described in (1.1) and the reaction–diffusion equation

(1.3) $$u_t - \Delta u + \alpha u = -f(u) - G .$$

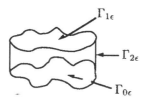

If we consider the equation (1.3) with homogeneous Neumann boundary conditions $\frac{\partial u}{\partial \nu_\epsilon} = 0$, then the limit equation is given by

(1.4) $$\begin{cases} \text{(i)} \ v_t + L_0 v + \alpha v = -f(v) - G_0 & \text{in} \Omega \\ \text{(ii)} \ \dfrac{\partial v}{\partial n} = 0 & \text{in} \ \partial\Omega, \end{cases}$$

where n is the outward normal to Ω, and L_0 and G_0 will be given later (see Chapter 3).

If we consider the equation (1.3) with the mixed conditions

$$\begin{cases} \dfrac{\partial u}{\partial \nu_\epsilon} = 0 & \text{on } \Gamma_{0\epsilon} \cup \Gamma_{1\epsilon} \ (\text{bottom and top of } Q_\epsilon) \\ u = 0 & \text{on } \Gamma_{2\epsilon} \ (\text{side of } Q_\epsilon), \end{cases}$$

then the limit equation will be (1.4)(i) with

(1.4)(ii bis) $$v = 0 \text{ in } \partial\Omega .$$

If $Q_\epsilon = \Omega \times (0, \epsilon)$, we can also put periodic boundary conditions in the vertical direction and Neumann or Dirichlet boundary conditions on the side $\Gamma_{2\epsilon} = \partial\Omega \times (0, \epsilon)$. Then the limit equation will be (1.4)(i) and (ii) or (1.4)(i) and (iibis).

If we consider (1.3) with homogeneous Dirichlet boundary conditions, then, in some sense, $u \to 0$ and the global attractor $\mathcal{A}_\epsilon \to \{0\}$. We shall make that more precise in Chapter 3.

Some oblique derivative boundary conditions could also be considered.

Non-homogeneous boundary conditions. In the examples of PDE above, we have chosen homogeneous boundary conditions, so that the solutions of these PDE do not exhibit a transition layer as $\epsilon \to 0$. On the other hand, if we take non-homogeneous boundary conditions, it is clear that, in general, transition layers are to be expected. These questions are very delicate and have not yet been studied in details. Let us give examples on the thin product domain $Q_\epsilon = (0, 1) \times (0, \epsilon)$. We consider the linear equation

(1.5)
$$\begin{cases} \text{(i)} & u_t - u_{xx} - u_{yy} + u = 0 \text{ in } Q_\epsilon \\ \text{(ii)} & u_y|_{y=0} = 0, \ u_y|_{y=\epsilon} = \frac{s}{\epsilon^a}, \ u_x|_{x=0} = u_x|_{x=1} = 0 \ . \end{cases}$$

where $s > 0$, $a > 0$. Then the unique stationary solution of (1.5) is given by

$$\widehat{u}_0(x, y) = \frac{s}{\epsilon^a} \frac{\text{ch} y}{\text{sh} y} \ .$$

We remark that $\|\widehat{u}_0\|_{L^2(Q_\epsilon)} \approx \epsilon^{-a-\frac{1}{2}}$. If we let $u = \widetilde{u} + \widehat{u}_0$, then $\widetilde{u}$ satisfies the equation (1.5)(i) with homogeneous Neumann boundary conditions.

We now consider the linear equation (1.5)(i) with the boundary conditions

(1.5)(iibis)
$$u_y|_{y=0} = u_y|_{y=\epsilon} = \frac{s}{\epsilon^a} , \ u_x|_{x=0} = u_x|_{x=1} = 0 \ .$$

Then the unique stationary solution of (1.5)(i), (iibis) is given by

$$\widehat{u}_1(x, y) = \frac{s}{\epsilon^a} \frac{(1 - e^\epsilon)}{e^{-\epsilon} - e^\epsilon}(e^{-\epsilon}e^y - e^{-y}) \ .$$

We remark that $\|\widehat{u}_1\|_{L^2(Q_\epsilon)} \approx \epsilon^{\frac{3}{2}-a}$ and that $\|\widehat{u}_1\|_{H^1(Q_\epsilon)} \approx \epsilon^{\frac{1}{2}-a}$. In particular, for $a = 1$, we get : $\|\widehat{u}_1\|_{L^2(Q_\epsilon)} \approx \epsilon^{\frac{1}{2}}$ and $\|\widehat{u}_1\|_{H^1(Q_\epsilon)} \approx \epsilon^{-\frac{1}{2}}$.

Consider now the general equation (1.3) on the thin domain Q_ϵ given by (1.1) with the boundary conditions

(1.6)
$$\frac{\partial u}{\partial \nu_\epsilon} = \chi_\epsilon \text{ in } \partial Q_\epsilon ,$$

where χ_ϵ is a regular function satisfying the adequate compatibility conditions. Let $u_0(x, y, \epsilon)$ be a solution of $-\Delta u + \alpha u = 0$ with the boundary conditions (1.6). If we let $u = \widetilde{u} + u_0$, then we obtain the equation

(1.7)
$$\begin{cases} \widetilde{u}_t - \Delta\widetilde{u} + \alpha\widetilde{u} = -f(\widetilde{u} + u_0) - G & \text{in } Q_\epsilon \\ \dfrac{\partial\widetilde{u}}{\partial\nu_\epsilon} = 0 & \text{in } \partial Q_\epsilon , \end{cases}$$

If χ_ϵ is "singular" in ϵ, then the behaviour of $-f(\tilde{u} + u_0)$ in ϵ can be also very singular, and transition layers appear.

1.3 Strategy of the analysis of PDE on thin domains. Abstract setting.

1.3.1 The case of simple thin domains. In order to explain the strategy, we consider a PDE (P), say a *parabolic* equation

$$(P) \qquad \begin{cases} \dfrac{du}{dt} + Au = F(u, \tilde{x}, \tilde{y}) \\[2mm] u_{|t=0} \in X \end{cases}$$

on a thin domain Q_ϵ "around" Ω, where A is a differential operator, $X(= D(A^s)$, $0 \le s \le 1)$ is a Banach space of functions defined on Q_ϵ.

We assume that (P) defines (at least) a local semigroup $T(t)$ on X. Sometimes we also need to use other Banach spaces H and Y with $Y \hookrightarrow X \hookrightarrow H$, $H = D(A^0)$. Y is usually a space of "more regular functions".

Example :

$$(1.8) \qquad \begin{cases} \dfrac{du}{dt} - \Delta u + \alpha u = -f(u) - G \\[2mm] \dfrac{\partial u}{\partial \nu_\epsilon} = 0 \end{cases}$$

with $Q_\epsilon = \Omega \times (0, \epsilon)$.

Here, $A = -\Delta + \alpha I$, with homogeneous Neumann boundary conditions, $X = D(A^{1/2}) = H^1(Q_\epsilon)$, $H = L^2(Q_\epsilon)$, $Y = H^2(Q_\epsilon)$.

Step 1. We perform a *change of variables* $(\tilde{x}, \tilde{y}) \in Q_\epsilon \mapsto (x, y)$ which transforms Q_ϵ into a fixed reference domain Q. Indeed it is more convenient to work on a fixed domain Q. Then the equation (P) is transformed into an equation $(P)_\epsilon$ on the domain Q:

$$(P)_\epsilon \qquad \begin{cases} \dfrac{du}{dt} + A_\epsilon u = F_\epsilon(u, \tilde{x}, \tilde{y}) \\[2mm] u_{|t=0} \in X_\epsilon \end{cases}$$

where X_ϵ is the Banach space of functions on Q, that we obtain from X by the above change of variables. Likewise, H_ϵ and Y_ϵ are the Banach spaces obtained from H and Y by the above change of variables. If $X = D(A^s)$, $H = D(A^0)$, then we define $X_\epsilon = D(A_\epsilon^s)$, $H_\epsilon = D(A_\epsilon^0)$ (and we introduce the right scaling into the norms of X_ϵ, H_ϵ, Y_ϵ). We define a local semigroup $T_\epsilon(t)$ on X_ϵ.

Example (continued) :

$$(1.8)_\epsilon \qquad \begin{cases} \dfrac{du}{dt} - \Delta_x u - \dfrac{1}{\epsilon^2}\dfrac{\partial^2 u}{\partial y^2} + \alpha u = -f(u) - G(x, \epsilon y) \\[2mm] \dfrac{\partial u}{\partial \nu} = 0 \end{cases}$$

where ν is the outward normal to ∂Q. Here,

$$A_\epsilon = -\Delta_x - \frac{1}{\epsilon^2}\frac{\partial^2}{\partial y^2} + \alpha I$$

with homogeneous Neumann boundary conditions,

$$X_\epsilon = D(A_\epsilon^{1/2}) = H_\epsilon^1(Q),$$

that is, the space $H^1(Q)$ with a norm $\| \cdot \|_{X_\epsilon}$, where

$$\|u\|_{X_\epsilon} \approx (\|u\|_{H^1(Q)}^2 + \frac{1}{\epsilon^2}\|\frac{\partial u}{\partial y}\|_{L^2(Q)}^2)^{1/2},$$

$$H_\epsilon = L^2(Q),$$
$$Y_\epsilon = H_\epsilon^2(Q)$$

with the norm

$$\|u\|_{Y_\epsilon} = (\|u\|_{H^2(Q)}^2 + \frac{1}{\epsilon^2}\|\frac{\partial u}{\partial y}\|_{H^1(Q)}^2 + \frac{1}{\epsilon^4}\|\frac{\partial^2 u}{\partial y^2}\|_{L^2(Q)}^2)^{1/2}.$$

Step 2. We try to define a *formal* limit operator A_0 of A_ϵ, where A_0 is defined on Ω. Then one can try to show that $A_\epsilon \to A_0$ or rather $A_\epsilon^{-1} \to A_0^{-1}$, if A_0 and A_ϵ are invertible in adequate spaces. To show this convergence, it is convenient to use variationnal forms, if it is possible (see the section 1.4 for some examples of limit operators A_0).

Example (continued) :

$$A_0 = -\Delta_x + \alpha I$$

with homogeneous Neumann boundary conditions on $\partial\Omega$.

Step 3. Having found the limit operator A_0, we define a *formal* limit problem

$$(\mathbf{P})_0 \qquad \begin{cases} \dfrac{dv}{dt} + A_0 v = F_0(v, x) \\[2mm] v_{|t=0} \in X_0, \end{cases}$$

where X_0 is a Banach space of functions defined on Ω, $X_0 = D(A_0^s)$; likewise we define $H_0 = D(A_0^0)$. We define a local dynamical system $T_0(t)$ on X_0.

Example (continued) : Here $(\mathbf{P})_0$ is given by

$(1.8)_0$
$$\begin{cases} \dfrac{dv}{dt} - \Delta_x v + \alpha v = -f(v) - G(x,0) \text{ in } \Omega \\ \dfrac{\partial v}{\partial n} = 0 \text{ in } \partial\Omega, \end{cases}$$

$X_0 = H^1(\Omega)$, $H_0 = L^2(\Omega)$, $Y_0 = H^2(\Omega)$. Note that, in this example, if $G(\widetilde{x},\widetilde{y}) = G(\widetilde{x})$, then every solution of $(\mathbf{P})_0$ is a solution of $(\mathbf{P})_\epsilon$.

Remarks :
• In most of our examples on simple thin domains, we have the inclusion properties

(1.9)
$$\begin{cases} X_0 \overset{I}{\hookrightarrow} X_\epsilon \\ H_0 \overset{I}{\hookrightarrow} H_\epsilon \end{cases}$$

where I is the canonical inclusion. This property simplifies some proofs. In the case of thin L–shaped domains, this property is replaced by $X_0 \xrightarrow{I_\epsilon} X_\epsilon$, where I_ϵ is a one–to–one mapping, close to the identity of X_0.
• In the study of parabolic equations, it is sometimes helpful to introduce a Banach space $\mathcal{X}$, independent of ϵ, such that $X_\epsilon \hookrightarrow \mathcal{X}$, for $\epsilon > 0$, $X_0 \overset{i}{\hookrightarrow} \mathcal{X}$ and $T_\epsilon(t)$ is defined on $\mathcal{X}$. In the example (1.8), $\mathcal{X} = H^1(Q)$.
• To compare the dynamical systems $T_\epsilon(t)$ and $T_0(t)$, we need to compare, for $u_0 \in X_\epsilon$, the orbit $T_\epsilon(t)u_0$ with some orbit $T_0(t)v_0$, where $v_0 \in X_0$ is well chosen and close to u_0 in some sense. This leads us to define a mapping M_ϵ from X_ϵ into X_0 (and H_ϵ into H_0). In the case of our simple thin domains, this mapping $M \equiv M_\epsilon$ is a projection from X_ϵ (resp. H_ϵ) onto X_0 (resp. H_0). In the case of non–simple thin domains, M_ϵ is no longer a projection, but it is close to a mean value operator. The introduction of the mapping M_ϵ helps also in the comparison of the operators A_ϵ and A_0.

Example (continued) : Let, for $u \in H^1(Q)$,

$$v(x) = (Mu)(x) = \int_0^1 u(x,y)dy .$$

M is an orthogonal projection in X_ϵ and H_ϵ and one has the Poincaré inequality :

$$\|u - Mu\|_{H_\epsilon} \le C\epsilon \|u\|_{X_\epsilon} .$$

Step 4. *In which sense is* $(\mathbf{P})_0$ *really a limit of* $(\mathbf{P})_\epsilon$? Since $(\mathbf{P})_0$ is defined on Ω, a lower–dimensional domain, the study of the properties of $(\mathbf{P})_0$ is simpler that the study of those of $(\mathbf{P})_\epsilon$. We want to compare the dynamics of $(\mathbf{P})_0$ and $(\mathbf{P})_\epsilon$ and compare their properties:
Qualitative properties.

- *local in time* : estimate $\|T_\epsilon(t)u_0 - T_0(t)Mu_0\|_{X_\epsilon}$ on finite time intervals.
- *local in time* : compare the equilibria, periodic orbits, ... of $(\mathbf{P})_0$ with those of $(\mathbf{P})_\epsilon$. This also leads to bifurcation problems.
- *global questions* : if $(\mathbf{P})_0$ (resp. $(\mathbf{P})_\epsilon$) has a global attractor $\mathcal{A}_0$ (resp. $\mathcal{A}_\epsilon$), we want to compare $\mathcal{A}_0$ and $\mathcal{A}_\epsilon$. Can we compare the sets $\mathcal{A}_0$ and $\mathcal{A}_\epsilon$ (u.s.c., l.s.c.)? Can we compare the flows? (Morse–Smale properties, determination of connecting orbits, comparison of inertial manifolds, ...).

After the steps 1, 2, 3, we are led to compare the dynamics of $T_\epsilon(t)$ and $T_0(t)$, where $T_\epsilon(t)$ (resp. $T_0(t)$) is a (local) semigroup on X_ϵ (resp. X_0), where

$$(1.9) \qquad X_0 \overset{I}{\hookrightarrow} X_\epsilon, \quad H_0 \overset{I}{\hookrightarrow} H_\epsilon,$$

where there exists a projection operator $M \in \mathcal{L}(X_\epsilon; X_0) \cap (H_\epsilon; H_0)$ and where we have good estimates of $\|T_\epsilon(t)u_0 - T_0(t)Mu_0\|_{X_\epsilon}$ and of $\|I - M\|_{\mathcal{L}(X_\epsilon; H_\epsilon)}$.

1.3.2 The case of non–simple thin domains : an example. Let us just give an example (see (Hale and Raugel (1993a)). We consider the equation

$$(1.10) \qquad \begin{cases} u_t - \Delta u = -f(u) - G & \text{in } Q_\epsilon \\ u = 0 & \text{in } \Gamma_\epsilon \\ \dfrac{\partial u}{\partial \nu_\epsilon} = 0 & \text{in } \partial Q_\epsilon \backslash \Gamma_\epsilon \end{cases}$$

where $\Gamma_\epsilon = \Gamma_\epsilon^1 \cup \Gamma_\epsilon^2$, Q_ϵ and Γ_ϵ^1, Γ_ϵ^2 are described below.

We assume that (1.10) defines a semigroup on $H^1(Q)$.

Step 1. Here the reference domain Q will be the product domain

$$Q = Q^1 \times Q^2$$

where

$$Q^i = \{(x,y) \in \mathbb{R}^2, \, 0 < x < 1, \, 0 < y < 1\}.$$

We make the changes of variables

$$(x,y) \in Q^1 \overset{\varphi_\epsilon^1}{\longmapsto} (x, \epsilon y) \in Q_\epsilon^1,$$

$$(x,y) \in Q^2 \overset{\varphi_\epsilon^2}{\longmapsto} (\epsilon x, y) \in Q_\epsilon^2,$$

and set :

$$J_\epsilon^1 = (\varphi_\epsilon^1)^{-1}(J_\epsilon), \quad J_\epsilon^2 = (\varphi_\epsilon^2)^{-1}(J_\epsilon).$$

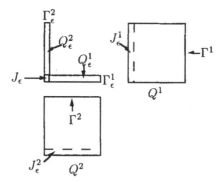

Here

$$H_\epsilon = \{u = (u_1, u_2) \in L^2(Q^1) \times L^2(Q^2) \ ;$$
$$u_1(\epsilon x, y) = u_2(x, \epsilon y) \text{ p.p.}, \ (x, y) \in (0, 1)^2\},$$

$$X_\epsilon = \{u = (u_1, u_2) \in H^1(Q^1) \times H^1(Q^2) \ ;$$
$$u_1|\Gamma^1 = u_2|\Gamma^2 = 0, \ u_1(\epsilon x, y) = u_2(x, \epsilon y) \text{ p.p.}, \ (x, y) \in (0, 1)^2\},$$

with appropriate norms. On X_ϵ, we put, for instance, the norm

(1.11) $$\|u\|_{X_\epsilon} = (\|u\|^2_{H^1(Q)} + \frac{1}{\epsilon^2}\|\frac{\partial u_1}{\partial y}\|^2_{L^2(Q^1)} + \frac{1}{\epsilon^2}\|\frac{\partial u_2}{\partial x}\|^2_{L^2(Q^2)})^{1/2} \ .$$

Using the above change of variables, we obtain a problem $(P)_\epsilon$, whose solutions u are considered in the "product" space X_ϵ, and which defines a semigroup $T_\epsilon(t)$ on X_ϵ.

Step 2. Here, A_0 is easily computed. It is given by

$$A_0((v_1, v_2)) = (-v_{1xx}, -v_{2yy})$$

with the boundary conditions $v_1(1) = v_2(1) = 0$ and the junction conditions $v_1(0) = v_2(0)$, $v_{1x}(0) = -v_{2y}(0)$. The spaces H_0 and X_0 are as follows

$$H_0 = \{v = (v_1, v_2) \in L^2(\Omega^1) \times L^2(\Omega^2)\},$$

$$X_0 = \{v = (v_1, v_2) \in H^1(\Omega^1) \times H^1(\Omega^2), \ v_1(1) = v_2(1) = 0, \ v_1(0) = v_2(0)\} \ .$$

Step 3. The formal limit problem $(P)_0$ is given by

$(\mathbf{P})_0$
$$\begin{cases} v_{1t} - v_{1xx} = -f(v_1) - G_{10} \text{ in } \Omega_1 \\ v_{2t} - v_{2yy} = -f(v_2) - G_{20} \text{ in } \Omega_2 \\ v_1(1, t) = v_2(1, t) = 0 \\ v_1(0, t) = v_2(0, t) = 0 \\ v_{1x}(0, t) = -v_{2y}(0, t) \end{cases}$$

and it defines a semigroup $T_0(t)$ on X_0.

Remarks : Clearly, X_0 (resp. H_0) is no longer included in X_ϵ (resp. H_ϵ). But we can define a one–to–one mapping $I_\epsilon : X_0 \to X_\epsilon$ which is close to the identity. We can also define a mapping $M_\epsilon : X_\epsilon \to X_0$ that is close to a partial mean value operator.

We are thus led to compare the dynamics of $T_\epsilon(t)$ and $T_0(t)$, where $T_\epsilon(t)$ (resp. $T_0(t)$) is a local semigroup on X_ϵ (resp. X_0) where $I_\epsilon : X_0 \to X_\epsilon$ is approximately the identity and $M_\epsilon : X_\epsilon \to X_0$ is approximately a partial mean value operator and where we have good estimates of $\|T_\epsilon(t)u_0 - T_0(t)M_\epsilon u_0\|_{\mathcal{X}_\epsilon}$, $\mathcal{X}_\epsilon$ being an adequate space containing X_ϵ and X_0. In our example, $\mathcal{X}_\epsilon$ is the space $H^1(Q) = H^1(Q^1) \times H^1(Q^2)$ equipped with the norm defined in (1.11).

1.4 Formal limit of some differential operators on thin domains.

In (Hale and Raugel (1992d)) we have shown how to compute the formal limit A_0 of a rather general family of operators A_ϵ. We do not reproduce the computations here and refer to this paper for more details. Here we consider the operator $A = -\Delta u + \alpha I$ with homogeneous Neumann boundary conditions. In the examples below, we give Q_ϵ, Q and the limit A_0.

- *First example.* This thin domain Q_ϵ is described in (1.1) and (1.2).

$g(x, \epsilon)$

The change of variables is given by $(\widetilde{x}, \widetilde{y}) \mapsto (x, y)$ where

(1.12)
$$\begin{cases} \widetilde{x} = x \\ \widetilde{y} = g(x, \epsilon)y \end{cases}$$

sends Q_ϵ onto Q. The mean value operator M is simply given by

(1.13)
$$(Mu)(x) = \int_0^1 u(x, y)dy \ .$$

The limit operator A_0 is :

$$A_0 v = -\frac{1}{g_0}(\sum_{i=1}^n g_0 v_{x_i})_{x_i}$$

with homogenenous Neumann boundary conditions.

- *Example 2.*

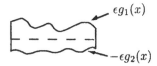

The limit operator is:

$$A_0 v = -\frac{1}{g_1 - g_2}\left(\sum_{i=1}^{n}(g_1 - g_2)v_{x_i}\right)_{x_i} .$$

- *Example 3.*

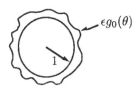

The limit operator is:

$$A_0 v = -\frac{1}{g_0}(g_0 v_\theta)_\theta .$$

- *Example* 4 (This example has also been considered by Yanagida (1988)).

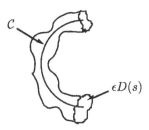

Here C is a portion of a rectifiable curve in $\mathbb{R}^3$, s is the arc–length, $a(s)$ is the area of the section $D(s)$; then the limit A_0 is :

$$A_0 v = -\frac{1}{a(s)}(a(s)v_s)_s$$

with homogeneous Neumann boundary conditions (see (Yanagida (1988)) and (Hale and Raugel (1992d)) for more details).

• *Example 5.*

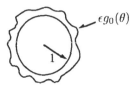

Here we consider the operator $A = \Delta^2$ with homogeneous Neumann boundary conditions

$$\frac{\partial u}{\partial n} = \frac{\partial \Delta u}{\partial n} = 0,$$

then the limit operator is :

$$A_0 v = -\frac{1}{g_0}(g_0(\frac{1}{g_0}(g_0 v_\theta)_\theta)_\theta)_\theta .$$

(see (Abounouh (1994)) for more details).

Chapter 2

Abstract comparison theorems.

2.1 Upper semicontinuity of attractors.

If the abstract semigroups $T_\epsilon(t)$ and $T_0(t)$, obtained in the chapter 1, are dissipative semigroups having global attractors $\mathcal{A}_\epsilon$ and $\mathcal{A}_0$ respectively, a first global question arises. Can we compare the sets $\mathcal{A}_\epsilon$ and $\mathcal{A}_0$?

To state precise results, we need to introduce some notation and definitions. For any Banach space Z and any subsets C, D of Z, we let

$$\delta_Z(C,D) = \sup_{c \in C} \inf_{d \in D} \|c - d\|_Z .$$

Remark that $\delta_Z(\cdot,\cdot)$ is a semi–distance. The Hausdorff distance between the sets C, D is given by $\sup(\delta_Z(C,D), \delta_Z(D,C))$.

Let $\mathcal{A}_\epsilon$, $0 \le \epsilon \le \epsilon_0$, be subsets of Z. The sets $\mathcal{A}_\epsilon$ are upper semicontinuous (u.s.c.) at $\epsilon = 0$ if

$$\delta_Z(\mathcal{A}_\epsilon, \mathcal{A}_0) \to 0 \text{ as } \epsilon \to 0 .$$

The sets $\mathcal{A}_\epsilon$ are lower semicontinuous (l.s.c.) at $\epsilon = 0$ if

$$\delta_Z(\mathcal{A}_0, \mathcal{A}_\epsilon) \to 0 \text{ as } \epsilon \to 0 .$$

Finally the sets $\mathcal{A}_\epsilon$ are continuous at $\epsilon = 0$ if they are both upper and lower semicontinuous at $\epsilon = 0$. In the above definitions, we can replace the space Z by a space Z_ϵ depending on ϵ, provided that, for each ϵ, $\mathcal{A}_\epsilon$ and $\mathcal{A}_0$ are included in Z_ϵ. Here we will use this setting.

Let $T(t)$ be a semigroup on Z. We recall that a global attractor $\mathcal{A}$ for $T(t)$, $t \ge 0$, is a compact set, which is invariant (i.e. $T(t)\mathcal{A} = \mathcal{A}$, for $t \ge 0$), such that, for any bounded set B in Z, we have $\delta_Z(T(t)B, \mathcal{A}) \to 0$ as $t \to +\infty$. Remark that, if a global attractor $\mathcal{A}$ exists, it is unique.

2.1.1 The case of simple thin domains. In this case, we had introduced Banach spaces X_0 and X_ϵ, $0 < \epsilon \le \epsilon_0$, such that

$$(2.1) \qquad\qquad X_0 \hookrightarrow X_\epsilon .$$

We had also introduced a continuous projection operator $M : X_\epsilon \to X_0$. (The operator M corresponds to a partial mean value operator). Here we assume that there are positive constants c_0, C_0, c_1 such that, for $0 < \epsilon \le \epsilon_0$,

$$(2.2) \qquad c_0\|\varphi\|_{X_\epsilon} \le \|\varphi\|_{X_0} \le C_0\|\varphi\|_{X_\epsilon} , \quad \forall \varphi \in X_0 ,$$

and

$$(2.3) \qquad \|M\varphi_\epsilon\|_{X_0} \le c_1\|\varphi_\epsilon\|_{X_\epsilon} , \quad \forall \varphi_\epsilon \in X_\epsilon .$$

We assume that $T_\epsilon(t)$, $0 < \epsilon \leq \epsilon_0$ (resp. $T_0(t)$) is a C^0–semigroup on X_ϵ (resp. on X_0) which has a global attractor $\mathcal{A}_\epsilon$ (resp. $\mathcal{A}_0$) in X_ϵ (resp. in X_0). Since $X_0 \hookrightarrow X_\epsilon$, we may consider $\mathcal{A}_0$ as a subset of X_ϵ. We make the following hypotheses:

(H1) *There exists a positive constant c_2 such that*

$$(2.4) \qquad \|\varphi_\epsilon\|_{X_\epsilon} \leq c_2, \quad \forall \varphi_\epsilon \in \mathcal{A}_\epsilon, \quad 0 \leq \epsilon \leq \epsilon_0 .$$

(H2) *For all $\eta > 0$, for all $\tau > 0$, there exists $\epsilon_1 = \epsilon_1(\tau, \eta)$ such that*

$$(2.5) \qquad \|T_\epsilon(\tau)\varphi_\epsilon - T_0(\tau)M\varphi_\epsilon\|_{X_\epsilon} \leq \eta, \quad \forall \varphi_\epsilon \in \mathcal{A}_\epsilon, \quad 0 < \epsilon \leq \epsilon_1 .$$

Actually it is sufficient to assume that the property (H2) holds for all $\tau \geq t_1$, where t_1 is some positive number.

PROPOSITION 2.1. — *If the hypotheses (H1) and (H2) are satisfied, then the attractors $\mathcal{A}_\epsilon$ are u.s.c. at $\epsilon = 0$, that is,*

$$\delta_{X_\epsilon}(\mathcal{A}_\epsilon, \mathcal{A}_0) \to 0 \quad \text{as} \quad \epsilon \to 0 .$$

Proof: Let $B_{X_0}(0, r)$ denote the ball in X_0 of center zero and radius r. Since $\mathcal{A}_0$ is the global attractor in X_0, for any $\eta > 0$, there is a time $\tau_\eta > 0$ such that

$$(2.6) \qquad T_0(t)B_{X_0}(0, c_1 c_2) \subset \mathcal{N}_{X_\epsilon}(\mathcal{A}_0, \frac{\eta}{2}) \text{ for } t \geq \tau_\eta ,$$

where $\mathcal{N}_{X_\epsilon}(\mathcal{A}_0, \frac{\eta}{2})$ is the $\frac{\eta}{2}$-neighbourhood of $\mathcal{A}_0$ in X_ϵ. Here we have used also the equivalence on X_0 of the norms of X_0 and X_ϵ, given in (2.2). Using the hypothesis (H1) and (2.3), we deduce from (2.6) that

$$(2.7) \qquad T_0(t)M\mathcal{A}_\epsilon \subset \mathcal{N}_{X_\epsilon}(\mathcal{A}_0, \frac{\eta}{2}) \text{ for } t \geq \tau_\eta , \ 0 < \epsilon \leq \epsilon_0 .$$

By Hypothesis (H2), there exists $\epsilon_1 = \epsilon_1(\tau_\eta, \frac{\eta}{2})$ such that

$$\|T_\epsilon(\tau_\eta)\varphi_\epsilon - T_0(\tau_\eta)M\varphi_\epsilon\|_{X_\epsilon} \leq \frac{\eta}{2}, \quad \varphi_\epsilon \in \mathcal{A}_\epsilon$$

which, together with (2.7), implies that $T_\epsilon(\tau_\eta)\mathcal{A}_\epsilon \subset \mathcal{N}_{X_\epsilon}(\mathcal{A}_0, \eta)$. Since $T_\epsilon(\tau_\eta)\mathcal{A}_\epsilon = \mathcal{A}_\epsilon$, this completes the proof.

$\square$

As we have just seen, the proposition 2.1 is elementary. The difficulty consists in showing the properties (H1) and (H2). Let us make some remarks about them.

REMARK 2.1. — In the parabolic case, that is, in the case where the solutions of the PDE become smoother in the spatial variables for $t > 0$, we can often show the following property:

(H2bis) *There are a positive increasing function $k(r)$, $r \geq 0$, and constants $a > 0$, $b \geq 0$ such that, for $0 < \epsilon \leq \epsilon_0$, $t \geq 0$,*

$$(2.5\text{bis}) \quad \|T_\epsilon(t)\varphi - T_0(t)M\varphi\|_{X_\epsilon} \leq \epsilon^a(1 + t^{-b})k(r)e^{k(r)t}, \quad \forall \varphi \in X_\epsilon, \ \|\varphi\|_{X_\epsilon} \leq r.$$

REMARK 2.2. — If the solutions of the PDE do not become smoother in the spatial variables for $t > 0$, then we must proceed differently to prove the hypothesis (H2). This happens, for instance, in the case of the *damped wave equation*. If $T_\epsilon(t)$ has a global attractor, then we have asymptotic smoothing. This implies, in some cases, that the elements in the attractor $\mathcal{A}_\epsilon$ are smoother in the spatial variables than the elements in X_ϵ. In these cases, one can introduce a Banach space Y_ϵ composed of smoother elements and compare the semigroups $T_\epsilon(t)\psi$ and $T_0(t)M\psi$ in the topology of X_ϵ for ψ in bounded sets of Y_ϵ. More precisely, we introduce, for each ϵ, $0 < \epsilon \leq \epsilon_0$, a Banach space Y_ϵ such that $Y_\epsilon \hookrightarrow X_\epsilon$, $T_\epsilon(t) : Y_\epsilon \to Y_\epsilon$ is a C^0–semigroup and we show that,

(H1ter) *There exists a positive constant c_2 such that*

$$(2.4\text{ter}) \quad \max(\|\varphi_\epsilon\|_{X_\epsilon}, \|\varphi_\epsilon\|_{Y_\epsilon}) \leq c_2, \quad \forall \varphi_\epsilon \in \mathcal{A}_\epsilon, \ 0 < \epsilon \leq \epsilon_0,$$

(H2ter) *There are a positive increasing function $k(r)$, $r \geq 0$, and a positive constant a such that, for $0 < \epsilon \leq \epsilon_0$, $t \geq 0$,*

$$(2.5\text{ter}) \quad \|T_\epsilon(t)\varphi - T_0(t)M\varphi\|_{X_\epsilon} \leq \epsilon^a k(r)e^{k(r)t}, \quad \forall \varphi \in Y_\epsilon, \ \|\varphi\|_{Y_\epsilon} \leq r.$$

In Section 3.3, we shall encounter a typical example of this case.

2.1.2 The case of non–simple thin domains.

We still consider Banach spaces X_0 and X_ϵ, $0 < \epsilon \leq \epsilon_0$; but, here, X_0 is no longer included in X_ϵ. We introduce a linear mapping $M_\epsilon : X_\epsilon \to X_0$ which is approximately a partial mean value operator. And, we assume that, for $0 < \epsilon \leq \epsilon_0$,

$$(2.8) \quad \|M_\epsilon\varphi_\epsilon\|_{X_0} \leq c_1\|\varphi_\epsilon\|_{X_\epsilon}, \quad \forall \varphi_\epsilon \in X_\epsilon.$$

We also introduce, for $0 < \epsilon \leq \epsilon_0$, a larger space $\mathcal{X}_\epsilon$ such that

$$X_0 \hookrightarrow \mathcal{X}_\epsilon \text{ and } X_\epsilon \hookrightarrow \mathcal{X}_\epsilon$$

and the following inequalities hold

$$(2.9) \quad c_0\|\varphi_0\|_{\mathcal{X}_\epsilon} \leq \|\varphi_0\|_{X_0} \leq C_0\|\varphi_0\|_{\mathcal{X}_\epsilon}, \quad \forall \varphi_0 \in X_0$$

and

$$(2.10) \quad c_3\|\varphi_\epsilon\|_{\mathcal{X}_\epsilon} \leq \|\varphi_\epsilon\|_{X_\epsilon} \leq C_3\|\varphi_\epsilon\|_{\mathcal{X}_\epsilon}, \quad \forall \varphi_\epsilon \in X_\epsilon.$$

We still assume that $T_\epsilon(t)$, $0 < \epsilon \leq \epsilon_0$ (resp. $T_0(t)$) is a C^0–semigroup on X_ϵ (resp. X_0). We suppose that (H1) holds and that (H2) is satisfied if X_ϵ is replaced by $\mathcal{X}_\epsilon$.

PROPOSITION 2.2. — *If the hypothesis* (H2) *is satisfied where* $|| \cdot ||_{X_\epsilon}$ *is replaced by* $|| \cdot ||_{X_\epsilon}$ *and if* (H1) *holds, then the attractors* $\mathcal{A}_\epsilon$ *are u.s.c.* *at* $\epsilon = 0$, *i.e.* $\delta_{X_\epsilon}(\mathcal{A}_\epsilon, \mathcal{A}_0) \to 0$ *as* $\epsilon \to 0$.

In the above framework, the simple thin domain case is a particular case of the non–simple thin domain case with $\mathcal{X}_\epsilon = X_\epsilon$ and $M_\epsilon = M$.

2.2 Lower semicontinuity of attractors in gradient systems and in gradient–like systems.

Without further hypotheses on the flow restricted to the attractor $\mathcal{A}_0$, there will be no lower semicontinuity of the sets $\mathcal{A}_\epsilon$ at $\epsilon = 0$. Let us consider the following ordinary differential equation depending on the real parameter ϵ:

$$\dot{x} = -(x+1)(x^2 - \epsilon) .$$

If $\epsilon < 0$, $\mathcal{A}_\epsilon = \{-1\}$; for $\epsilon = 0$, $\mathcal{A}_0 = [-1, 0]$ and if $0 < \epsilon \leq 1$, $\mathcal{A}_\epsilon = [-1, \sqrt{\epsilon}]$. Clearly, the sets $\mathcal{A}_\epsilon$ are not lower semicontinuous at $\epsilon = 0$, for $\epsilon \leq 0$. This drastic change in the size of $\mathcal{A}_\epsilon$ is caused here by the fact that zero is not a hyperbolic equilibrium point.

2.2.1 Lower semicontinuity results. Roughly speaking, lower semicontinuity results are only known in the case where the limit semigroup $T_0(t)$ is a gradient system or is gradient–like. *We keep here the hypotheses of Section* 2.1. We denote by E_0 (resp. E_ϵ) the set of equilibrium points of $T_0(t)$ (resp. $T_\epsilon(t)$). Recall that φ_j^0 is an equilibrium point of $T_0(t)$ if $T_0(t)\varphi_j^0 = \varphi_j^0$, for all $t \geq 0$. If U_j^0 is a neighbourhood of φ_j^0 in X_0, we define the local unstable set

$$W_0^u(\varphi_j^0, U_j^0) \equiv W_{\text{loc},0}^u(\varphi_j^0)$$
$$= \{v \in U_j^0 \; ; \; T_0(-t)v \in U_j^0, \; t \geq 0, \; T_0(-t)v \to \varphi_j^0 \text{ in } X_0 \text{ as } t \to +\infty\} .$$

Likewise, if φ_j^ϵ belongs to E_ϵ and if U_j^ϵ is a neighbourhood of φ_j^ϵ in X_ϵ, we define the local unstable set

$$W_\epsilon^u(\varphi_j^\epsilon, U_j^\epsilon) \equiv W_{\text{loc},\epsilon}^u(\varphi_j^\epsilon)$$
$$= \{u \in U_j^\epsilon \; ; \; T_\epsilon(-t)u \in U_j^\epsilon, \; t \geq 0, \; T_\epsilon(-t)u \to \varphi_j^\epsilon \text{ in } X_\epsilon \text{ as } t \to +\infty\} .$$

We also define the global unstable sets $W_0^u(\varphi_j^0)$ and $W_\epsilon^u(\varphi_j^\epsilon)$ by

$$W_0^u(\varphi_j^0) = \{v \in X_0 \; ; \; T_0(-t)v \text{ exists}, \; t \geq 0, \; T_0(-t)v \to \varphi_j^0 \text{ in } X_0 \text{ as } t \to +\infty\} .$$

and

$$W_\epsilon^u(\varphi_j^\epsilon) = \{u \in X_\epsilon \; ; \; T_\epsilon(-t)u \text{ exists}, \; t \geq 0, \; T_\epsilon(-t)u \to \varphi_j^\epsilon \text{ in } X_\epsilon \text{ as } t \to +\infty\} .$$

We consider the case of simple thin domains and, besides the hypotheses (H1) and (H2), we assume that:

(H3) *The set E_0 is finite ; $E_0 = \{\varphi_1^0, \ldots, \varphi_{N_0}^0\}$,*

(H4) $\mathcal{A}_0 = \bigcup_{j=1}^{N_0} W_0^u(\varphi_j^0)$, *that is, $T_0(t)$ is gradient-like* (then, for every neighbourhood U_j^0 of φ_j^0, $j = 1, \ldots, N_0$,

$$\mathcal{A}_0 = \bigcup_{t \geq 0} \bigcup_{j=1}^{N_0} T_0(t) W_0^u(\varphi_j^0, U_j^0))$$

and

(H5) *for $j = 1, \ldots, N_0$, there exists a neighbourhood $\widetilde{U}_j^0$ of φ_j^0 in X_0 such that*

$$\lim_{\epsilon \to 0} \delta_{X_\epsilon}(W_0^u(\varphi_j^0, \widetilde{U}_j^0), \mathcal{A}_\epsilon) = 0 .$$

PROPOSITION 2.3. — *If the hypotheses (H1), (H2), (H3), (H4) and (H5) hold, then the attractors $\mathcal{A}_\epsilon$ are l.s.c. at $\epsilon = 0$, i.e.*

$$\delta_{X_\epsilon}(\mathcal{A}_0, \mathcal{A}_\epsilon) \to 0 \text{ as } \epsilon \to 0$$

which, with Proposition 2.2, implies that the attractors $\mathcal{A}_\epsilon$ are continuous at $\epsilon = 0$.

Proof: Let $\varphi \in \mathcal{A}_0$, $\varphi \notin W_0^u(\varphi_j^0, \widetilde{U}_j^0)$, $j = 1, \ldots, N_0$, then there exists $t_\varphi > 0$ and $v_i \in W_0^u(\varphi_i^0, \widetilde{U}_i^0)$, for some i, such that

$$\varphi = T_0(t_\varphi) v_i .$$

Since $T_0(t)$ is a C^0–semigroup and $\mathcal{A}_0$ is compact, for any $\eta > 0$, there exists $\delta > 0$, depending only on $\mathcal{A}_0$, such that, if $\|v - v_i\|_{X_0} \leq \delta$, then

$$(2.11) \qquad \|T_0(t_\varphi)v - T_0(t_\varphi)v_i\|_{X_0} \leq c_0 \frac{\eta}{2} .$$

By the hypotheses (H2) and (H5), we can choose ϵ_1 such that, for $0 < \epsilon \leq \epsilon_1$, there exists $w_\epsilon \in \mathcal{A}_\epsilon$ so that

$$(2.12) \qquad \|v_i - w_\epsilon\|_{X_\epsilon} \leq \frac{\delta}{c_1}$$

and, for $0 < \epsilon \leq \epsilon_1$,

$$(2.13) \qquad \|T_\epsilon(t_\varphi)w_\epsilon - T_0(t_\varphi)Mw_\epsilon\|_{X_\epsilon} \leq \frac{\eta}{2} .$$

Since $Mv_i = v_i$, we deduce from (2.11), (2.12), (2.13) and (2.3) that, for $0 < \epsilon \leq \epsilon_1$,

$$\|T_\epsilon(t_\varphi)w_\epsilon - \varphi\|_{X_\epsilon}$$
$$\leq \|T_\epsilon(t_\varphi)w_\epsilon - T_0(t_\varphi)Mw_\epsilon\|_{X_\epsilon} + \|T_0(t_\varphi)Mw_\epsilon - T_0(t_\varphi)Mv_i\|_{X_\epsilon} \leq \eta .$$

Assume that the sets $\mathcal{A}_\epsilon$ are not l.s.c. at $\epsilon = 0$. Then there exists $\eta > 0$ and, for any $n \in \mathbb{N}$, there exists $\epsilon_n > 0$, with $\epsilon_n < \frac{1}{n}$, such that $\delta_{X_{\epsilon_n}}(\mathcal{A}_0, \mathcal{A}_{\epsilon_n}) > \eta$. Thus there exists a sequence $\varphi_n \in \mathcal{A}_0$ with $\delta_{X_{\epsilon_n}}(\varphi_n, \mathcal{A}_{\epsilon_n}) > \frac{\eta}{2}$. Since $\mathcal{A}_0$ is compact, this sequence φ_n converges to an element $\psi \in \mathcal{A}_0$. One easily sees that there exists $n_0 \in \mathbb{N}$ such that, for $n \geq n_0$, $\delta_{X_{\epsilon_n}}(\psi, \mathcal{A}_{\epsilon_n}) \geq \frac{\eta}{4}$, which contradicts the first part of the proof.

$\square$

REMARK 2.3. — Usually the condition (H5) is shown by proving that, for ϵ small enough, $T_\epsilon(t)$ has, at least, N_0 equilibrium points φ_j^ϵ and that there exists, for $j = 1, \ldots, N_0$, a bounded neighbourhood $\tilde{U}_j^\epsilon$ of φ_j^ϵ in X_ϵ such that

$$\lim_{\epsilon \to 0} \delta_{X_\epsilon}(W_0^u(\varphi_j^0, \tilde{U}_j^0), W_\epsilon^u(\varphi_j^\epsilon, \tilde{U}_j^\epsilon)) = 0 .$$

Such a property can be proved if φ_j^0 is hyperbolic.

In the general case, it is difficult to prove the hypothesis (H5) if φ_j^0 is not hyperbolic. However, in a particular regular perturbation of the Chafee–Infante equation, Kostin (1994a, 1994b) proved the condition (H5) by using local strongly unstable sets and local center unstable sets.

REMARK 2.4. — In this proposition we could replace the equilibrium points by more general invariant sets. But the difficulty is to find exemples satisfying the corresponding hypotheses (H4) and (H5).

In the case of non–simple thin domains (as we can see in the proof of Proposition 2.3), we need to show that $\|M_\epsilon w_\epsilon - v_i\|_{X_0}$ is small, if $\|w_\epsilon - v_i\|_{X_\epsilon}$ is small. We write

$$M_\epsilon w_\epsilon - v_i = M_\epsilon w_\epsilon - M_\epsilon I_\epsilon v_i + M_\epsilon I_\epsilon v_i - v_i$$

and

$$\|M_\epsilon w_\epsilon - v_i\|_{X_0} \leq C_3 c_1 \|w_\epsilon - I_\epsilon v_i\|_{X_\epsilon} + \|M_\epsilon I_\epsilon v_i - v_i\|_{X_0}$$
$$\leq C_3 c_1 (\|w_\epsilon - v_i\|_{X_\epsilon} + \|v_i - I_\epsilon v_i\|_{X_\epsilon}) + \|M_\epsilon I_\epsilon v_i - v_i\|_{X_0}$$

where $I_\epsilon \in \mathcal{L}(X_0, X_\epsilon)$ is approximately the identity.

Let us now assume:

(H6) For any $\eta > 0$, $\exists \epsilon_2 = \epsilon_2(\eta)$ such that, for $0 < \epsilon \leq \epsilon_2$,

(2.15) $$\|\varphi - I_\epsilon \varphi\|_{X_\epsilon} + \|\varphi - M_\epsilon I_\epsilon \varphi\|_{X_0} \leq \eta, \quad \forall \varphi \in \mathcal{A}_0 .$$

This condition is easy to show since usually the attractor $\mathcal{A}_0$ is more regular in the spatial variables than the functions in X_0.

PROPOSITION 2.4. — In the case of non–simple thin domains, if the hypotheses (H1), (H2)(with X_ϵ replaced by $\mathcal{X}_\epsilon$), (H3), (H4), (H5) and (H6) hold, then the attractors $\mathcal{A}_\epsilon$ are l.s.c. at $\epsilon = 0$, i.e. $\delta_{\mathcal{X}_\epsilon}(\mathcal{A}_0, \mathcal{A}_\epsilon) \to 0$ as $\epsilon \to 0$.

2.2.2 "Error" estimates. If we want to estimate $\delta_{\mathcal{X}_\epsilon}(\mathcal{A}_\epsilon, \mathcal{A}_0)$ and $\delta_{\mathcal{X}_\epsilon}(\mathcal{A}_0, \mathcal{A}_\epsilon)$, we need to make more precise hypotheses. We restrict our study to the case where $T_0(t)$ is a gradient system.

DEFINITION 2.5. — *A continuous function $V : X_0 \to \mathbb{R}$ is a Lyapunov function for $T_0(t)$ if:*

(i) $V(T_0(t)\varphi)$ *is non–increasing in t, for every φ in X_0,*

(ii) *if φ is such that $V(T_0(t)\varphi) = V(\varphi)$, $\forall t \in \mathbb{R}$, then φ is an equilibrium point (i.e. $T_0(t)\varphi = \varphi$, $\forall t \in \mathbb{R}$).*

REMARK 2.5. — Assume that $\bigcup_{t \geq 0} T_0(t)\varphi$ is relatively compact in X_0, and that V is a Lyapunov function for $T_0(t)$. Then $V(T_0(t)\varphi)$, $t \geq 0$, is a non–increasing and bounded function, since $\bigcup_{t \geq 0} V(T_0(t)\varphi)$ is relatively compact. Therefore $c_\varphi \equiv \lim_{t \to +\infty} V(T_0(t)\varphi)$ exists. If $\psi \in \omega(\varphi)$, the ω–limit set of φ, there exists a sequence $t_n \to +\infty$ such that $T_0(t_n)\varphi \to \psi$. By continuity of V we have

$$(2.16) \qquad V(\psi) = c_\varphi , \quad \forall \psi \in \omega(\varphi) .$$

Since $T_0(t)$ is continuous, the set E_0 of equilibrium points of $T_0(t)$ is closed. By a general theorem, $T_0(t)(\omega(\varphi)) = \omega(\varphi) \neq \emptyset$. Since, by (2.16), $V(T_0(t)\psi) = V(\psi) = c_\varphi$, $\forall \psi \in \omega(\varphi)$, ψ is an equilibrium point and $E_0 \neq \emptyset$. Since $\delta_{X_0}(T_0(t)\varphi, \omega(\varphi)) \to 0$ as $t \to +\infty$ we conclude that $\delta_{X_0}(T_0(t)\varphi, E_0) \to 0$ as $t \to +\infty$.

Assume that $\bigcup_{t \leq 0} T_0(t)\varphi$ is relatively compact in X_0 and $\varphi \notin E_0$. Then the α–limit set $\alpha(\varphi)$ is compact. If $\psi \in \alpha(\varphi)$, there exists a sequence $t_n \to -\infty$ as $n \to +\infty$ such that $T_0(t_n)\varphi \to \psi$ as $n \to +\infty$ and $t_{n-1} - t_n \geq 1$, $\forall n$. Then, for any $t \in (0,1)$, we have:

$$V(T_0(t_{n-1})\varphi) \leq V(T_0(t_n + t)\varphi) \leq V(T_0(t_n)\varphi), \quad \forall n$$

and $V(T_0(t_n + t)\varphi) \to V(\psi)$ as $n \to +\infty$. Since $V(T_0(t_n + t)\varphi)$ also converges to $V(T_0(t)\psi)$ as $n \to +\infty$, it follows that $V(T_0(t)\psi) = V(\psi)$, $\forall t \in (0,1)$, and thus, for all $t \in \mathbb{R}$. Hence, ψ belongs to E_0.

DEFINITION 2.6. — *A C^r–semigroup $T_0(t) : X_0 \to X_0$, $t \geq 0$, $r \geq 1$, is said to be a gradient system if:*

(i) *each bounded positive orbit is precompact,*

(ii) *there exists a Lyapunov function V for $T_0(t)$ such that $V(\varphi) \to +\infty$ as $\|\varphi\|_{X_0} \to +\infty$.*

The gradient systems satisfy the following properties (see Hale (1988) p.51-52 and also Babin and Vishik (1989)).

THEOREM 2.7. — *If $T_0(t)$, $t \geq 0$, is an asymptotically smooth gradient system on X_0 and E_0 is bounded in X_0, then there is a global attractor $\mathcal{A}_0$ for $T_0(t)$ and*

$$\mathcal{A}_0 = W_0^u(E_0) = \{v \in X_0 \; ; \; T(-t)v \to E_0 \text{ as } t \to +\infty\} .$$

And $\mathcal{A}_0$ is connected. If, moreover, E_0 is a discrete set, then E_0 is a finite set and

$$\mathcal{A}_0 = \bigcup_{\varphi_j^0 \in E_0} W_0^u(\varphi_j^0) .$$

In particular, if each element of E_0 is hyperbolic (see the definition below), then E_0 is finite.

The existence of $\mathcal{A}_0$ is a consequence of the facts that $T_0(t)$ is point dissipative (see Remark 2.5), asymptotically smooth and that orbits of bounded sets are bounded. The remark 2.5 implies then that $\mathcal{A}_0 = W_0^u(E_0)$.

Assume that $E_0 = \{\varphi_j^0, \ 1 \leq j \leq N_0\}$ and let $e_1 > e_2 > \cdots > e_{m_0}$ be the distinct points of the set $\{V(\varphi_1^0), \ldots, V(\varphi_{N_0}^0)\}$. We define the set

$$E_0^j = \{\varphi \in E_0 \ ; \ V(\varphi) = e_j\}, \quad \text{for } 1 \leq j \leq m_0 \ .$$

The sets $\{E_0^1, \ldots, E_0^{m_0}\}$ form a Morse decomposition of the attractor $\mathcal{A}_0$, that is, each E_0^j is compact, invariant and, for any $v \in \mathcal{A}_0 \backslash \bigcup_{j=1}^{m_0} E_0^j$, there are integers k and ℓ, $k < \ell$, such that $\alpha(v) \in E_0^k$ and $\omega(v) \in E_0^\ell$. We now introduce, for $1 \leq k \leq m_0$ and for $\epsilon > 0$ small enough, the sets

$$\mathcal{A}_0^k = \bigcup\{W_0^u(\varphi) \ ; \ \varphi \in \bigcup_{j=k}^{m_0} E_0^j\}$$

$$\mathcal{O}_\epsilon^{k-1} = \{\psi \in X_0 \ ; \ V(\psi) \leq e_{k-1} - \epsilon\} \ .$$

COROLLARY 2.8. — Assume that the conditions of the Theorem 2.7 hold and that $E_0 = \{\varphi_j^0, \ 1 \leq j \leq N_0\}$. Then $\mathcal{A}_0^1 = \mathcal{A}_0$, the global attractor for $T_0(t)$, and, for $k > 1$, $\mathcal{A}_0^k$ is the global attractor for $T_0(t)$, restricted to the set $\mathcal{O}_\epsilon^{k-1}$.

In (Babin and Vishik (1983, 1989a)) and (Vishik (1992)), the following definition of a hyperbolic equilibrium point φ of a semigroup $T_0(t)$ is given.

DEFINITION 2.9. — An equilibrium point φ of the semigroup $T_0(t)$, $t \geq 0$, on X_0 is said to be hyperbolic (in the sense of Babin and Vishik) if:

(i) the operators $T_0(t)$, $t \geq 0$, are of class $C^{1+k}(O_\varphi; X_0)$, where O_φ is a neighbourhood of φ in X_0, k is a positive real number. Moreover, for any $\tau > 0$, the Hölder constant of $DT_0(t)$ on O_φ is bounded by a positive constant C_τ, for $0 \leq t \leq \tau$.

(ii) The operators $DT_0(t)$, $t \geq 0$, form a linear semigroup on X_0.

(ii) The spectrum of $DT_0(t)\varphi$, $t \geq 0$, does not intersect the unit circle $\{\xi \ : \ |\xi| = 1\}$ in $\mathbb{C}$. Moreover the invariant subspace X_{0+} of $DT_0(t)\varphi$ corresponding to the part $\sigma_+(\varphi)$ of the spectrum of $DT_0(t)\varphi$ lying outside the unit circle does not depend on t and is finite-dimensional (Its dimension is called the index of φ).

In Chapter 1, we have only considered evolution equations of the form $u_t + Au = f(x, u)$, where A is a differential operator. Under some smoothness conditions on f, one easily shows that the corresponding semigroup $T_0(t)$ satisfies the conditions (i) and (ii) of the definition 2.9. Usually, in these cases, $T_0(t)$ belongs to the space $C^{1,1}(X_0; X_0)$. Thus, in our framework, the hypotheses (i) and (ii) are not restrictive at all. We will let to the reader to check that they are satisfied.

In the proof of the estimate of the distance between $\mathcal{A}_0$ and $\mathcal{A}_\epsilon$, we widely use the following property of hyperbolic equilibrium points (see, for instance. Babin and Vishik (1983, 1989a, b), Vishik (1992), Wells (1976)).

LEMMA 2.10. — *Let $T_0(t)$ be a C^0-semigroup on X_0 and φ be a hyperbolic equilibrium point of $T_0(t)$. Then:*

(i) *there exists a neighbourhood U_φ of φ in X_0 such that the unstable set $W_0^u(\varphi, U_\varphi)$ is a C^1-submanifold of X_0 of dimension* indφ. *More precisely, $W_0^u(\varphi, U_\varphi)$ is the graph of a function g_+ of class C^{1+k} over a neighbourhood of 0 in X_{0+}.*

(ii) *There exist three positive constants $\rho = \rho_\varphi$, $K = K_\varphi$, $\beta = \beta_\varphi$, such that, if $T_0(t)v_0$ is in the ball $B_{X_0}(\varphi, \rho)$ for $[0, \tau]$, there exists $v_1 \in W_0^u(\varphi, U_\varphi)$ such that,*

$$(2.17) \qquad T_0(t)v_1 \in W_0^u(\varphi, U_\varphi), \qquad for\ 0 \le t \le \tau,$$

and

$$(2.18) \qquad \|T_0(t)v_1 - T_0(t)v_0\|_{X_0} \le K \exp(-\beta t), \qquad for\ 0 \le t \le \tau + 1 .$$

REMARK 2.6. — We can also define the local unstable set

$$\widetilde{W}_0^u(\varphi, U_\varphi) \equiv \widetilde{W}_{loc,0}^u(\varphi) = \{v \in U_\varphi\ ;\ T_0(-n)v \in U_\varphi\ ,\ n \in \mathbb{N}\ ;$$
$$T_0(-n)v \to \varphi \text{ in } X_0,\ n \to +\infty\} .$$

The lemma 2.10 still holds with $W_0^u(\varphi, U_\varphi)$ replaced by $\widetilde{W}_0^u(\varphi, U_\varphi)$ (see the references given before Lemma 2.10). Likewise, the theorems 2.11 and 2.12 below still hold if $W_\epsilon^u(\varphi_j^\epsilon, U_j^\epsilon)$ is replaced by $\widetilde{W}_\epsilon^u(\varphi_j^\epsilon, U_j^\epsilon)$, for $0 \le \epsilon \le \epsilon_0$.

Now we are able to state the theorems giving the estimates between $\mathcal{A}_0$ and $\mathcal{A}_\epsilon$. We will state theorems which include the cases of simple and non–simple thin domains. Besides the hypotheses (H1) and (H2) and the conditions on the spaces X_0, X_ϵ, $\mathcal{X}_\epsilon$ given in Section 2.1, we make the following hypotheses

(H7) $T_0(t)$, $t \ge 0$, *is a C^r-gradient system, with $r \ge 1$, which is asymptotically smooth. Moreover, the set E_0 of equilibrium points of $T_0(t)$ is bounded.*

Then, by Theorem 2.7, $T_0(t)$ has a global attractor $\mathcal{A}_0$.

(H8) *There exist two positive constants C_4 and α such that, for any v_1, v_2 in $\mathcal{A}_0$, for $t \ge 0$,*

$$(2.19) \qquad \|T_0(t)v_1 - T_0(t)v_2\|_{X_0} \le C_4 \exp(\alpha t)\|v_1 - v_2\|_{X_0} .$$

(H9) *Each element φ_j^0 of E_0 is hyperbolic (in the above sense).*

This implies that E_0 is a finite set of N_0 elements and $\mathcal{A}_0$ is the union of the unstable sets $W_0^u(\varphi_j^0)$, $1 \le j \le N_0$. Moreover, for $1 \le j \le N_0$, there exist a neighbourhood U_j^0 of φ_j^0 and positive constants ρ_j, K_j, β_j such that the statements (i) and (ii) of Lemma 2.10 hold.

(H10) *For $0 < \epsilon \le \epsilon_0$, $T_\epsilon(t)$ is a C^0-semigroup and has a global attractor $\mathcal{A}_\epsilon$. The set E_ϵ of equilibrium points of $T_\epsilon(t)$ contains, at least, N_0 elements φ_j^ϵ.*

Moreover, for $1 \leq j \leq N_0$, φ_j^ϵ is a hyperbolic equilibrium point and φ_j^ϵ converges to φ_j^0 in $\mathcal{X}_\epsilon$ as ϵ goes to 0.

(H11) *There exist a positive number a_0 and, for $1 \leq j \leq N_0$, a neighbourhood U_j^ϵ of φ_j^ϵ in X_ϵ such that*

$$(2.20) \qquad \delta_{\mathcal{X}_\epsilon}(W_0^u(\varphi_j^0, U_j^0), W_\epsilon^u(\varphi_j^\epsilon, U_j^\epsilon)) \leq C_4 \epsilon^{a_0} .$$

(H12) *For any positive number t_0, there exists a positive constant $C_0^* \equiv C_0^*(t_0)$ such that, for $0 < \epsilon \leq \epsilon_0$, for $\psi_\epsilon \in \mathcal{A}_\epsilon$, $\psi_0 \in \mathcal{A}_0$, for $t \geq t_0$,*

$$(2.21) \qquad \|T_0(t)\psi_0 - T_\epsilon(t)\psi_\epsilon\|_{\mathcal{X}_\epsilon} \leq C_0^* \exp(\alpha t)(\epsilon^{a_0} + \|\psi_0 - \psi_\epsilon\|_{\mathcal{X}_\epsilon}) .$$

Note that, since the hypotheses (H1) and (H2) hold, the attractors $\mathcal{A}_\epsilon$ are u.s.c in $\mathcal{X}_\epsilon$ at $\epsilon = 0$. Let us make some remarks about the hypotheses (H8) and (H12).

REMARK 2.7. — In most of the examples given in Chapter 1, the hypothesis (H8) is replaced by the stronger property:

(H8bis) *There exists a positive non–decreasing function $K(\cdot)$ such that, if v_1, v_2 belong to the ball $B_{X_0}(0, r)$, then, for $t \geq 0$,*

$$(2.19\text{bis}) \qquad \|T_0(t)v_1 - T_0(t)v_2\|_{X_0} \leq K(r)(\exp K(r)t)\|v_1 - v_2\|_{X_0} .$$

In most of these examples, one can also show that the following hypothesis is satisfied:

(H13) *For any positive number t_0, there exists a positive constant $C_1^* \equiv C_1^*(t_0)$ such that, for $0 < \epsilon \leq \epsilon_0$, for $\psi_\epsilon \in \mathcal{A}_\epsilon$, for $t \geq t_0$,*

$$(2.22) \qquad \|T_\epsilon(t)\psi_\epsilon - T_0(t)M_\epsilon\psi_\epsilon\|_{\mathcal{X}_\epsilon} \leq C_1^* \epsilon^{a_0} \exp(\alpha t) .$$

Since the attractors $\mathcal{A}_\epsilon$ are u.s.c. in $\mathcal{X}_\epsilon$ at $\epsilon = 0$, the conditions (2.19bis) and (2.22) imply that, for $\psi_\epsilon \in \mathcal{A}_\epsilon$, $\psi_0 \in \mathcal{A}_0$, for $t \geq t_0$,
(2.23)
$$\|T_0(t)\psi_0 - T_\epsilon(t)\psi_\epsilon\|_{\mathcal{X}_\epsilon} \leq C_1^* \epsilon^{a_0} \exp(\alpha t)$$
$$+ c_0^{-1} K((1 + c_1)c_2)(\exp K((1 + c_1)c_2)t)\|\psi_0 - M_\epsilon\psi_\epsilon\|_{X_0} .$$

Then, in the case of simple thin domains, (2.21) is a direct consequence of (2.23) since $M_\epsilon\psi_0 = \psi_0$. In the case of non–simple thin domains, (2.23) implies (2.21) if one shows in addition that the following condition holds:

(H14) *There exists a positive constant C_5 such that, for $\psi_0 \in \mathcal{A}_0$,*

$$(2.24) \qquad \|\psi_0 - M_\epsilon I_\epsilon \psi_0\|_{\mathcal{X}_\epsilon} + \|\psi_0 - I_\epsilon \psi_0\|_{\mathcal{X}_\epsilon} \leq C_5 \epsilon^{a_0} .$$

Now one can state the following result, whose proof in a slightly different setting has been first given in (Hale and Raugel (1989), Theorem 2.4). The proof of this theorem 2.4 can be easily adapted to our frame and can be even simplified (for details, see Raugel (1993)).

THEOREM 2.11. — *Assume that the hypotheses* (H1), (H2), (H7) *to* (H12) *as well as the conditions on the spaces* X_0, X_ϵ, $\mathcal{X}_\epsilon$ *given in Section 2.1 hold. Then we have the following estimate, for* $0 < \epsilon \leq \epsilon_0$,

$$(2.25) \qquad \delta_{\mathcal{X}_\epsilon}(\mathcal{A}_0, \mathcal{A}_\epsilon) \leq C\epsilon^q,$$

where C *and* q *are positive constants. For instance, we can choose* q *as follows*

$$(2.26) \qquad q = (1 + \frac{\alpha}{\beta})^{1-m_0} a_0,$$

where $\beta = \inf_{1 \leq j \leq N_0} \beta_j$.

The number q appears in the course of the proof. If one looks more closely at the proof, one can even choose larger values of q than the one given in (2.26).

To give an estimate of $\delta_{\mathcal{X}_\epsilon}(\mathcal{A}_\epsilon, \mathcal{A}_0)$, we need the following additional hypotheses, for $0 < \epsilon \leq \epsilon_0$,

(H15) The attractor $\mathcal{A}_\epsilon$ is the union of the unstable sets $W_\epsilon^u(\varphi_j^\epsilon)$, $1 \leq j \leq N_0$.

(H16) For $1 \leq j \leq N_0$, there exists a neighbourhood $\tilde{U}_j^\epsilon$ of φ_j^ϵ in X_ϵ such that

$$(2.27) \qquad \delta_{\mathcal{X}_\epsilon}(W_\epsilon^u(\varphi_j^\epsilon, \tilde{U}_j^\epsilon), W_0^u(\varphi_j^0, U_j^0)) \leq C_4 \epsilon^{a_0}.$$

As for the theorem 2.11, the proof of the following result has been first given in a slightly different setting in (Hale and Raugel (1989), Theorem 2.5). The proof of this Theorem 2.5 can be easily adapted to our frame and can be even simplified (for more details, see Raugel (1993)).

THEOREM 2.12. — *Assume that the hypotheses* (H1), (H2), (H7), (H8bis), (H9) *to* (H11), (H13) *to* (H16) *hold. Then, we have the following estimate, for* $0 < \epsilon \leq \epsilon_0$,

$$(2.28) \qquad \delta_{\mathcal{X}_\epsilon}(\mathcal{A}_\epsilon, \mathcal{A}_0) \leq C\epsilon^q,$$

where q *is given in* (2.26).

Similar and even stronger results have been given by Babin and Vishik (1989a, b) and by Vishik (1992), in the case where the semigroups $T_\epsilon(t)$ are all defined on a same space X and $T_\epsilon(t)$ is continuous in ϵ.

The proofs of the theorems 2.11 and 2.12 consist in a recursion argument, which contains at most m_0 steps. In these proofs, we widely use the Lyapunov function V of $T_0(t)$, the lemma 2.10 and the corollary 2.8. In the proof of Theorem 2.12, we also use the following lemma. Let d be a positive number satisfying

$$d < \inf(\inf_{1 \leq j, \ell \leq m_0} |e_j - e_\ell|, \inf_{1 \leq j \leq m_0} |e_j|)$$

and define the set

$$\mathcal{M}_{\epsilon,d}^k = \{u_\epsilon \in \mathcal{A}_\epsilon ; V(M_\epsilon u_\epsilon) \leq e^k - d\}.$$

LEMMA 2.13. — *Assume that the hypotheses of Theorem 2.12 hold. Then, for any positive number η, there exist two positive numbers $\epsilon_1 = \epsilon_1(\eta, d)$ and $\tau = \tau(\eta, d)$ such that, for $0 \leq \epsilon \leq \epsilon_1$, for $t \geq \tau$,*

$$(2.29) \qquad T_\epsilon(t)\mathcal{M}_{\epsilon,d}^{k-1} \subset \mathcal{N}_{\mathcal{X}_\epsilon}(\mathcal{A}_0^k, \eta), \qquad \text{for } 2 \leq k \leq m_0 .$$

In the next sections, we will show the validity of only a part of the hypotheses of Theorems 2.11 and 2.12. We will mainly compare the equilibrium points of $T_\epsilon(t)$ and $T_0(t)$ and study some spectral properties. The comparison of the local unstable manifolds of the equilibrium points φ_j^0 and φ_j^ϵ has been studied in (Hale and Raugel (1992d)), in a general setting. In Chapter 3, we compare these local unstable manifolds in the case of the reaction diffusion equation on a simple thin domain and give a direct short proof of it.

2.3 Comparison of equilibrium points.

In the previous section, we saw the importance of the study of the equilibrium points. In (Hale and Raugel (1992d)), we studied the general case of approximation of hyperbolic equilibrium points of the semigroup $T_0(t)$ by those of $T_\epsilon(t)$. There, we also compared periodic orbits γ_0 of $T_0(t)$ with those of $T_\epsilon(t)$, if the periodic orbits γ_0 are hyperbolic and $T_0(t)$ and $T_\epsilon(t)$ are defined by parabolic equations.

Here we restrict our study to some "particular case" and refer to (Hale and Raugel (1992d)) for the general case. Often, in the applications, in the study of the equilibria of $T_\epsilon(t)$ and $T_0(t)$, one is led to the study of the solutions in X_0 (resp. X_ϵ) of the equations

$$(2.30)_0 \qquad A_0 v = f_0(v)$$

(resp.

$$(2.30)_\epsilon \qquad A_\epsilon u = f_\epsilon(u)) ,$$

where A_0 (resp. A_ϵ) is a sectorial operator on H_0 (resp. H_ϵ), $X_0 = D(A_0^\beta)$ (resp. $X_\epsilon = D(A_\epsilon^\beta)$), $0 < \beta < 1$, and f_0 (resp. f_ϵ) is a C^1-mapping from X_0 to H_0 (resp. X_ϵ to H_ϵ). To simplify, we assume that $\text{Re}\sigma(A_\epsilon) > 0$, for $0 \leq \epsilon \leq \epsilon_0$. We are thus led to study the solutions v (resp. u) in X_0 (resp. X_ϵ) of the equation

$$(2.31)_0 \qquad F_0(v) \equiv v - A_0^{-1} f_0(v)$$

(resp.

$$(2.31)_\epsilon \qquad F_\epsilon(u) \equiv u - A_\epsilon^{-1} f_\epsilon(u)) .$$

We shall work in the framework of non-simple thin domains. To this end, we need the following additional notation and properties. Besides the spaces X_0, X_ϵ, $\mathcal{X}_\epsilon$, we introduce the spaces Y_0, H_0, H_ϵ, $\mathcal{H}_\epsilon$ such that

$$Y_0 \hookrightarrow X_0 \hookrightarrow H_0 , \quad X_\epsilon \hookrightarrow H_\epsilon , \quad H_0 \hookrightarrow \mathcal{H}_\epsilon , \quad H_\epsilon \hookrightarrow \mathcal{H}_\epsilon .$$

We can choose the constants c_0, C_0, c_3, C_3 in (2.9) and (2.10) so that, for $0 < \epsilon \leq \epsilon_0$,

(2.32)
$$c_0\|\varphi\|_{\mathcal{H}_\epsilon} \leq \|\varphi\|_{H_0} \leq C_0\|\varphi\|_{\mathcal{H}_\epsilon}, \forall \varphi \in H_0,$$

(2.33)
$$c_3\|\varphi_\epsilon\|_{\mathcal{H}_\epsilon} \leq \|\varphi_\epsilon\|_{H_\epsilon} \leq C_3\|\varphi_\epsilon\|_{\mathcal{H}_\epsilon}, \forall \varphi_\epsilon \in H_\epsilon.$$

We also assume that there exists a constant $c_4 > 0$, independent of ϵ, such that, for $0 < \epsilon \leq \epsilon_0$,

(2.34)
$$\|\varphi_\epsilon\|_{H_\epsilon} \leq c_4\|\varphi_\epsilon\|_{X_\epsilon}, \forall \varphi_\epsilon \in X_\epsilon.$$

Before we had already introduced the mappings $M_\epsilon \in \mathcal{L}(X_\epsilon, X_0)$ and $I_\epsilon \in \mathcal{L}(X_0, X_\epsilon)$. We assume that these mappings satisfy the following estimates, for $0 < \epsilon \leq \epsilon_0$,

(2.35)
$$\|I - M_\epsilon\|_{\mathcal{L}(X_\epsilon, \mathcal{H}_\epsilon)} \leq c_5\epsilon^{p_0},$$

(2.36)
$$\begin{cases} \text{(i)} & \|I - I_\epsilon\|_{\mathcal{L}(X_0, \mathcal{H}_\epsilon)} \leq c_5\epsilon^{p_0}, \\ \text{(ii)} & \|I - I_\epsilon\|_{\mathcal{L}(Y_0, X_\epsilon)} \leq c_5\epsilon^{p_1}, \\ \text{(iii)} & \|I - M_\epsilon I_\epsilon\|_{\mathcal{L}(Y_0, X_0)} \leq c_5\epsilon^{p_1}, \\ \text{(iv)} & \|I_\epsilon\|_{\mathcal{L}(X_0, X_\epsilon)} \leq c_5. \end{cases}$$

where c_5, p_0, p_1 are positive real numbers. In our examples of thin domains, the above properties hold.

Concerning our problem $(2.31)_\epsilon$, we make the following assumptions:

(A1) *There exist positive constants a_0, C such that, for $0 < \epsilon \leq \epsilon_0$,*

(2.37)
$$\|A_\epsilon^{-1}\|_{\mathcal{L}(H_\epsilon, X_\epsilon)} \leq C, \quad \|A_0^{-1}\|_{\mathcal{L}(H_0, Y_0)} \leq C,$$

(2.38)
$$\|A_0^{-1}h_0 - A_\epsilon^{-1}h_\epsilon\|_{X_\epsilon} \leq C(\|h_0 - h_\epsilon\|_{\mathcal{H}_\epsilon} + \epsilon^{a_0}(\|h_0\|_{\mathcal{H}_\epsilon} + \|h_\epsilon\|_{\mathcal{H}_\epsilon}),$$
$$\forall h_\epsilon \in H_\epsilon, \ h_0 \in H_0.$$

(A2) *There is a positive increasing function $k_0(\cdot)$ such that, for $0 \leq \epsilon \leq \epsilon_0$, $\varphi_\epsilon \in X_\epsilon$, $\varphi_{i\epsilon} \in X_\epsilon$, $i = 1, 2$, with $\|\varphi_\epsilon\|_{X_\epsilon} \leq r$, $\|\varphi_{i\epsilon}\|_{X_\epsilon} \leq r$,*

(2.39)
$$\|Df_\epsilon(\varphi_\epsilon)\|_{\mathcal{L}(X_\epsilon; H_\epsilon)} + \|f_\epsilon(\varphi_\epsilon)\|_{H_\epsilon} \leq k_0(r),$$

(2.40)
$$\|Df_\epsilon(\varphi_{1\epsilon}) - Df_\epsilon(\varphi_{2\epsilon})\|_{\mathcal{L}(X_\epsilon; H_\epsilon)} \leq k_0(r)\|\varphi_{1\epsilon} - \varphi_{2\epsilon}\|_{X_\epsilon}.$$

(A3) *There are positive constants a_1, a_2 and a positive increasing function $k_1(\cdot)$ such that, for $0 < \epsilon \leq \epsilon_0$, for $\varphi_\epsilon \in X_\epsilon$, $\varphi_0 \in X_0$, with $\|\varphi_\epsilon\|_{X_\epsilon} \leq r$, $\|\varphi_0\|_{X_\epsilon} \leq r$, for $\psi_\epsilon \in X_\epsilon$,*

(2.41)
$$\|f_\epsilon(\varphi_\epsilon) - f_0(\varphi_0)\|_{\mathcal{H}_\epsilon} \leq k_1(r)(\epsilon^{a_1} + \|\varphi_\epsilon - \varphi_0\|_{X_\epsilon}),$$

$$(2.42) \quad \|Df_\epsilon(\varphi_\epsilon)\psi_\epsilon - Df_0(\varphi_0)M_\epsilon\psi_\epsilon\|_{\mathcal{H}_\epsilon} \leq k_1(r)(\epsilon^{a_1} + \epsilon^{a_2} + \|\varphi_\epsilon - \varphi_0\|_{\mathcal{X}_\epsilon})\|\psi_\epsilon\|_{\mathcal{X}_\epsilon} .$$

REMARK 2.8. — In the case of simple thin domains, the hypotheses become much simpler. In this case, $I_\epsilon = I$ and $\mathcal{X}_\epsilon$ (resp. $\mathcal{H}_\epsilon$) are replaced by X_ϵ (resp. H_ϵ).

REMARK 2.9. — In most of our examples, we can show that f_ϵ is actually defined on $\mathcal{X}_\epsilon$ and $f_\epsilon \in C^2(\mathcal{X}_\epsilon, \mathcal{H}_\epsilon)$. The above conditions on f_ϵ and f_0 say that f_ϵ and Df_ϵ are locally Lipschitz continuous functions, uniformly in ϵ, and that f_ϵ is close to f_0 in $C^1(X_0, \mathcal{H}_\epsilon)$. Then the last condition (2.42) reduces to

$$(2.42\text{bis}) \qquad \|Df_\epsilon(\varphi_0)(\psi_\epsilon - M_\epsilon\psi_\epsilon)\|_{\mathcal{H}_\epsilon} \leq k_1(r)\epsilon^{a_2}\|\psi_\epsilon\|_{\mathcal{X}_\epsilon} ,$$

which is satisfied in our examples and comes in part from the property (2.35).

We now can show the following properties.

THEOREM 2.14. — *Assume that $v_0 \in X_0$ is a hyperbolic solution of $(2.31)_0$ and that the conditions* (A1), (A2) *and* (A3) *hold. Then there exist positive constants ϵ_1, r_1, K_1 such that, for $0 < \epsilon \leq \epsilon_1$, the equation $(2.31)_\epsilon$ has a unique solution u_0^ϵ in $B_{\mathcal{X}_\epsilon}(v_0; r_1) \cap X_\epsilon$. Moreover, we have*

$$(2.43) \qquad \|(DF_\epsilon(u_0^\epsilon))^{-1}\|_{\mathcal{L}(X_\epsilon; X_\epsilon)} \leq 2K_1$$

and

$$(2.44) \qquad \|u_0^\epsilon - v_0\|_{\mathcal{X}_\epsilon} \leq C\epsilon^\beta ,$$

where $\beta = \min(p_0, a_0, a_1)$.

Proof: Let us recall that $v_0 \in X_0$ is a hyperbolic solution of $(2.31)_0$ if and only if v_0 is a solution of $(2.31)_0$ and $I - A_0^{-1}Df_0(v_0) \in \mathcal{L}(X_0; X_0)$ is an isomorphism of X_0. Let $r_0 > 0$ be such that $(\|v_0\|_{\mathcal{X}_\epsilon}, \|v_0\|_{X_0}) \leq r_0$, $0 < \epsilon \leq \epsilon_0$.

In order to prove the above result, we shall apply the strict contraction fixed point theorem in the form given by (Crouzeix and Rappaz (1990)). We set, for $0 < \epsilon \leq \epsilon_0$,

$$\delta_\epsilon = \|F_\epsilon(I_\epsilon v_0)\|_{\mathcal{X}_\epsilon} ,$$

and, for $\theta > 0$,

$$\ell_\epsilon(\theta) = \sup_{u \in B_{X_\epsilon}(I_\epsilon v_0; \theta)} \|DF_\epsilon(I_\epsilon v_0) - DF_\epsilon(u)\|_{\mathcal{L}(X_\epsilon; X_\epsilon)} .$$

We assume that there exists a positive constant K_1 such that, for $0 < \epsilon \leq \widetilde{\epsilon}_0$,

$$(2.45) \qquad \|(DF_\epsilon(I_\epsilon v_0))^{-1}\|_{\mathcal{L}(X_\epsilon; X_\epsilon)} \leq K_1 .$$

Then the strict contraction fixed point theorem says: if $2K_1\ell_\epsilon(2K_1\delta_\epsilon) < 1$, then, for any $\theta \geq 2K_1\delta_\epsilon$ such that $K_1\ell_\epsilon(\theta) < 1$, the equation $(2.31)_\epsilon$ has a unique solution u_0^ϵ in $B_{\mathcal{X}_\epsilon}(I_\epsilon v_0; \theta)$. Moreover,

$$(2.46) \qquad \|(DF_\epsilon(u_0^\epsilon))^{-1}\|_{\mathcal{L}(X_\epsilon; X_\epsilon)} \leq 2K_1 .$$

and

(2.47)
$$\|u_0^\epsilon - I_\epsilon v_0\|_{X_\epsilon} \le K_1 \delta_\epsilon .$$

We are going to estimate δ_ϵ, $\ell_\epsilon(\theta)$ and show that (2.45) holds. By $(2.31)_0$ and the hypotheses (A1) and (A3), we can write

$$
\begin{aligned}
\|F_\epsilon(I_\epsilon v_0)\|_{X_\epsilon} &= \|I_\epsilon v_0 - A_\epsilon^{-1} f_\epsilon(I_\epsilon v_0) - v_0 + A_0^{-1} f_0(v_0)\|_{X_\epsilon} \\
&\le \|v_0 - I_\epsilon v_0\|_{X_\epsilon} + C\big[(\|v_0 - I_\epsilon v_0\|_{X_\epsilon} + \epsilon^{a_1}) k_1(r_2) \\
&\quad + \epsilon^{a_0}(\|f_0(v_0)\|_{\mathcal{H}_\epsilon} + \|f_\epsilon(I_\epsilon v_0)\|_{\mathcal{H}_\epsilon})\big] ,
\end{aligned}
$$

where $r_2 = r_0(1 + \frac{c_5}{c_3})$. From these inequalities and from (2.10), we deduce that, for $0 < \epsilon \le \epsilon_0$,

(2.48)
$$\delta_\epsilon \le C(\epsilon^{p_0} + \epsilon^{a_0} + \epsilon^{a_1}) .$$

By the assumptions (A1) and (A2), we obtain,

(2.49)
$$\ell_\epsilon(\theta) \le C\theta(k_0(r_2) + k_0(r_2 + \theta)) .$$

We now show that $DF_\epsilon(I_\epsilon v_0)$ is invertible on X_ϵ and obtain an estimate of the quantity $\|(DF_\epsilon(I_\epsilon v_0))^{-1}\|_{\mathcal{L}(X_\epsilon;X_\epsilon)}$. We can write
(2.50)
$$DF_\epsilon(I_\epsilon v_0)w = w - I_\epsilon A_0^{-1} Df_0(v_0) M_\epsilon w + (I_\epsilon A_0^{-1} Df_0(v_0) M_\epsilon w - A_\epsilon^{-1} Df_\epsilon(I_\epsilon v_0)w) .$$

Using the hypotheses (A1), (A2), (A3) as well as (2.36), (2.8), we have, for $0 < \epsilon \le \epsilon_0$,

$$
\begin{aligned}
\|A_\epsilon^{-1} Df_\epsilon(I_\epsilon v_0)w &- I_\epsilon A_0^{-1} Df_0(v_0) M_\epsilon w\|_{X_\epsilon} \\
&\le \|(I - I_\epsilon) A_0^{-1} Df_0(v_0) M_\epsilon w\|_{X_\epsilon} \\
&\quad + \|A_\epsilon^{-1} Df_\epsilon(I_\epsilon v_0)w - A_0^{-1} Df_0(v_0) M_\epsilon w\|_{X_\epsilon} \\
&\le C\epsilon^{p_1} k_0(r_0)\|w\|_{X_\epsilon} + C\big(\|Df_\epsilon(I_\epsilon v_0)w - Df_0(v_0) M_\epsilon w\|_{\mathcal{H}_\epsilon} \\
&\quad + \epsilon^{a_0}(\|Df_0(v_0) M_\epsilon w\|_{\mathcal{H}_\epsilon} + \|Df_\epsilon(I_\epsilon v_0)w\|_{\mathcal{H}_\epsilon})\big)
\end{aligned}
$$

which implies

(2.51)
$$
\begin{aligned}
\|A_\epsilon^{-1} Df_\epsilon(I_\epsilon v_0)w &- I_\epsilon A_0^{-1} Df_0(v_0) M_\epsilon w\|_{X_\epsilon} \\
&\le C(\epsilon^{p_1} k_0(r_0) + \epsilon^{a_0} k_0(r_1) + k_1(r_1)(\epsilon^{a_1} + \epsilon^{a_2} + \epsilon^{p_1}))\|w\|_{X_\epsilon} .
\end{aligned}
$$

It remains to show that $I - I_\epsilon A_0^{-1} Df_0(v_0) M_\epsilon$ is an isomorphism from X_ϵ onto X_ϵ. At first, we remark that, for any v in X_0,

$$(I - M_\epsilon I_\epsilon A_0^{-1} Df_0(v_0))v = (I - A_0^{-1} Df_0(v_0))v - ((M_\epsilon I_\epsilon - I) A_0^{-1} Df_0(v_0))v .$$

Since, by (2.36) and assumption (A1),

$$\|((M_\epsilon I_\epsilon - I) A_0^{-1} Df_0(v_0))v\|_{X_0} \le Cc_5 \epsilon^{p_1} k_0(r_0),$$

we deduce from the above equality that, for $0 < \epsilon \le \epsilon_2$, with $\epsilon_2 > 0$, $I - M_\epsilon I_\epsilon A_0^{-1} Df_0(v_0)$ is an isomorphism of X_0 and

$$(2.52) \qquad \|(I - M_\epsilon I_\epsilon A_0^{-1} Df_0(v_0))^{-1}\|_{\mathcal{L}(X_0; X_0)} \le K_0,$$

where K_0 is a positive constant.

Let now $U_\epsilon \in X_\epsilon$. We want to show that there exists a unique element w_ϵ in X_ϵ such that

$$(2.53) \qquad U_\epsilon = w_\epsilon - I_\epsilon A_0^{-1} Df_0(v_0) M_\epsilon w_\epsilon .$$

We note that (2.53) is equivalent to the system

$$\begin{cases} (i) & M_\epsilon U_\epsilon = M_\epsilon w_\epsilon - M_\epsilon I_\epsilon A_0^{-1} Df_0(v_0) M_\epsilon w_\epsilon \\ (ii) & (I - M_\epsilon) U_\epsilon = (I - M_\epsilon) w_\epsilon - (I - M_\epsilon) I_\epsilon A_0^{-1} Df_0(v_0) M_\epsilon w_\epsilon \end{cases}$$

Since $I - M_\epsilon I_\epsilon A_0^{-1} Df_0(v_0)$ is an isomorphism of X_0, there exists a unique v_1 in X_0 such that

$$M_\epsilon U_\epsilon = v_1 - M_\epsilon I_\epsilon A_0^{-1} Df_0(v_0) v_1$$

and, by (2.52),

$$(2.54) \qquad \|v_1\|_{X_0} \le K_0 \|M_\epsilon U_\epsilon\|_{X_0} \le C K_0 \|U_\epsilon\|_{X_\epsilon} .$$

We set

$$(2.55) \qquad w_\epsilon = U_\epsilon + I_\epsilon A_0^{-1} Df_0(v_0) v_1 .$$

We have

$$M_\epsilon w_\epsilon = M_\epsilon U_\epsilon + M_\epsilon I_\epsilon A_0^{-1} Df_0(v_0) v_1 = v_1$$

which implies

$$U_\epsilon = w_\epsilon - I_\epsilon A_0^{-1} Df_0(v_0) M_\epsilon w_\epsilon .$$

From (2.54) and (2.55), we at once infer that

$$\|w_\epsilon\|_{X_\epsilon} \le (C K_0 k_0(r_0) + 1) \|U_\epsilon\|_{X_\epsilon} .$$

In a similar way, one shows that, if $U_\epsilon = 0$, then $w_\epsilon = 0$. We thus have shown that $DF_\epsilon(I_\epsilon v_0)$ is an isomorphism of X_ϵ and, for $0 < \epsilon \le \epsilon_3$, with $\epsilon_3 > 0$,

$$(2.56) \qquad \|(DF_\epsilon(I_\epsilon v_0))^{-1}\|_{\mathcal{L}(X_\epsilon; X_\epsilon)} \le K_1,$$

where K_1 is a positive constant, independent of ϵ.

Now the theorem is a direct consequence of (2.46), (2.47), (2.48) and (2.56).

$\square$

REMARK 2.10. — One can state a global version of Theorem 2.14. Assume now that the equations

$$\frac{du}{dt} + A_\epsilon u = f_\epsilon(u)$$

and

$$\frac{dv}{dt} + A_0 v = f_0(v)$$

generate $\mathcal{C}^0$-semigroups $T_\epsilon(t)$ and $T_0(t)$ on X_ϵ and X_0, that $T_\epsilon(t)$ and $T_0(t)$ have global attractors $\mathcal{A}_\epsilon$ and $\mathcal{A}_0$ and that the hypotheses (H1) and (H2) (with X_ϵ replaced by $\mathcal{X}_\epsilon$) hold. Then, due to Proposition 2.2, the attractors $\mathcal{A}_\epsilon$ are upper semicontinuous at $\epsilon = 0$. Assume also that all the equilibrium points of $T_0(t)$ are hyperbolic in the sense of this section. Then the set E_0 of equilibrium points of $T_0(t)$ is a finite set of say N_0 points φ_j^0, $1 \leq j \leq N_0$.

In the neighbourhood of every equilibrium point φ_j^0, we can apply the theorem 2.14, which shows that the set E_ϵ of equilibrium points of $T_\epsilon(t)$ contains at least N_0 hyperbolic points φ_j^ϵ, $1 \leq j \leq N_0$. These points φ_j^ϵ satisfy the estimate (2.44). Using the compactness of $\mathcal{A}_0$ and the uppersemicontinuity property of the sets $\mathcal{A}_\epsilon$, one shows that E_ϵ contains exactly N_0 points φ_j^ϵ, $1 \leq j \leq N_0$.

REMARK 2.11. — Using similar techniques and a Lyapunov–Schmidt method, we can also seek the solutions of

$$F_\epsilon(u, \lambda) \equiv u - A_\epsilon^{-1} f_\epsilon(u, \lambda) = 0$$

in a neighbourhood of a solution (v_0, λ_0) of

$$F_0(v, \lambda) \equiv v - A_0^{-1} f_0(v, \lambda) = 0,$$

even if $I - A_0^{-1} D_v f_0(v, \lambda)$ is no longer an isomorphism of X_0 (see (Hale, Raugel (1992d)).

2.4 Comparison of eigenvalues.

In section 2.2, we saw that, in order to obtain lowersemicontinuity results, we need to compare the local unstable manifolds $W_{loc,0}^u(\varphi_j^0)$ and $W_{loc,\epsilon}^u(\varphi_j^\epsilon)$ of the equilibria φ_j^0 and φ_j^ϵ of $T_0(t)$ and $T_\epsilon(t)$. To achieve that, we need to compare some spectral properties of $DT_0(t)\varphi_j^0$ and $DT_\epsilon(t)\varphi_j^\epsilon$. Such results can also be useful for other purposes (see Chapter 4). If $T_0(t)$ and $T_\epsilon(t)$ are, for instance, generated by the equations

$$\frac{dv}{dt} + A_0 v = f_0(v)$$

and

$$\frac{du}{dt} + A_\epsilon u = f_\epsilon(u),$$

where A_0 and A_ϵ are sectorial operators on H_0 and H_ϵ, we are led to compare some spectral properties of the operators $A_0 - Df_0(\varphi_j^0)$ and $A_\epsilon - Df_\epsilon(\varphi_j^\epsilon)$. Actually, it is more convenient to compare the spectra of the two operators $B_0 = (A_0 - Df_0(\varphi_j^0) + \alpha_0 I)^{-1}$ and $B_\epsilon = (A_\epsilon - Df_\epsilon(\varphi_j^\epsilon) + \alpha_0 I)^{-1}$ where $\alpha_0 > 0$ is large enough.

Here we shall compare the spectra of B_0 and B_ϵ in the framework of simple thin domains. The results below has been first given in (Hale and Raugel (1992b)). In the case of non–simple thin domains, we obtain the same results as below, but the proofs are much longer and more technical. The details in the non–simple thin domains case are given in (Hale, Raugel (1993a)). The problem of eigenvalues of operators on thin domains had been considered before by Ramm (1985), but the proofs do not seem to be complete. Another approach, using Rayleigh quotients, is given in (Iosif'yian, Oleinik and Shamaev (1989)). For eigenvalue problems in more complicated junctions between elastic structure problems, we refer to (Bourquin and Ciarlet (1989)), where Rayleigh quotients are also used.

Like in Section 2.3, we introduce the spaces X_0, X_ϵ, H_0, H_ϵ such that

$$(2.57) \qquad X_0 \hookrightarrow H_0\,, \quad X_\epsilon \hookrightarrow H_\epsilon\,, \quad H_0 \hookrightarrow H_\epsilon\,, \quad X_0 \hookrightarrow X_\epsilon$$

and the properties (2.2) and (2.32) hold. We also introduce a continuous operator M:

$$M : H_\epsilon \to H_0\,, \quad MX_\epsilon = X_0\,, \quad MH_\epsilon = H_0\,,$$

satisfying the properties (2.3) and (2.35) (we recall that here $\mathcal{H}_\epsilon = H_\epsilon$). Without loss of generality, we can suppose that

$$(2.58) \qquad \|M\varphi_\epsilon\|_{H_0} \le c_1 \|\varphi_\epsilon\|_{H_\epsilon}\,, \quad \forall \varphi_\epsilon \in H_\epsilon\,.$$

Finally we introduce the operators B_0 and B_ϵ, $0 < \epsilon \le \epsilon_0$, such that

$$(2.59) \quad \begin{cases} \text{(i)} & B_0 \in \mathcal{L}(X_0; X_0) \cap \mathcal{L}(H_0; X_0)\,, \quad B_\epsilon \in \mathcal{L}(X_\epsilon; X_\epsilon) \cap \mathcal{L}(H_\epsilon; X_\epsilon)\,, \\ \text{(ii)} & \|B_\epsilon\|_{\mathcal{L}(H_\epsilon; X_\epsilon)} \le C\,, \quad \text{for } 0 \le \epsilon \le \epsilon_0\,, \\ \text{(iii)} & \|(B_\epsilon - B_0)h\|_{X_\epsilon} \le C\epsilon^{a_0} \|h\|_{H_0}\,, \quad \forall h \in H_0\,, \ 0 \le \epsilon \le \epsilon_0\,, \end{cases}$$

where a_0, C are positive constants.

Arguing as in Section 2.3, we prove the following result.

LEMMA 2.15. — *There exists a positive constant c and, for any compact set K in the resolvent set of B_0 with $0 \notin K$, there are positive constants c_K, ϵ_K such that, for $0 < \epsilon \le \epsilon_K$, we have*

$$(2.60) \qquad \|B_\epsilon - B_0 M\|_{\mathcal{L}(X_\epsilon; X_\epsilon)} \le C(\epsilon^{a_0} + \epsilon^{p_0})\,,$$

$$(2.61) \qquad \sup_{\lambda \in K} \|(\lambda I - B_0 M)^{-1}\|_{\mathcal{L}(X_\epsilon; X_\epsilon)} \le c_K\,,$$

$$(2.62) \qquad \sup_{\lambda \in K} \|(\lambda I - B_\epsilon)^{-1}\|_{\mathcal{L}(X_\epsilon; X_\epsilon)} \le c_K$$

and

$$(2.63) \qquad \sup_{\lambda \in K} \|(\lambda I - B_\epsilon)^{-1} - (\lambda I - B_0 M)^{-1}\|_{\mathcal{L}(X_\epsilon; X_\epsilon)} \le c_K(\epsilon^{a_0} + \epsilon^{p_0})\,,$$

Proof: Due to the hypotheses (2.35) and (2.59), we can write:

$$\|B_\epsilon - B_0 M\|_{\mathcal{L}(X_\epsilon;X_\epsilon)} \leq \|B_\epsilon(I - M)\|_{\mathcal{L}(X_\epsilon;X_\epsilon)} + \|B_\epsilon M - B_0 M\|_{\mathcal{L}(X_\epsilon;X_\epsilon)}$$
$$\leq \|B_\epsilon\|_{\mathcal{L}(H_\epsilon;X_\epsilon)}\|I - M\|_{\mathcal{L}(X_\epsilon;H_\epsilon)}$$
$$+ \|B_\epsilon M - B_0 M\|_{\mathcal{L}(X_\epsilon;X_\epsilon)}$$
$$\leq c(\epsilon^{a_0} + \epsilon^{p_0}) .$$

which proves (2.60).

To prove (2.61), we solve the equation

$$(\lambda I - B_0 M)\overline{u} = u, \quad \overline{u}, \ u \in X_\epsilon .$$

If we set

$$u = Mu + (I - M)u = v + w ,$$
$$\overline{u} = M\overline{u} + (I - M)\overline{u} = \overline{v} + \overline{w} ,$$

then this equation is equivalent to

$$\begin{cases} \lambda\overline{w} = w \\ (\lambda - B_0 M)\overline{v} = (\lambda - B_0)\overline{v} = v . \end{cases}$$

Since $0 \notin K$ and there is a constant $\overline{c}_K$ such that

$$\sup_{\lambda \in K} \|(\lambda I - B_0)^{-1}\|_{\mathcal{L}(X_0;X_0)} \leq \overline{c}_K ,$$

this proves (2.61).

The estimate (2.62) is a direct consequence of (2.60) and (2.61) and the following equality

$$\lambda I - B_\epsilon = (\lambda I - B_0 M)(I - (\lambda I - B_0 M)^{-1}(B_\epsilon - B_0 M)) .$$

To prove (2.63), we note that

$$(\lambda I - B_\epsilon)^{-1} - (\lambda I - B_0 M)^{-1} = -(\lambda I - B_\epsilon)^{-1}(B_\epsilon - B_0 M)(\lambda I - B_0 M)^{-1}$$

and use (2.60) to (2.62). $\quad\square$

Now we assume that

(2.64) $\qquad B_0 \in \mathcal{L}(X_0 : X_0)$ is a compact operator.

This hypothesis is not necessary, but simplifies the statements below. For any $\eta > 0$, $\lambda \in \mathbb{C}$, let $B(\lambda, \eta)$ be the disk of radius η and center λ. Since B_0 is compact, for any $\eta > 0$, there are an integer $p = p(\eta)$ and complex numbers $\lambda_1, \ldots, \lambda_p$ such that

(2.65) $\qquad \sigma_\eta(B_0) \equiv \sigma(B_0) \cap (\mathbb{C}\backslash B(0,\eta)) = \{\lambda_1, \ldots, \lambda_p\}$

where $\sigma(B_0)$ denotes the spectrum of B_0. Moreover there exist a compact set $\widetilde{K}$ with $0 \in \widetilde{K}$ and a positive constant C such that $\sigma(B_0) \cap (\mathbb{C}\backslash\widetilde{K}) = \emptyset$ and

(2.66) $\qquad \sup_{\lambda \in (\mathbb{C}\backslash\widetilde{K})} \|(\lambda I - B_0)^{-1}\|_{\mathcal{L}(X_0;X_0)} \leq C .$

LEMMA 2.15bis. — *There exist positive constants $\widetilde{C}$ and $\widetilde{\epsilon}$ such that, for $0 < \epsilon \leq \widetilde{\epsilon}$, the estimates (2.61) to (2.63) hold, where $\sup_{\lambda \in K}$ is replaced by $\sup_{\lambda \in \widetilde{U}}$, with $\widetilde{U} = \mathbb{C} \backslash \widetilde{K}$.*

The proof of Lemma 2.15bis is similar to the one of Lemma 2.15 and uses (2.66).

Remark that we have:

$$\sigma(B_0 M) \cap (\mathbb{C} \backslash B(0, \eta)) = \sigma(B_0) \cap (\mathbb{C} \backslash B(0, \eta)), \quad \text{for any } \eta > 0 .$$

From this relation and the lemmas 2.15 and 2.15bis, we deduce that, for any $\eta > 0$, for any $\nu > 0$, there is an $\epsilon_0(\nu, \eta) > 0$ such that, for $0 < \epsilon \leq \epsilon_0(\nu, \eta)$,

$$(2.67) \qquad \sigma(B_\epsilon) \cap (\mathbb{C} \backslash B(0, \eta)) \subset \bigcup_{j=1}^{p} B(\lambda_j, \nu) .$$

If $\sigma_1 \equiv \sigma_1(B_0)$, with $0 \notin \sigma_1$, is any spectral set of B_0, then it induces continuous projections $P_0 \equiv P_0(\sigma_1)$, $Q_0 \equiv Q_0(\sigma_1) = I - P_0$ on X_0 for which the subspaces $P_0 X_0$ and $Q_0 X_0$ are invariant under B_0 and $\sigma(B_0/P_0 X_0) = \sigma_1(B_0)$. If γ is a closed curve in $\mathbb{C}$ which encloses $\sigma_1(B_0)$ and no other points of $\sigma(B_0)$ (in particular, γ does not enclose 0), then, for $v \in X_0$, the projection P_0 is given by

$$(2.68) \qquad P_0 v \equiv P_0(\sigma_1) v = \frac{1}{2\pi i} \int_\gamma (\lambda I - B_0)^{-1} v d\lambda .$$

There exists $\epsilon_\gamma > 0$ such that, for $0 < \epsilon \leq \epsilon_\gamma$, we can define the projection operators $\mathcal{P}_0$ and P_ϵ on X_ϵ by

$$(2.69) \qquad \mathcal{P}_0 u \equiv \mathcal{P}_0(\sigma_1) u = \frac{1}{2\pi i} \int_\gamma (\lambda I - B_0 M)^{-1} u d\lambda$$

and

$$(2.70) \qquad P_\epsilon u \equiv P_\epsilon(\sigma_1) u = \frac{1}{2\pi i} \int_\gamma (\lambda I - B_\epsilon)^{-1} u d\lambda .$$

LEMMA 2.16. — *With the notation above and assuming that $\sigma_1(B_0)$ does not contain 0, we have*

$$\mathcal{P}_0(\sigma_1) = \mathcal{P}_0(\sigma_1) M .$$

Proof: Letting $u = Mu + (I - M)u = v + w$, we can write:

$$\mathcal{P}_0(\sigma_1) u - \mathcal{P}_0(\sigma_1) Mu = \frac{1}{2\pi i} \int_\gamma ((\lambda I - B_0 M)^{-1} u - (\lambda I - B_0)^{-1} Mu) d\lambda$$

$$= \frac{1}{2\pi i} \int_\gamma ((\lambda I - B_0)^{-1} v + \lambda^{-1} w - (\lambda I - B_0)^{-1} v) d\lambda$$

$$= \frac{1}{2\pi i} (\int_\gamma \lambda^{-1} d\lambda) w = 0 .$$

$\square$

Arguing like in (Kato (1966),p.33), we prove the following result.

THEOREM 2.17. — *Fix $\eta > 0$ and suppose that $\lambda_0 \in \sigma(B_0) \cap (\mathbb{C}\backslash B(0,\eta))$ has multiplicity d_0. Choose $\nu > 0$ so that $B(0,\eta) \cap B(\lambda_0, 2\nu) = \emptyset$, $\sigma(B_0) \cap B(\lambda_0, 2\nu) = \{\lambda_0\}$. Then there exist positive constants $\epsilon_1 = \epsilon_1(\nu, \eta)$, $c_1 = c_1(\nu, \eta)$ such that, for $0 < \epsilon \leq \epsilon_1$, the projection P_ϵ given by (2.70) is well defined, where γ is any closed curve in $B(\lambda_0, \nu)\backslash B^\circ(\lambda_0, \frac{3\nu}{4})$ enclosing λ_0. Moreover $\dim P_\epsilon X_\epsilon = \dim P_0 X_\epsilon = d_0$ and*

$$(2.71) \qquad \|P_\epsilon - P_0 M\|_{\mathcal{L}(X_\epsilon; X_\epsilon)} \leq C(\epsilon^{a_0} + \epsilon^{p_0}) .$$

Proof: Let γ be a closed curve in $B(\lambda_0, \nu)\backslash B^\circ(\lambda_0, \frac{3\nu}{4})$ enclosing λ_0. Using Lemma 2.15, we at once show that, for $0 < \epsilon \leq \epsilon_2$, where $\epsilon_2 > 0$, the projections P_0, $\mathcal{P}_0$ and P_ϵ are well defined and (2.71) is a direct consequence of (2.63). It remains to show that $\dim P_\epsilon X_\epsilon = d_0$. We set

$$\mathcal{Q}_0 = I - \mathcal{P}_0 , \quad Q_\epsilon = I - P_\epsilon$$

and we follow (Kato (1966),p.33). We define $R_\epsilon = (P_\epsilon - \mathcal{P}_0)^2$ and observe that there exists $\epsilon_3 > 0$ such that, for $0 < \epsilon \leq \epsilon_3$, the operator $(I - R_\epsilon)^{-1/2}$ is a well defined operator commuting with $\mathcal{P}_0$ and P_ϵ. If we define the operators

$$U_\epsilon^1 = \mathcal{P}_0 P_\epsilon + \mathcal{Q}_0 Q_\epsilon , \quad U_\epsilon = U_\epsilon^1 (I - R_\epsilon)^{-1/2} = (I - R_\epsilon)^{-1/2} U_\epsilon^1 ,$$

$$V_\epsilon^1 = P_\epsilon \mathcal{P}_0 + Q_\epsilon \mathcal{Q}_0 , \quad V_\epsilon = V_\epsilon^1 (I - R_\epsilon)^{-1/2} = (I - R_\epsilon)^{-1/2} V_\epsilon^1 ,$$

then $U_\epsilon V_\epsilon = V_\epsilon U_\epsilon = I$, $U_\epsilon = V_\epsilon^{-1}$, $V_\epsilon = U_\epsilon^{-1}$ and $U_\epsilon P_\epsilon = \mathcal{P}_0 U_\epsilon$ and $P_\epsilon V_\epsilon = V_\epsilon \mathcal{P}_0$. Therefore the operators P_ϵ and $\mathcal{P}_0$ are conjugate: $\mathcal{P}_0 = U_\epsilon P_\epsilon (U_\epsilon)^{-1}$. As a consequence, the subspaces $\mathcal{P}_0 X_\epsilon$ and $P_\epsilon X_\epsilon$ are isomorphic and have the same dimension d_0. □

REMARK 2.12. — Like in (Hale, Raugel (1993a), Lemma 9.3 and Remark 9.1), we can show the following estimates:

$$(2.72) \qquad \begin{cases} \text{(i)} \quad \|U_\epsilon - I\|_{\mathcal{L}(X_\epsilon; X_\epsilon)} \leq C(\epsilon^{a_0} + \epsilon^{p_0}) \\ \text{(ii)} \quad \|V_\epsilon - I\|_{\mathcal{L}(X_\epsilon; X_\epsilon)} \leq C(\epsilon^{a_0} + \epsilon^{p_0}) . \end{cases}$$

COROLLARY 2.18. — *If the hypotheses of Theorem 2.17 are satisfied, then, for $0 < \epsilon \leq \epsilon_1$, $B_\epsilon \in \mathcal{L}(X_\epsilon; X_\epsilon)$ has d_0 eigenvalues λ_j^ϵ, $1 \leq j \leq d_0$, counted with their multiplicities in $B(\lambda_0, \nu)$. If $\mathcal{X}_0 = \mathcal{P}_0 X_0 = P_0 X_0$ and if $\mathcal{B}_\epsilon \in \mathcal{L}(\mathcal{X}_0, \mathcal{X}_0)$ is given by $\mathcal{B}_\epsilon = U_\epsilon B_\epsilon (U_\epsilon)^{-1}$, then the eigenvalues of B_ϵ in $B(\lambda_0, \nu)$ coincide with the eigenvalues of the operator $\mathcal{B}_\epsilon$. Moreover, we have*

$$(2.73) \qquad \|\mathcal{B}_\epsilon - B_0\|_{\mathcal{L}(\mathcal{X}_0; \mathcal{X}_0)} \leq C(\epsilon^{a_0} + \epsilon^{p_0}) .$$

In particular, the eigenvalues λ_j^ϵ converge to λ_0 as $\epsilon \to 0$.

The proof of Corollary 2.18 is a particular case of the proof of (Hale,Raugel (1993a), Corollary 9.7).

REMARK 2.13. — We can also define the space $\mathcal{X}_\epsilon = (U_\epsilon)^{-1} \mathcal{X}_0$ and we check that $\mathcal{X}_\epsilon = P_\epsilon X_\epsilon$. If $\Phi_0 = (\varphi_{01}, \ldots, \varphi_{0d_0})$ is a basis of $\mathcal{X}_0$, then $\Phi_\epsilon = (\varphi_{\epsilon 1}, \ldots, \varphi_{\epsilon d_0})$, where $\varphi_{\epsilon j} = (U_\epsilon)^{-1} \varphi_{0j}$, is a basis of $\mathcal{X}_\epsilon$. Arguing as in the proof of the corollary 2.18 above, we obtain

$$(2.74) \qquad \|\varphi_{\epsilon j} - \varphi_{\epsilon 0}\|_{X_\epsilon} \leq C(\epsilon^{a_0} + \epsilon^{p_0}) .$$

□

Assume now that there is a positive number ℓ_0 such that

$$(2.75) \qquad \left| \|\varphi\|_{X_\epsilon} - \|\varphi\|_{X_0} \right| \leq C \epsilon^{\ell_0} \|\varphi\|_{X_0}, \quad \forall \varphi \in X_0,$$

then we can show the following corollary.

COROLLARY 2.19. — *Assume that the hypotheses of Theorem 2.17 hold, but now that $d_0 = 1$; that is, λ_0 is a simple eigenvalue of B_0. Then, for $0 < \epsilon \leq \epsilon_1$, B_ϵ has one and only one eigenvalue λ_ϵ in $B_0(\lambda_0, \nu)$; this eigenvalue is simple and satisfies*

$$(2.76) \qquad |\lambda_\epsilon - \lambda_0| \leq C(\epsilon^{a_0} + \epsilon^{p_0}) .$$

We can also choose an eigenfunction $\varphi_\epsilon \in X_\epsilon$ (resp. $\varphi_0 \in X_0$) corresponding to λ_ϵ (resp. λ_0) such that $\|\varphi_\epsilon\|_{X_\epsilon} = \|\varphi_0\|_{X_0} = 1$ and

$$(2.77) \qquad \|\varphi_\epsilon - \varphi_0\|_{X_\epsilon} \leq C(\epsilon^{a_0} + \epsilon^{p_0} + \epsilon^{\ell_0}) .$$

COROLLARY 2.20. — *If the operator B_0 has only simple eigenvalues, then, for any integer $N > 0$, there is a positive number $\widetilde{\epsilon}_1 = \widetilde{\epsilon}_1(N)$ such that, for $0 < \epsilon \leq \widetilde{\epsilon}_1$, the operator B_ϵ has the first N eigenvalues simple with the ordering being according to nonincreasing modulus.*

REMARK 2.14. — Let Z be a Banach space and let $\mu_0 \in Z$. We introduce the operator $B_0 = B_0(\mu_0)$. If r_0 is a positive number, we denote the ball in Z of center μ_0 and radius r_0 by $B_Z(\mu_0, r_0)$. For each ϵ, $0 < \epsilon \leq \epsilon_0$, we introduce a subset $\mathcal{M}_\epsilon$ of $B_Z(\mu_0, r_0)$ such that $\mu_0 \in \overline{\mathcal{M}}_\epsilon$ and, for $\mu_\epsilon \in \mathcal{M}_\epsilon$, we define an operator $B_\epsilon(\mu_\epsilon)$. We assume that the operators $B_0(\mu_0)$ and $B_\epsilon(\mu_\epsilon)$ satisfy the hypotheses (2.59)(i), (ii) and

$$(2.59)(iii) \qquad \|(B_\epsilon(\mu_\epsilon) - B_0(\mu_0))h\|_{X_\epsilon} \leq C(\epsilon^{a_0} + \|\mu_\epsilon - \mu_0\|_Z^{a_3})\|h\|_{H_0},$$

where $a_3 > 0$.

Then, obviously, all the estimates and results of Section 2.4 are still valid if we replace the operators B_0, B_ϵ by $B_0(\mu_0)$, $B_\epsilon(\mu_\epsilon)$ and if we replace the term $(\epsilon^{a_0} + \epsilon^{p_0})$ by $(\epsilon^{a_0} + \epsilon^{p_0} + \|\mu_\epsilon - \mu_0\|_Z^{a_3})$ in the estimates.

Chapter 3

Description of the results
in specific examples of P.D.E.

*In this chapter we apply the abstract results of Chapter 2 to some of the examples
described in the first chapter and give some additional properties of them. Sometimes
we also give alternate proofs of certain results without using the results of Chapter
2.*

In all the following, C denotes a generic positive constant independent of ϵ and
$K(\cdot)$ denotes a generic positive non–decreasing function of a real variable.

3.1 Reaction–diffusion equations.

3.1.1 A reaction–diffusion equation on a simple thin domain (our model
simple thin domain). Let $\Omega \subset \mathbb{R}^n$, $n \leq 2$, be a bounded domain, which is regular
or a C^1–polygonal domain in the sense of (Grisvard, (1985)). Let $\epsilon_0 > 0$ and
$g : \overline{\Omega} \times [0, \epsilon_0] \to \mathbb{R}$ be a function of class C^3 satisfying

(3.1)
$$\begin{cases} g(\widetilde{x}, 0) = 0, \quad g_0(\widetilde{x}) = \dfrac{\partial g}{\partial \epsilon}(\widetilde{x}, 0) > 0, \quad \widetilde{x} \in \overline{\Omega}, \\ g(\widetilde{x}, \epsilon) > 0, \quad \widetilde{x} \in \overline{\Omega}, \quad 0 < \epsilon \leq \epsilon_0. \end{cases}$$

For $0 < \epsilon \leq \epsilon_0$, let Q_ϵ be the domain

$$Q_\epsilon = \{(\widetilde{x}, \widetilde{y}) \in \mathbb{R}^{n+1} \ ; \ 0 < \widetilde{y} < g(\widetilde{x}, \epsilon), \widetilde{x} \in \Omega\}$$

and denote by ν_ϵ the outward normal to ∂Q_ϵ. We choose $\widetilde{\delta} > 0$ so that $\widetilde{Q} = \Omega \times (0, \widetilde{\delta})$
contains Q_ϵ for $0 < \epsilon \leq \epsilon_0$.

For α a positive constant and G a function belonging to $C^0(\widetilde{Q})$, we consider the
equation

(3.2)
$$\begin{cases} \text{(i)} \quad u_t - \Delta u + \alpha u = -f(u) - G \quad \text{in } Q_\epsilon \\ \text{(ii)} \quad \dfrac{\partial u}{\partial \nu_\epsilon} = 0 \quad\quad\quad\quad\quad \text{in } \partial Q_\epsilon, \end{cases}$$

where $f : \mathbb{R} \to \mathbb{R}$ is a C^2–function satisfying

(3.3)
$$\limsup_{|s| \to +\infty} -\frac{f(s)}{s} \leq \alpha_0 < \alpha,$$

(3.4)
$$\begin{cases} |f''(s)| \leq c(1 + |s|^\gamma) \text{ for } s \in \mathbb{R}, \\ 0 \leq \gamma < +\infty \text{ if } n = 1, \ 0 \leq \gamma \leq 1 \text{ if } n = 2, \end{cases}$$

and c, α_0 are constants.

REMARKS 3.1. — In the case $n = 1$, we could choose $f \in C^2(\mathbb{R}, \mathbb{R})$ satisfying the condition

(3.5) $$|f''(s)| \leq ce^{|s|^\gamma}, \quad 0 < \gamma < 2,$$

then the proofs are a little more complicated.

• All the results below are still true if $f(u)$ is replaced by $f(x, u)$ which satisfies (3.3), (3.4) in a uniform way.

If we transform coordinates to the canonical domain $Q = \Omega \times (0, 1)$ by letting

$$\widetilde{x} = x, \quad \widetilde{y} = g(x, \epsilon)y,$$

we obtain the system

$(3.6)_\epsilon$
$$\begin{cases} \text{(i)} \quad u_t + L_\epsilon u + \alpha u = -f(u) - G_\epsilon & \text{in } Q \\ \text{(ii)} \quad \dfrac{\partial u}{\partial \nu_{B_\epsilon}} = B_\epsilon u \cdot \nu = 0 & \text{in } \partial Q, \end{cases}$$

where ν is the unit outward normal to ∂Q, L_ϵ is the operator

(3.7)(i) $$L_\epsilon u = -\frac{1}{g}\text{div} B_\epsilon u$$

with

(3.7)(ii) $$B_\epsilon u = \begin{bmatrix} gu_{x_1} - g_{x_1}yu_y \\ gu_{x_2} - g_{x_2}yu_y \\ -g_{x_1}yu_{x_1} - g_{x_2}yu_{x_2} + \frac{1}{g}(1 + (g_{x_1}y)^2 + (g_{x_2}y)^2)u_y \end{bmatrix},$$

and where

(3.8) $$G_\epsilon(x, y) = G(x, g(x, \epsilon)y), \quad G_0(x) = G(x, 0).$$

Hereafter, we assume that

(3.9) $$\|G_\epsilon - G_0\|_{0,Q} \leq C\epsilon.$$

(The property (3.9) holds if, for instance, G is in $W^{1,\infty}(\widetilde{Q})$.)

We want to write the equation $(3.6)_\epsilon$ as an abstract evolutionary equation. Let $\| \cdot \|_{0,Q}$, $\| \cdot \|_{1,Q}$, $\| \cdot \|_{2,Q}$ denote respectively the classical norms in $L^2(Q)$, $H^1(Q)$, $H^2(Q)$. For $0 < \epsilon \leq \epsilon_0$, we let H_ϵ be the space $L^2(Q)$ endowed with the norm $\| \cdot \|_{H_\epsilon}$ induced by the inner product

$$(u_1, u_2)_{H_\epsilon} = \int_Q \frac{g}{\epsilon} u_1 u_2 dx dy.$$

The hypothesis (3.1) implies that there exist positive constants c_1, C_1 such that

$$c_1\|u\|_{0,Q} \le \|u\|_{H_\epsilon} \le C_1\|u\|_{0,Q}, \quad \forall u \in L^2(Q).$$

The bilinear form $(u_1, u_2) \mapsto \int_{Q_\epsilon}(\nabla u_1 \cdot \nabla u_2 + \alpha u_1 u_2)d\tilde{x}d\tilde{y}$ is transformed by the above change of variables into the symmetric bilinear form on $(H^1(Q))^2$

$$a_\epsilon(u_1, u_2) = (\mathcal{L}_\epsilon^{1/2}u_1, \mathcal{L}_\epsilon^{1/2}u_2)_{H_\epsilon} + \alpha(u_1, u_2)_{H_\epsilon}$$

where $\mathcal{L}_\epsilon^{1/2}$ is the "gradient operator" on $H^1(Q)$:

$$\mathcal{L}_\epsilon^{1/2}u = \left(u_{x_1} - \frac{g_{x_1}}{g}yu_y, u_{x_2} - \frac{g_{x_2}}{g}yu_y, \frac{1}{g}u_y\right).$$

It is wellknown that $a_\epsilon(\cdot, \cdot)$ defines an unbounded linear operator A_ϵ on $H^1(Q)$ which is *selfadjoint* on H_ϵ, positive, $A_\epsilon = L_\epsilon + \alpha I$ with homogeneous Neumann boundary conditions and $D(A_\epsilon^{1/2}) \cong H^1(Q)$. We have:

$$a_\epsilon(u, u)^{1/2} = \|A_\epsilon^{1/2}u\|_{H_\epsilon}.$$

The conditions (3.1) imply that there are positive constants $\tilde{c}_1$, $\tilde{C}_1$ such that

$$(3.10) \quad \tilde{c}_1(\|u\|_{1,Q}^2 + \frac{1}{\epsilon^2}\|u_y\|_{0,Q}^2)^{1/2} \le \|A_\epsilon^{1/2}u\|_{H_\epsilon} \le \tilde{C}_1(\|u\|_{1,Q}^2 + \frac{1}{\epsilon^2}\|u_y\|_{0,Q}^2)^{1/2}.$$

For $0 < \epsilon \le \epsilon_0$, we let $X_\epsilon(= X_\epsilon^1)$ be the space $D(A_\epsilon^{1/2})$ endowed with the norm $\|u\|_{X_\epsilon} = \|A_\epsilon^{1/2}u\|_{H_\epsilon}$. We also let X_ϵ^2 be the space $D(A_\epsilon)$ endowed with the norm $\|u\|_{X_\epsilon^2} = \|A_\epsilon u\|_{H_\epsilon}$. If Ω is a regular or a C^1-polygonal domain with *convex angles* in the sense of (Grisvard (1985)), we have:

$$D(A_\epsilon) = \{u \in H^2(Q) ; \frac{\partial u}{\partial \nu_{B_\epsilon}} = 0 \text{ in } \partial Q\}.$$

With this notation, the equation $(3.6)_\epsilon$ is equivalent to the abstract evolutionary equation

$$(3.11)_\epsilon \qquad\qquad u_t + A_\epsilon u = -f(u) - G_\epsilon.$$

Let us remark that $f : u \in H^1(Q) \mapsto f(u) \in L^2(Q)$ is a C^1-mapping.

We want to compare the problem $(3.6)_\epsilon$ or $(3.11)_\epsilon$ with the limit problem

$$(3.6)_0 \qquad \begin{cases} \text{(i)} \quad v_t - \frac{1}{g_0}\sum_{i=1}^{n}(g_0 v_{x_i})_{x_i} + \alpha v = -f(v) - G_0 \quad \text{in } \Omega \\[2ex] \text{(ii)} \quad \dfrac{\partial v}{\partial n} = 0 \qquad\qquad\qquad\qquad\qquad\qquad \text{in } \partial\Omega \end{cases}$$

where n is the unit outward normal to $\partial\Omega$. As above, we let H_0 be the space $L^2(\Omega)$ endowed with the norm $||\cdot||_{H_0}$ induced by the inner product

$$(v_1, v_2)_{H_0} = \int_\Omega g_0 v_1 v_2 dx \ .$$

By (3.1), the norm $||\cdot||_{H_0}$ is equivalent to the classical norm of $L^2(\Omega)$. We introduce the bilinear form $a_0(\cdot, \cdot)$ on $(H^1(\Omega))^2$:

$$a_0(v_1, v_2) = (\nabla_x v_1, \nabla_x v_2)_{H_0} + \alpha(v_1, v_2)_{H_0} \ .$$

The form $a_0(\cdot, \cdot)$ defines an unbounded linear operator A_0 on $H^1(\Omega)$ which is *selfadjoint* on H_0 and positive. We set $X_0 (\equiv X_0^1) = D(A_0^{1/2})$ endowed with the norm $||v||_{X_0} = ||A_0^{1/2} v||_{H_0}$. We also let X_0^2 be the space $D(A_0)$ endowed with the norm $||v||_{X_0^2} = ||A_0 v||_{H_0}$. Clearly $X_0 \cong H^1(\Omega)$. If Ω is a regular or a $\mathcal{C}^1$-polygonal domain with convex angles, then $X_0^2 = D(A_0) = \{v \in H^2(\Omega) \ ; \ \frac{\partial v}{\partial n} = 0 \text{ on } \partial Q\}$. With the above notation, the abstract evolutionary equation corresponding to the problem $(3.6)_0$ is

$$(3.11)_0 \qquad\qquad v_t + A_0 v = -f(v) - G_0 \ .$$

In order to compare the problems $(3.11)_\epsilon$ and $(3.11)_0$, we introduce the projection $M \in \mathcal{L}(L^2(Q); L^2(\Omega))$ given by

$$Mu(x) = \int_0^1 u(x, y) dy \ .$$

LEMMA 3.1. — *If u belongs to $H^j(Q)$, $j \geq 0$, then Mu belongs to $H^j(\Omega)$ and*

$$(3.12) \qquad\qquad ||Mu||_{j,\Omega} \leq ||u||_{j,Q} \ .$$

Moreover, there exists a positive constant C such that, for $0 \leq \epsilon \leq \epsilon_0$,
    (i) *for any $u \in H^1(Q)$,*

$$(3.13) \qquad\qquad ||u - Mu||_{H_\epsilon} \leq C\epsilon||\frac{1}{\epsilon}\frac{\partial u}{\partial y}||_{0,Q} \leq C\epsilon||u||_{X_\epsilon}$$

(Poincaré inequality),
    (ii) *for any $u \in H^2(Q)$ with $\frac{\partial u}{\partial y}(x, 0) = 0$*

$$(3.14) \qquad\qquad ||u - Mu||_{H_\epsilon} + \epsilon||u - Mu||_{X_\epsilon} \leq C\epsilon^2||u||_{X_\epsilon^2} \ .$$

Finally, for any $u \in H^1(Q)$, we have the Sobolev inequalities,

(i) *if* $n = 1$, *for* $2 < p < \infty$,

(3.15)
$$\|u - Mu\|_{L^p(Q)} \le C\left(\frac{1}{2-s}\right)^{\frac{\gamma_s(p-2)}{2p}} \epsilon^{1-\frac{\gamma_s(p-2)}{4p}} \|u\|_{X_\epsilon}$$

where $\gamma_s = \frac{s}{s-1}$, $1 < s < 2$, $s \ge \frac{2p}{2+p}$,

(ii) *if* $n = 2$, *for* $2 \le p \le 6$,

(3.16)
$$\|u - Mu\|_{L^p(Q)} \le C\epsilon^{\frac{2}{p}} \|u\|_{X_\epsilon} .$$

See (Hale and Raugel (1992a, c,d)).

Since $f : X_\epsilon \to H_\epsilon$ is a C^1-mapping, $(3.11)_\epsilon$ defines a (local) semigroup $T_\epsilon(t)$ on X_ϵ (see (Henry (1981))). It is easily seen that $T_\epsilon(t)$ is compact for $t > 0$. We define the energy functional, for $\varphi \in X_\epsilon$, $0 \le \epsilon \le \epsilon_0$,

(3.17)
$$V_\epsilon(\varphi) = \frac{1}{2}\|A_\epsilon^{1/2}\varphi\|_{H_\epsilon}^2 + (F(\varphi), 1)_{H_\epsilon} + (G_\epsilon, \varphi)_{H_\epsilon} ,$$

where $F(s) = \int_0^s f(\sigma)d\sigma$. The hypothesis (3.3) implies that, for any $\eta > 0$, there exists a positive constant c_η such that

(3.18)
$$\begin{cases} \text{(i)} & \forall s \in \mathbb{R}, \quad -f(s)s \le (\alpha_0 + \eta)s^2 + c_\eta , \\ \text{(ii)} & \forall s \in \mathbb{R}, \quad -F(s) \le \frac{1}{2}(\alpha_0 + \eta)s^2 + c_\eta . \end{cases}$$

Using the properties (3.18) as well as the hypotheses (3.4), we show that there are positive constants c_2 to c_6 such that, for $\varphi \in X_\epsilon$,

(3.19)
$$c_2\|A_\epsilon^{1/2}\varphi\|_{H_\epsilon}^2 - c_3 \le V_\epsilon(\varphi) \le c_4\|A_\epsilon^{1/2}\varphi\|_{H_\epsilon}^2 + c_5\|A_\epsilon^{1/2}\varphi\|_{H_\epsilon}^{\gamma+3} + c_6 .$$

Moreover one easily shows that, if $u_\epsilon(t)$ is a solution of $(3.11)_\epsilon$, then, for $t \ge 0$,

(3.20)
$$\frac{d}{dt}V_\epsilon(u_\epsilon(t)) = -\|\frac{du_\epsilon}{dt}(t)\|_{H_\epsilon}^2 .$$

Let $\overline{\varphi}_\epsilon$ be an equilibrium point of $(3.11)_\epsilon$; using the properties (3.1) of $g(x,\epsilon)$ and the property (3.18)(i) with $\eta = \frac{\alpha - \alpha_0}{4}$, as well as the Young inequality, we obtain, for $0 < \epsilon \le \epsilon_0$,

$$c\int_Q \frac{g}{\epsilon}((\overline{\varphi}_{\epsilon x_1})^2 + (\overline{\varphi}_{\epsilon x_2})^2 + \frac{1}{\epsilon^2}(\overline{\varphi}_{\epsilon y})^2)dxdy + \alpha\int_Q \frac{g}{\epsilon}(\overline{\varphi}_\epsilon)^2 dxdy$$

$$\le c_\eta\int_Q \frac{g}{\epsilon}dxdy + \frac{1}{\alpha - \alpha_0}\int_Q \frac{g}{\epsilon}(G_\epsilon)^2 dxdy + \frac{\alpha + \alpha_0}{2}\int_Q \frac{g}{\epsilon}(\overline{\varphi}_\epsilon)^2 dxdy$$

which becomes,

(3.21)
$$\|A_\epsilon^{1/2}\overline{\varphi}_\epsilon\|_{H_\epsilon}^2 \le C\left(c_\eta + \frac{1}{\alpha - \alpha_0}\|G\|_{C^0(\widetilde{Q})}^2\right)\text{mes}(\Omega) \le C .$$

A similar estimate holds for $\epsilon = 0$. Therefore the set E_ϵ of equilibrium points is uniformly bounded in X_ϵ, with respect to ϵ. The above properties imply that $T_\epsilon(t)$ is a gradient system. Since E_ϵ is bounded, the theorem 2.7 implies that $T_\epsilon(t)$, $0 \le \epsilon \le \epsilon_0$, has a global attractor $\mathcal{A}_\epsilon$ and

(3.22)
$$\mathcal{A}_\epsilon = W_\epsilon^u(E_\epsilon) .$$

From the properties (3.19), (3.20) and (3.22), we at once deduce the statement (i) from Theorem 3.2 below.

THEOREM 3.2. — (i) *There is a positive constant C_0 such that, for $0 \leq \epsilon \leq \epsilon_0$,*

$$(3.23) \qquad \|\varphi\|_{X_\epsilon} \leq C_0 , \quad \forall \varphi \in \mathcal{A}_\epsilon .$$

(ii) *There exists a positive, non–decreasing function $K_0(\cdot)$ such that, for any positive number r_0, for any u_0 in $B_{H^1(Q)}(0, r_0)$, we have, for $t \geq 0$,*

$$(3.24) \qquad \|t(T_\epsilon(t)u_0 - T_0(t)Mu_0)\|^2_{X_\epsilon} + \int_0^t \|\frac{d}{ds}(s(T_\epsilon(s)u_0 - T_0(s)Mu_0))\|^2_{H_\epsilon} ds$$

$$\leq \epsilon^2 K_0(r_0)\exp(K_0(r_0)t) .$$

From Theorem 3.2 and Proposition 2.1, we at once deduce the following corollary.

COROLLARY 3.3. — *Under the above hypotheses, the attractors $\mathcal{A}_\epsilon$ are u.s.c. at $\epsilon = 0$, i.e. $\delta_{X_\epsilon}(\mathcal{A}_\epsilon, \mathcal{A}_0) \to 0$ as $\epsilon \to 0$.*

Proof of the statement (ii) *of Theorem 3.2* : In this proof, we use energy type estimates.

1) Let $u_\epsilon(t) = T_\epsilon(t)u_0$, $v(t) = T_0(t)Mu_0$.

Taking the inner product in H_ϵ of $(3.11)_\epsilon$ with u_ϵ and using again (3.18) with $\eta = \frac{\alpha - \alpha_0}{4}$, we obtain, for $t \geq 0$,

$$\frac{d}{dt}\|u_\epsilon(t)\|^2_{H_\epsilon} + \|A_\epsilon^{1/2}u_\epsilon(t)\|^2_{H_\epsilon} \leq C(1 + \|G\|^2_{C^0(\widetilde{Q})})$$

and,

$$(3.25)(i) \qquad \|u_\epsilon(t)\|^2_{H_\epsilon} + \int_0^t \|A_\epsilon^{1/2}u_\epsilon(s)\|^2_{H_\epsilon} ds \leq \|u_0\|^2_{H_\epsilon} + Ct(1 + \|G\|^2_{C^0(\widetilde{Q})}) .$$

Since

$$\alpha\|u\|^2_{H_\epsilon} \leq \|A_\epsilon^{1/2}u\|^2_{H_\epsilon} ,$$

we also deduce from the first inequality, after having applied the Gronwall lemma that,

$$(3.26) \qquad \|u_\epsilon(t)\|^2_{H_\epsilon} \leq e^{-\alpha t}\|u_0\|^2_{H_\epsilon} + C(1 + \|G\|^2_{C^0(\widetilde{Q})}) .$$

Integrating now the first inequality from τ to t and using (3.26), we get, for $0 \leq \tau \leq t$,

$$(3.25)(ii)$$

$$\|u_\epsilon(t)\|^2_{H_\epsilon} + \int_\tau^t \|A_\epsilon^{1/2}u_\epsilon(s)\|^2_{H_\epsilon} ds \leq e^{-\alpha\tau}\|u_0\|^2_{H_\epsilon} + C(t - \tau + 1)(1 + \|G\|^2_{C^0(\widetilde{Q})}) .$$

Taking the inner product in H_ϵ of $(3.11)_\epsilon$ with $u_\epsilon^{2\gamma+3}$ and using (3.18), with $\eta_0 = \frac{\alpha-\alpha_0}{8}$, the Hölder inequality together with the Young inequality, we obtain, for $t \geq 0$,

$$\frac{1}{2(\gamma+2)}\frac{d}{dt}\left(\int_Q \frac{g}{\epsilon}u_\epsilon^{2(\gamma+2)}dxdy\right) + \int_Q (2\gamma+3)\frac{g}{\epsilon}\left[\left(\sum_{i=1}^n u_\epsilon^{2\gamma+2}(u_{\epsilon x_i} - \frac{g_{x_i}}{g}yu_{\epsilon y})^2 + \right.\right.$$

$$\left.\frac{1}{g^2}u_\epsilon^{2\gamma+2}(u_{\epsilon y})^2\right]dxdy + \alpha\int_Q \frac{g}{\epsilon}u_\epsilon^{2(\gamma+2)}dxdy$$

$$\leq (\alpha_0+\eta_0)\int_Q \frac{g}{\epsilon}u_\epsilon^{2(\gamma+2)}dxdy + c_{\eta_0}\int_Q \frac{g}{\epsilon}u_\epsilon^{2\gamma+2}dxdy + \int_Q \frac{g}{\epsilon}G_\epsilon u_\epsilon^{2\gamma+3}dxdy$$

$$\leq (\alpha_0 + \frac{\alpha-\alpha_0}{4})\int_Q \frac{g}{\epsilon}u_\epsilon^{2(\gamma+2)}dxdy + C(1+||G||^{2(\gamma+2)}_{C^0(\widetilde{Q})}) \ .$$

Integrating the above inequality from 0 to t, we deduce, for $t \geq 0$,

$$(3.27) \quad \int_Q \frac{g}{\epsilon}u_\epsilon^{2(\gamma+2)}dxdy \leq C(e^{-\frac{3(\gamma+2)}{2}(\alpha-\alpha_0)t}||u_0||^{2(\gamma+2)}_{L^{2(\gamma+2)}(Q)} + (1+||G||^{2(\gamma+2)}_{C^0(\widetilde{Q})})) \ .$$

Since $\frac{du_\epsilon}{dt}(t)$ belongs to $H^1(Q)$ for $t > 0$, we can take the inner product in H_ϵ of $(3.11)_\epsilon$ with $\frac{du_\epsilon}{dt}$. We thus obtain

$$||\frac{du_\epsilon}{dt}||^2_{H_\epsilon} + \frac{d}{dt}||A_\epsilon^{1/2}u_\epsilon||^2_{H_\epsilon} \leq C\int_Q \frac{g}{\epsilon}u_\epsilon^{2(\gamma+2)}dxdy + C(1+||G||^2_{C^0(\widetilde{Q})}),$$

which becomes, with (3.27),

$$(3.28) \quad ||\frac{du_\epsilon}{dt}||^2_{H_\epsilon} + \frac{d}{dt}||A_\epsilon^{1/2}u_\epsilon||^2_{H_\epsilon}$$

$$\leq C\left[(1+||G||^{2(\gamma+2)}_{C^0(\widetilde{Q})}) + e^{-\frac{3(\gamma+2)}{2}(\alpha-\alpha_0)t}||u_0||^{2(\gamma+2)}_{1,Q}\right] \ .$$

Applying the uniform Gronwall lemma to the estimate (3.28) and using the inequalities (3.25), we get, for $0 \leq t \leq 1$,

$$(3.29)(i) \quad ||A_\epsilon^{1/2}u_\epsilon||^2_{H_\epsilon} \leq C[\frac{1}{t}||u_0||^2_{0,Q} + (1+||G||^{2(\gamma+2)}_{C^0(\widetilde{Q})}) + ||u_0||^{2(\gamma+2)}_{1,Q}] \ ,$$

and, for $t \geq 1$,
$$(3.29)(ii)$$
$$||A_\epsilon^{1/2}u_\epsilon||^2_{H_\epsilon} \leq C[e^{-\frac{3(\gamma+2)}{2}(\alpha-\alpha_0)t}||u_0||^{2(\gamma+2)}_{1,Q} + e^{-\alpha t}||u_0||^2_{0,Q} + (1+||G||^{2(\gamma+2)}_{C^0(\widetilde{Q})})] \ .$$

We do not use the estimates (3.29) here. But we gave them by sake of completeness.

2) *Estimate of* $||T_\epsilon(t)u_0 - T_0(t)Mu_0||_{X_\epsilon}$.

The function $Mu_\epsilon - v$ satisfies the following equation, for any $\tilde{v}$ in $H^1(\Omega)$,

$$(3.30) \quad (\frac{d}{dt}(Mu_\epsilon - v), \tilde{v})_{H_0} + a_0(Mu_\epsilon - v, \tilde{v}) =$$
$$- (Mf(u_\epsilon) - f(v), \tilde{v})_{H_0} - (MG_\epsilon - G_0, \tilde{v})_{H_0} + \sum_{i=1}^{n}(\frac{g_{x_i}}{g}M(yu_{\epsilon y}), \tilde{v}_{x_i})_{H_0}$$
$$+ \sum_{i=1}^{n}((\frac{g_{x_i}}{g} - \frac{g_{0x_i}}{g_0})(Mu_{\epsilon x_i} - \frac{g_{x_i}}{g}M(yu_{\epsilon y})), \tilde{v})_{H_0}$$

Replacing $\tilde{v}$ by $Mu_\epsilon - v$ in (3.30) and using the fact that M is an orthogonal projection, we obtain the following inequality, for $t \geq 0$,

$$(3.31) \quad \frac{d}{dt}\|(Mu_\epsilon - v)(t)\|^2_{H_0} + \|A_0^{1/2}(Mu_\epsilon - v)(t)\|^2_{H_0}$$
$$\leq C[(Mf(u_\epsilon) - f(v), Mu_\epsilon - v)_{H_0} + \|G_\epsilon - G_0\|^2_{H_\epsilon} + \epsilon^2\|A_\epsilon^{1/2}u_\epsilon\|^2_{H_\epsilon}$$
$$+ \sum_{i=1}^{n}(\sup_{\bar{\Omega}}|\frac{g_{x_i}}{g} - \frac{g_{0x_i}}{g_0}|^2)\|u_\epsilon\|^2_{1,Q}] .$$

But, due to the hypothesis (3.4) and to Lemma 3.1, we can write, by using the estimate (3.27) for both functions u_ϵ and v, for $t \geq 0$,

$$(3.32) \quad |(Mf(u_\epsilon) - f(v), Mu_\epsilon - v)_{H_0}|$$
$$\leq C[(1 + \|v\|^{\gamma+1}_{L^{2(\gamma+2)}(\Omega)} + \|u_\epsilon\|^{\gamma+1}_{L^{2(\gamma+2)}(Q)})\|Mu_\epsilon - v\|_{1,Q}$$
$$(\|Mu_\epsilon - v\|_{H_0} + \|u_\epsilon - Mu_\epsilon\|_{H_\epsilon})]$$
$$\leq CK_1(\|u_0\|_{1,Q})(\|Mu_\epsilon - v\|^2_{H_\epsilon} + \epsilon^2\|A_\epsilon^{1/2}u_\epsilon\|^2_{H_\epsilon})$$
$$+ \frac{1}{2}\|A_0^{1/2}(Mu_\epsilon - v)\|^2_{H_0} ,$$

where $K_1(\cdot)$ is a positive non-decreasing function. The hypotheses (3.1) made on g imply that, for $0 \leq \epsilon \leq \epsilon_0$,

$$(3.33) \quad \sum_{i=1}^{n}\sup_{\bar{\Omega}}|\frac{g_{x_i}}{g} - \frac{g_{0x_i}}{g_0}| \leq C\epsilon .$$

From (3.31), (3.32), (3.33) and the hypothesis (3.9), we deduce that, for $t \geq 0$,

$$\frac{d}{dt}\|(Mu_\epsilon - v)(t)\|^2_{H_0} + \|A_0^{1/2}(Mu_\epsilon - v)(t)\|^2_{H_0}$$
$$\leq C[K_1(\|u_0\|_{1,Q})\|Mu_\epsilon - v\|^2_{H_0} + \epsilon^2(1 + \|u_\epsilon\|^2_{1,Q})$$
$$+ \epsilon^2(1 + K_1(\|u_0\|_{1,Q}))\|A_\epsilon^{1/2}u_\epsilon\|^2_{H_\epsilon}] .$$

Integrating this inequality from 0 to t and using the estimate (3.25)(i) for both functions u_ϵ and v, we obtain, after an application of Gronwall lemma,

$$(3.34) \quad \|(Mu_\epsilon - v)(t)\|^2_{H_0} + \int_0^t \|A_0^{1/2}(Mu_\epsilon - v)(s)\|^2_{H_0} ds$$

$$\leq C\epsilon^2 K(\|u_0\|_{1,Q}) \exp(K(\|u_0\|_{1,Q})t) \ .$$

Applying again the lemma 3.1 and the inequality (3.25)(i), we infer from the above estimate that, for $t \geq 0$,

$$(3.35) \quad \int_0^t \|(u_\epsilon - v)(s)\|^2_{H_\epsilon} ds \leq C\epsilon^2 K_2(\|u_0\|_{1,Q}) \exp(K_2(\|u_0\|_{1,Q})t) \ .$$

where $K_2(\cdot)$ is a positive non–decreasing function.

The function $u_\epsilon - v$ satisfies the following equation, for any u in $H^1(Q)$,

$$(3.36) \quad (\frac{d}{dt}(u_\epsilon - v), u)_{H_\epsilon} + a_\epsilon(u_\epsilon - v, u)$$

$$= -(f(u_\epsilon) - f(v), u)_{H_\epsilon} - (G_\epsilon - G_0, u)_{H_\epsilon} + \sum_{i=1}^n ((\frac{g_{x_i}}{g} - \frac{g_{0x_i}}{g_0})v_{x_i}, u)_{H_\epsilon}$$

$$+ \sum_{i=1}^n (\frac{g_{x_i}}{g} v_{x_i}, y u_y)_{H_\epsilon} \ .$$

We set $Z_\epsilon(t) = t(u_\epsilon(t) - v(t))$. Multiplying the equality (3.36) by t and replacing u by $\frac{dZ_\epsilon}{dt}$, we obtain, for $t \geq 0$,

$$(3.37) \quad \|\frac{dZ_\epsilon}{dt}\|^2_{H_\epsilon} + \frac{d}{dt}(a_\epsilon(Z_\epsilon, Z_\epsilon))$$

$$\leq C \Big[t^2 \|f(u_\epsilon) - f(v)\|^2_{H_\epsilon} + t^2 \|G_\epsilon - G_0\|^2_{H_\epsilon} + \|u_\epsilon - v\|^2_{H_\epsilon}$$

$$+ \sum_{i=1}^n \sup_{\overline{\Omega}} |\frac{g_{x_i}}{g} - \frac{g_{0x_i}}{g_0}|^2 \|t\frac{\partial v}{\partial x_i}\|^2_{H_\epsilon} + t \sum_{i=1}^n (\frac{g_{x_i}}{g} v_{x_i}, y\frac{\partial}{\partial y}(\frac{dZ_\epsilon}{dt}))_{H_\epsilon} \Big] \ .$$

But,

$$(3.38) \quad t \sum_{i=1}^n (\frac{g_{x_i}}{g} v_{x_i}, y\frac{\partial}{\partial y}(\frac{dZ_\epsilon}{dt}))_{H_\epsilon}$$

$$\leq \sum_{i=1}^n \frac{d}{dt}(t\frac{g_{x_i}}{g} v_{x_i}, y\frac{\partial}{\partial y} Z_\epsilon)_{H_\epsilon} + 2\|Z_\epsilon\|^2_{X_\epsilon}$$

$$+ 2C\epsilon^2 (\|v\|^2_{1,\Omega} + t^2 \|\frac{dv}{dt}\|^2_{1,\Omega}) \ .$$

Integrating the inequality (3.37) from 0 to t, using the estimates (3.25)(i) and (3.27) for both functions v and u_ϵ, together with the estimates (3.33), (3.35), (3.38) and the condition (3.9), we obtain, for $t \geq 0$,

$$(3.39) \quad \|Z_\epsilon(t)\|_{X_\epsilon}^2 + \int_0^t \|\frac{dZ_\epsilon}{dt}(s)\|_{H_\epsilon}^2 ds$$

$$\leq C\Big[(1 + K_1(\|u_0\|_{1,Q})) \int_0^t \|Z_\epsilon(s)\|_{X_\epsilon}^2 ds$$

$$+ \epsilon^2 (t^2 + K_2(\|u_0\|_{1,Q})) \exp(K_2(\|u_0\|_{1,Q})t)$$

$$+ (t^2 + 1) \int_0^t \|v(s)\|_{1,\Omega}^2 ds$$

$$+ \int_0^t s^2 \|\frac{dv}{dt}\|_{1,\Omega}^2 ds) + \epsilon^2 t^2 \|v(t)\|_{1,\Omega}^2 \Big] .$$

From the properties (3.19) and (3.20) for $\epsilon = 0$ and the fact that $\|v(0)\|_{1,\Omega} \leq \|u_0\|_{1,Q}$, we deduce that, for $t \geq 0$,

$$(3.40) \qquad\qquad \|v(t)\|_{1,\Omega} \leq K(\|u_0\|_{1,Q}) .$$

Moreover, by (Henry (1981), page 71), the function $t\frac{dv}{dt}(t)$ belongs to $L^\infty((0,\infty); X_0)$ and one easily shows that

$$(3.41) \qquad\qquad \int_0^t s^2 \|\frac{dv}{dt}\|_{1,\Omega}^2 ds \leq C(1 + t^2) K(\|u_0\|_{1,Q}) .$$

Finally, taking into account the estimates (3.40) and (3.41) and applying the Gronwall lemma to (3.39), we obtain the estimate (3.24).

$\square$

PROPOSITION 3.4. — *If $Q_\epsilon = \Omega \times (0,\epsilon)$ and $G(\widetilde{x}, \widetilde{y}) = G(\widetilde{x})$, then there exists a positive constant ϵ_1 such that, for $0 \leq \epsilon \leq \epsilon_1$,*

$$\mathcal{A}_\epsilon = \mathcal{A}_0 .$$

In this case, the equations $(3.6)_\epsilon$ or $(3.11)_\epsilon$ are simply written,

$$(3.42)_\epsilon \quad \begin{cases} (i) & \begin{cases} v_t - \Delta_x v + \alpha v = -f(v) - M(f(u) - f(v)) - G_0 \\ \frac{\partial v}{\partial n} = 0 \end{cases} \\ (ii) & \begin{cases} w_t - \Delta_x w - \frac{1}{\epsilon^2} w_{yy} + \alpha w = -(I - M)f(v + w) \\ \frac{\partial w}{\partial \nu} = 0 \end{cases} \end{cases}$$

where $u \equiv v + w \equiv Mu + (I - M)u$. Clearly each solution of $(3.42)_\epsilon$(i) is a solution of the whole system $(3.42)_\epsilon$, which implies that $\mathcal{A}_0 \subset \mathcal{A}_\epsilon$. To show that $\mathcal{A}_\epsilon \subset \mathcal{A}_0$, we show that, for $t \geq 0$,

$$||w(t)||_{X_\epsilon} \leq K(r_0) \exp(-\frac{tK(r_0)}{\epsilon^2})||w(0)||_{X_\epsilon} ,$$

if $||u(0)||_{X_\epsilon} \leq r_0$. For more details, see (Hale, Raugel (1992a)) and also Chapter 5.

If we want further comparison results between $(3.11)_\epsilon$ and $(3.11)_0$, we also need to study the mappings $DT_\epsilon(t)$. Let u_0, $\tilde{u}_0$ be elements of $H^1(Q)$. We set $\tilde{u}_\epsilon(t) = (DT_\epsilon(t)u_0)\tilde{u}_0$ and $\tilde{v}(t) = (DT_0(t)Mu_0)M\tilde{u}_0$. The functions $\tilde{u}_\epsilon(t)$ and $\tilde{v}(t)$ satisfy the non–autonomous equations

$(3.43)_\epsilon$
$$\begin{cases} \tilde{u}_{\epsilon t} + A_\epsilon \tilde{u}_\epsilon = -Df(u_\epsilon)\tilde{u}_\epsilon \\ \tilde{u}_\epsilon(0) = \tilde{u}_0 \end{cases}$$

and

$(3.43)_0$
$$\begin{cases} \tilde{v}_t + A_0 \tilde{v} = -Df(v)\tilde{v} \\ \tilde{v}(0) = M\tilde{u}_0 \end{cases}$$

where $u_\epsilon(t) = T_\epsilon(t)u_0$, $v(t) = T_0(t)Mu_0$. Arguing as in the proof of statement (ii) of Theorem 3.2, we can prove the following results. For more details, see (Raugel (1993)).

LEMMA 3.5. — (i) *There exists a positive, non–decreasing function $K_3(\cdot)$, independent of ϵ, such that, for any positive number r_0, for any u_{01}, u_{02} in $B_{H^1(Q)}(0, r_0)$, we have, for $t \geq 0$, for $0 \leq \epsilon \leq \epsilon_0$,*
$(3.44)_\epsilon$
$$t||DT_\epsilon(t)u_{01} - DT_\epsilon(t)u_{02}||_{\mathcal{L}(L^2(Q);X_\epsilon)} \leq K_3(r_0) \exp(K_3(r_0)t)||u_{01} - u_{02}||_{0,Q} .$$

(ii) *There exists a positive, non–decreasing function $K_4(\cdot)$, independent of ϵ, such that, for any positive number r_0, for any u_0 in $B_{H^1(Q)}(0, r_0)$, we have, for $t \geq 0$, for $0 \leq \epsilon \leq \epsilon_0$,*

(3.45)
$$t||DT_\epsilon(t)u_0 - D(T_0(t)M)u_0||_{\mathcal{L}(H^1(Q);X_\epsilon)} \leq \epsilon K_4(r_0) \exp(K_4(r_0)t) .$$

Assume now that all the equilibrium points of $T_0(t)$ are hyperbolic. Since $T_0(t)$ has a global (compact) attractor, $T_0(t)$ has only a finite number N_0 of equilibrium points φ_j^0, $1 \leq j \leq N_0$. According to the results of Section 2.3, $T_\epsilon(t)$ has exactly N_0 equilibrium points φ_j^ϵ, $1 \leq j \leq N_0$, which are all hyperbolic. Arguing as in Section 2.3, one can show that there exist positive constants ϵ_1, r_1 such that, for $0 < \epsilon \leq \epsilon_1$, $T_\epsilon(t)$ has a unique equilibrium point φ_j^ϵ in $B_{X_\epsilon \cap H^1(Q)}(\varphi_j^0, r_1)$ and

(3.46)
$$\sup(||\varphi_j^\epsilon - \varphi_j^0||_{X_\epsilon}, ||\varphi_j^\epsilon - \varphi_j^0||_{H^1(Q)}) \leq C\epsilon .$$

We want to show that, if all the equilibrium points of $T_0(t)$ are hyperbolic, then the attractors $\mathcal{A}_\epsilon$ are l.s.c. at $\epsilon = 0$. To this end, as we have seen it in Section 2.2, we need to compare the local unstable manifolds of the equilibria φ_j^0 and φ_j^ϵ. We will compare the local unstable manifolds at φ_j^0 and φ_j^ϵ of the mappings S_0 and S_ϵ from $H^1(Q)$ into $H^1(Q)$, introduced in (3.47) below.

Let φ_j^0 be a given equilibrium point of $T_0(t)$ and φ_j^ϵ be the corresponding equilibrium point of $T_\epsilon(t)$. We set $\varphi_0 = \varphi_j^0$, $\varphi_\epsilon = \varphi_j^\epsilon$. We introduce the following mappings from $H^1(Q)$ into $H^1(Q)$:

$$(3.47) \qquad \begin{cases} S_0(u) = T_0(1)M(u + \varphi_0) - \varphi_0 \\ S_\epsilon(u) = T_\epsilon(1)(u + \varphi_\epsilon) - \varphi_\epsilon \ . \end{cases}$$

For $\epsilon \geq 0$, $S_\epsilon(0) = 0$. Moreover, 0 is a hyperbolic equilibrium point of S_0. We set $L = DS_0(0)$. There is a decomposition of $H^1(Q) = X_1 \oplus X_2$ with associated continuous operators P_1, P_2 which commute with L, such that X_1 is finite dimensional and such that the spectrum $\sigma(L)$ can be written as $\sigma(L) = \sigma(L_1) \cup \sigma(L_2)$ with $L_1 = LP_1$, $L_2 = LP_2$, where $\sigma(L_1)$ and $\sigma(L_2)$ lie respectively outside and inside the unit circle with centre 0 in $\mathbb{C}$. Since 0 is a hyperbolic fixed point, there exist two small positive numbers δ_1, δ_2 such that the spectral radii $r(L_2)$ and $r(L_1^{-1})$ satisfy

$$r(L_2) = 1 - 2\delta_2, \quad r(L_1^{-1}) = 1 - 2\delta_1 \ .$$

For any $u_1 \in X_1$, $u_2 \in X_2$, we set:

$$\|u_1\|_0 = \sup_{n \geq 0} \frac{\|L_1^{-n} u_1\|_{1,Q}}{(1 - \delta_1)^n} \ ,$$

$$\|u_2\|_0 = \sup_{n \geq 0} \frac{\|L_2^n u_2\|_{1,Q}}{(1 - \delta_2)^n}$$

and, for $u \in H^1(Q)$, we set:

$$\|u\|_0 = \sup(\|P_1 u\|_0, \|P_2 u\|_0) \ .$$

In (Babin and Vishik (1983), (1989a)), it is proved that

(i) $\quad \|L_1^{-1} u_1\|_0 \leq (1 - \delta_1) \|u_1\|_0, \quad \forall u_1 \in X_1$

(ii) $\quad \|L_2 u_2\|_0 \leq (1 - \delta_2) \|u_2\|_0, \quad \forall u_2 \in X_2$

and, for any $u \in H^1(Q)$,

$$(3.48) \qquad C_0^{-1} \|u\|_{1,Q} \leq \|u\|_0 \leq C_0 \|u\|_{1,Q}$$

where $C_0 > 1$ is a constant.

We denote by $\widetilde{X}$ (resp. $\widetilde{X}_1$, $\widetilde{X}_2$) the space $H^1(Q)$ (resp. X_1, X_2) equipped with the norm $\| \cdot \|_0$. We can show that, for $R > 0$ small enough and for $\epsilon > 0$ small enough, the mappings S_0 and S_ϵ belong to the space $\mathcal{L}_R(\Lambda_R; \widetilde{X})$, that is, to the set of Lipschitz-continuous mappings S from Λ_R into X satisfying

$$S(0) = 0$$

and

$$\|(S(u) - Lu) - (S(\widetilde{u}) - L\widetilde{u})\|_0 < \sigma\|u - \widetilde{u}\|_0 , \quad \forall u, \widetilde{u} \in \Lambda_R ,$$

where

$$\sigma < \min(\delta_1, \delta_2) ,$$

and

$$\Lambda_R = \{u \in \widetilde{X} \; ; \; \|u\|_0 \le R\} .$$

(Here, $\mathcal{L}_R(\Lambda_R; \widetilde{X})$ is endowed with the classical topology of $\mathcal{C}^0(\Lambda_R; \widetilde{X})$.)

Indeed, by Lemma 3.5, we can write, for $0 < \epsilon \le \epsilon_0$,

$$(3.49) \quad \|(S_\epsilon(u) - Lu) - (S_\epsilon(\widetilde{u}) - L\widetilde{u})\|_0$$

$$\le C_0 \| \int_0^1 [DS_\epsilon(\widetilde{u} + s(u - \widetilde{u})) - DS_0(\widetilde{u} + s(u - \widetilde{u}))](u - \widetilde{u})ds\|_{1,Q}$$

$$+ C_0\|(S_0(u) - Lu) - (S_0(\widetilde{u}) - L\widetilde{u})\|_{1,Q}$$

$$\le C\epsilon\|u - \widetilde{u}\|_{1,Q} + C(\|\widetilde{u}\|_{1,Q} + \|u\|_{1,Q})\|u - \widetilde{u}\|_{1,Q}$$

$$\le (C\epsilon + CR)\|u - \widetilde{u}\|_0 .$$

Thus there exists $\widetilde{R}$ and $\widetilde{\epsilon}_1$ such that S_0 and S_ϵ, $0 < \epsilon \le \widetilde{\epsilon}_1$, belong to $\mathcal{L}_{\widetilde{R}}(\Lambda_{\widetilde{R}}; \widetilde{X})$. We now introduce the following subspace of $\mathcal{C}^1(\Lambda_{\widetilde{R}}; \widetilde{X})$

$$\mathcal{F}_{\widetilde{R}}(\Lambda_{\widetilde{R}}; \widetilde{X}) = \{S \in \mathcal{L}_{\widetilde{R}}(\Lambda_{\widetilde{R}}; \widetilde{X}) \cap \mathcal{C}^1(\Lambda_{\widetilde{R}}; \widetilde{X}) \; ;$$

$$\|DS(u) - DS(\widetilde{u})\|_{\mathcal{L}(\widetilde{X}; \widetilde{X})} \le L\|u - \widetilde{u}\|_0 , \; \forall u, \widetilde{u} \in \Lambda_{\widetilde{R}}\}$$

endowed with the classical topology of $\mathcal{C}^1(\Lambda_{\widetilde{R}}; \widetilde{X})$, where L is a fixed adequate positive constant. Applying one more time the lemma 3.5(i), we see that the mappings S_ϵ, $0 \le \epsilon \le \widetilde{\epsilon}_1$, belong also to $\mathcal{F}_{\widetilde{R}}(\Lambda_{\widetilde{R}}; \widetilde{X})$. Hence we can apply the result of (Wells (1976)).

Let $\widetilde{W}_\epsilon^u(S_\epsilon, 0, \Lambda_{\widetilde{R}})$ be the set of all $u \in \Lambda_{\widetilde{R}}$ such that there exist $u_1, u_2, \ldots \in \Lambda_{\widetilde{R}}$ with $S_\epsilon(u_{n+1}) = u_n$, for $n \ge 1$, and $\|u_n\|_0 \to 0$ as $n \to +\infty$. For $0 \le \epsilon \le \widetilde{\epsilon}_1$, there exists a $\mathcal{C}^1$–mapping

$$g_1^{S_\epsilon} : B_{\widetilde{X}_1}(0, \widetilde{R}) \to \widetilde{X}_2$$

such that

$$(3.50) \quad \widetilde{W}_\epsilon^u(S_\epsilon, 0, \Lambda_{\widetilde{R}}) = \{u \in \Lambda_{\widetilde{R}} \; ; \; u = u_1 + g_1^{S_\epsilon}(u_1), \; u_1 \in B_{\widetilde{X}_1}(0, \widetilde{R})\}$$

and

$$(3.51) \quad \|g_1^{S_\epsilon} - g_1^{S_0}\|_{\mathcal{C}^1(B_{\widetilde{X}_1}(0, \widetilde{R}); \widetilde{X}_2)} \le K_1\|S_\epsilon - S_0\|_{\mathcal{C}^1(\Lambda_{\widetilde{R}}; \widetilde{X})} .$$

But, by Theorem 3.2, by Lemma 3.5 and by (3.46), we have, for $u \in \Lambda_{\widetilde{R}}$,

$$\|S_\epsilon(u) - S_0(u)\|_0 \le C_0(\|T_0(1)M(u + \varphi_0) - T_\epsilon(1)(u + \varphi_0)\|_{1,Q}$$

$$(3.52) \qquad\qquad + \|T_\epsilon(1)\varphi_\epsilon - T_\epsilon(1)\varphi_0\|_{1,Q} + \|\varphi_\epsilon - \varphi_0\|_{1,Q})$$

$$\le C\epsilon ,$$

and

(3.53) $\|DS_\epsilon(u) - DS_0(u)\|_{\mathcal{L}(\widetilde{X};\widetilde{X})}$

$$\leq C_0 \|D(T_0(1)M)u - DT_\epsilon(1)u\|_{\mathcal{L}(H^1(Q);H^1(Q))} \leq C\epsilon .$$

Finally, from the estimates (3.51), (3.52) and (3.53), we deduce that, for $0 \leq \epsilon \leq \widetilde{\epsilon}_1$,

(3.54) $$\|g_1^{S_\epsilon} - g_1^{S_0}\|_{C^1(B_{\widetilde{X}_1}(0,\widetilde{R});\widetilde{X}_2)} \leq C\epsilon .$$

Arguing as above, for any equilibrium point φ_j^0 of $T_0(t)$, we obtain the following lemma.

LEMMA 3.6. — *Assume that all the equilibrium points of $T_0(t)$ are hyperbolic, then $T_0(t)$ has N_0 equilibrium points φ_j^0, $1 \leq j \leq N_0$. There exists a positive constant $\widetilde{\epsilon}_1$ such that, for $0 < \epsilon \leq \widetilde{\epsilon}_1$, $T_\epsilon(t)$ has exactly N_0 equilibrium points, which are all hyperbolic. Moreover, for $1 \leq j \leq N_0$, there exists a neighbourhood Λ_j of φ_j^0 in $H^1(Q)$ such that, for $0 \leq \epsilon \leq \widetilde{\epsilon}_1$,*

(3.55) $\sup(\delta_{H^1(Q)}(\widetilde{W}_\epsilon^u(T_\epsilon(1), \varphi_j^\epsilon, \Lambda_j), \widetilde{W}_0^u(T_0(1)M, \varphi_j^0, \Lambda_j)),$

$$\delta_{H^1(Q)}(\widetilde{W}_0^u(T_0(1)M, \varphi_j^0, \Lambda_j), \widetilde{W}_\epsilon^u(T_\epsilon(1), \varphi_j^\epsilon, \Lambda_j))) \leq C\epsilon .$$

Since the local unstable manifolds are negatively invariant, one at once deduces from (3.55) and Theorem 3.2 that the distance in X_ϵ between the local unstable manifolds is of order ϵ (that is, (3.55) holds with $H^1(Q)$ replaced by X_ϵ).

Now we can apply the theorems 2.11 and 2.12 (see also Remark 2.6).

THEOREM 3.7. — *Assume that all the equilibrium points of $T_0(t)$ are hyperbolic, then the attractors $\mathcal{A}_\epsilon$ are l.s.c. at $\epsilon = 0$ and there exist positive numbers ϵ_1, C and p, $0 < p \leq 1$, such that, for $0 < \epsilon \leq \epsilon_1$,*

(3.56) $$\delta_{X_\epsilon}(\mathcal{A}_\epsilon, \mathcal{A}_0) + \delta_{X_\epsilon}(\mathcal{A}_0, \mathcal{A}_\epsilon) \leq C\epsilon^p .$$

REMARK 3.2. — The hyperbolicity of all equilibrium points of $(3.11)_0$ is generic in G_0 (see (Babin and Vishik (1983), (1989a)). If $n = 1$, this property is also generic in $f \in C^1(\mathbb{R}; \mathbb{R})$ (see (Brunovsky and Chow (1984)), (Smoller and Wasserman (1984)).

We may also obtain more information by imposing some further conditions on the flow defined by $T_0(t)$. We say that the gradient system $T_0(t)$ is a *Morse-Smale system* if all the equilibrium points are hyperbolic and the stable and unstable manifolds of the equilibrium points are transversal. In the case $n = 1$, a result of Henry (1985) or Angenent (1986) implies that the stable and unstable manifolds of the equilibrium points of $T_0(t)$ are transversal as soon as the equilibrium points are hyperbolic.

THEOREM 3.8. — *Suppose that the semigroup $T_0(t) : H^1(\Omega) \to H^1(\Omega)$ is a Morse–Smale system. Then there exist positive constants ϵ_1 and t_0 such that, for $0 \leq \epsilon \leq \epsilon_1$, the semigroups $T_\epsilon(t)$ are Morse–Smale systems and there is a homeomorphism $h_\epsilon : \mathcal{A}_0 \to \mathcal{A}_\epsilon$ satisfying the conjugacy condition: $h_\epsilon \circ T_0(t_0) = T_\epsilon(t_0) \circ h_\epsilon$.*

Proof : Since $T_0(t)$ is bounded dissipative and the convergence result (3.24) holds, we can choose a bounded set B_0 in $H^1(Q)$ and a constant $t_0 > 0$ such that $MB_0 \subset B_0$, $T_0(t)MB_0 \subset B_0$, for $t \geq t_0$, $T_\epsilon(t_0)B_0 \subset B_0$ and $\mathcal{A}_\epsilon \subset B_0$, for $0 \leq \epsilon \leq \epsilon_0$. We consider the mappings $S_0 = T_0(t_0)M$ and $S_\epsilon = T_\epsilon(t_0)$ in $C_b^{1,1}(B_0, B_0)$. By Theorem 3.2 and Lemma 3.5, S_ϵ converges to S_0 in $C_b^{1,1}(B_0, B_0)$. By the backward uniqueness results of (Bardos and Tartar (1973)) or of (Ghidaglia (1986)), the mappings $S_\epsilon \in C_b^1(B_0, B_0)$, for $0 \leq \epsilon \leq \epsilon_0$, and the mapping $T_0(t_0) \in C_b^1(MB_0, MB_0)$ are reversible in the sense of (Hale, Margalhães and Oliva (1984)). It is also easy to see that S_0, considered as a mapping from B_0 into B_0, is reversible. By Corollary 3.3, the attractors $\mathcal{A}_\epsilon$ are u.s.c. at $\epsilon = 0$. It is a classical matter to show that the attractors $\mathcal{A}_\epsilon$ are u.s.c. at $\epsilon = \tilde{\epsilon}_0$, for $0 < \tilde{\epsilon}_0 \leq \epsilon_0$. Hence the set $\{S_\epsilon \; ; \; 0 \leq \epsilon \leq \epsilon_0\}$ can be chosen as a $KC_b^1(B_0, B_0)$ space (for the definition of a $KC_b^1(B_0, B_0)$ space, see (Hale, Margalhães and Oliva (1984), Chapter 10)). Moreover, if $T_0(t) : X_0 \to X_0$ is a Morse–Smale system, it is straightforward to show that $S_0 \equiv T_0(t_0)M$ is a Morse–Smale map. Thus we can apply Theorem 10.27 of (Hale, Magalhães and Oliva (1884)) which gives us the theorem 3.8. (For more details, see (Raugel (1993))).

$\square$

If $T_0(t)$ is no longer a Morse–Smale system, we can have some information about the different connecting orbits of $T_\epsilon(t)$ by using the Conley index theory (see (Mischaikow and Raugel (1994))).

REMARK 3.3. — We can also study mixed boundary conditions or Dirichlet ones. Let us denote by $\Gamma_{j,\epsilon}$ (resp. Γ_j), $j = 0, 1, 2$, the portions of the boundary of Q_ϵ (resp. Q) given by

$$\Gamma_{0,\epsilon} = \Omega \times \{0\} \quad (\text{resp. } \Gamma_0 = \Omega \times \{0\}),$$

$$\Gamma_{1,\epsilon} = \{(\tilde{x}, \tilde{y}) \in \mathbb{R}^{n+1} \; ; \; \tilde{x} \in \Omega, \; \tilde{y} = g(\tilde{x}, \epsilon)\} \quad (\text{resp. } \Gamma_1 = \Omega \times \{1\}),$$

$$\Gamma_{2,\epsilon} = \{(\tilde{x}, \tilde{y}) \in \mathbb{R}^{n+1} \; ; \; \tilde{x} \in \partial\Omega, \; 0 < \tilde{y} < g(\tilde{x}, \epsilon)\} \quad (\text{resp. } \Gamma_2 = \partial\Omega \times (0, 1)).$$

We may define the corresponding unit outward normals $\nu_{j,\epsilon}$ on $\Gamma_{j,\epsilon}$ (resp. ν_j on Γ_j). We consider the equation (3.2)(i) with the boundary conditions

(3.2)(iibis)
$$\begin{cases} u = 0 \text{ in } \Gamma_{2,\epsilon} \\ \dfrac{\partial u}{\partial \nu_{j,\epsilon}} = 0 \text{ in } \Gamma_{j,\epsilon}, \; j = 0, 1 . \end{cases}$$

In the new variables (x, y) of the fixed domain Q, this boundary value problem becomes the equation $(3.6)_\epsilon(i)$ with the boundary conditions

$(3.6)_\epsilon$(iibis)
$$\begin{cases} u = 0 \text{ in } \Gamma_2 \\ \dfrac{\partial u}{\partial \nu_{B_\epsilon, j}} \equiv B_\epsilon u \cdot \nu_j = 0 \text{ in } \Gamma_{j,\epsilon}, \ j = 0, 1 . \end{cases}$$

The limit equation is $(3.6)_0(i)$ with the boundary conditions

$(3.6)_0$(iibis)
$$v = 0 \text{ in } \partial\Omega .$$

The bilinear form $a_\epsilon(\cdot, \cdot)$ defines an unbounded linear operator A_ϵ on $V_0 \equiv \{u \in H^1(Q) \ ; \ v = 0 \text{ in } \Gamma_2\}$ and we set $X_\epsilon = D(A_\epsilon^{1/2})$. Likewise we define the space X_0. Since we did not use any $H^2(Q)$–regularity results in the proof of the above results, the theorem 3.2, the corollary 3.3, the proposition 3.4, the lemma 3.5, the lemma 3.6 and the theorems 3.7 and 3.8 still hold. Finally, we can also consider the equation $(3.2)(i)$ or $(3.6)_\epsilon(i)$ with homogeneous Dirichlet boundary conditions. The semigroup $T_\epsilon(t)$ defined by $(3.6)_\epsilon(i)$ on $X_\epsilon \approx H_0^1(Q)$ has a global attractor $\mathcal{A}_\epsilon$. One shows that there are positive constants ϵ_1, C, such that, for $0 < \epsilon \le \epsilon_1$,

$$(3.57) \qquad \|\varphi_\epsilon\|_{X_\epsilon} \le C\epsilon, \quad \forall \varphi_\epsilon \in \mathcal{A}_\epsilon .$$

We now consider the particular case $n = 1$.

If $n = 1$, that is, if Q_ϵ is a thin domain "around" a line segment, we can show that the equation (3.2) or $(3.6)_\epsilon$ has an inertial manifold. To state a precise result, it is convenient to modify the equation $(3.11)_\epsilon$ (or $(3.6)_\epsilon$) outside a bounded set containing the attractor $\mathcal{A}_\epsilon$ in such a way that the flow of the new equation is linear outside this region and thus is contracting. Since, in this particular case, the sets $\mathcal{A}_\epsilon$ are contained in a ball $B_{H^2(Q)}(0, R)$ we can do that in a uniform way with respect to ϵ. (For the details, see (Hale,Raugel (1992a)). We denote by $\widetilde{T}_\epsilon(t)$ the semigroup corresponding to the modified equation.

We say that $\mathcal{M}^\epsilon$ is an inertial manifold for the equation $(3.6)_\epsilon$ or $(3.11)_\epsilon$ if $\mathcal{M}^\epsilon$ is a *finite dimensional submanifold* of $H^1(Q)$ of class $\mathcal{C}^1$ which contains the attractor $\mathcal{A}_\epsilon$ and is positively invariant under the flow $\widetilde{T}_\epsilon(t)$, that is, $\widetilde{T}_\epsilon(t)\mathcal{M}^\epsilon \subset \mathcal{M}^\epsilon$ for $t \ge 0$. In a similar manner, we define a modified equation of the equation $(3.6)_0$ or $(3.11)_0$, a corresponding semigroup $\widetilde{T}_0(t)$ and an inertial manifold $\mathcal{M}^0$ for this flow $\widetilde{T}_0(t)$. In (Hale, Raugel (1992a)), we have proved the following result.

THEOREM 3.9. — (i) *If $n = 1$, there exist an integer m and a positive number ϵ_1 such that there are an inertial manifold $\mathcal{M}^0 \subset H^1(Q)$ of $(3.11)_0$ of dimension m and, for each $0 < \epsilon \le \epsilon_1$, an inertial manifold $\mathcal{M}^\epsilon \subset H^1(Q)$ of $(3.11)_\epsilon$ of dimension m. Furthermore, for $0 \le \epsilon \le \epsilon_1$, the flow $\widetilde{T}_\epsilon(t)$ on the inertial manifold $\mathcal{M}^\epsilon$ is given by the ordinary differential equation*

$$(3.58)_\epsilon \qquad \frac{d\rho}{dt} = q_\epsilon(\rho), \quad \rho \in \mathbb{R}^m$$

and the functions q_ϵ converge to q_0 in $C^1(U; \mathbb{R}^m)$ as $\epsilon \to 0$, for every bounded open set U in $\mathbb{R}^m$. Moreover there exists a positive number p_0, $0 < p_0 \leq 1$, such that

$$\|q_\epsilon - q_0\|_{C^1(U;\mathbb{R}^m)} \leq C\epsilon^{p_0} .$$

(ii) If $Q_\epsilon = \Omega \times (0, \epsilon)$, $G(\tilde{x}, \tilde{y}) = G(\tilde{x})$, then, for $0 \leq \epsilon \leq \epsilon_1$, $\mathcal{M}^0$ is also an inertial manifold for $(3.11)_\epsilon$.

The proof of Theorem 3.9 uses the gap property of the eigenvalues of A_0 and the comparison results of eigenvalues of Section 2.4. In (Hale, Raugel (1992a)), we used the method of graphs to construct the inertial manifolds, as described in (Mallet–Paret, Sell (1988)). It would also have been possible to construct them by using the method of integrals.

REMARK 3.4. — Since, by Theorem 3.9, we are reduced to systems of ordinary differential equations, we can use the classical Morse–Smale results to show that, if $n = 1$ and the equilibrium points of $(3.6)_0$ are hyperbolic, then, for $0 \leq \epsilon \leq \epsilon_1$, the flows defined by $T_\epsilon(t)$ and $T_0(t)$ are *topologically equivalent* on the attractors, that is, there exist homeomorphisms $h_\epsilon : \mathcal{A}_0 \to \mathcal{A}_\epsilon$ which preserve orbits and the sense of direction time. (See (Hale, Raugel (1992a)).

Let us now give an example which shows the effect of the domain shape on the dynamics of $T_\epsilon(t)$ and $T_0(t)$ and uses Theorem 3.9. This example is described with all the details in (Hale, Raugel (1992d)).

Let $g_0(x, \mu)$ be a smooth positive function of (x, μ) in $[0, 1] \times K$ where K is a compact interval in $\mathbb{R}$. We also write: $g_{0\mu}(x) = g_0(x, \mu)$. We now consider the equation (3.2) with $G = 0$ on the thin domain

$$Q_\epsilon = Q_{\epsilon,\mu} = \{(\tilde{x}, \tilde{y}) \in \mathbb{R}^2 ; \ 0 < \tilde{y} < \epsilon g_{0\mu}(\tilde{x}), \ 0 < \tilde{x} < 1\} .$$

After change of variables, we obtain the equation $(3.6)_\epsilon$ or $(3.11)_\epsilon$ with $G_\epsilon = 0$. The limit equation is $(3.6)_0$ with g_0 replaced by $g_{0\mu}$, $G_0 = 0$ and $n = 1$. Without loss of generality, we can assume that $\int_0^1 g_{0\mu}(s)ds = 1$. If we make the change of variables

$$z = \int_0^x g_{0\mu}(s)ds$$

and set

$$v(x) = \tilde{v}(z), \quad \tilde{g}_{0\mu}(z) = g_{0\mu}(x(z)),$$

then $\tilde{v}$ is a solution of the equation

$(3.59)_\mu$
$$\begin{cases} \text{(i)} \quad \tilde{v}_t - (\tilde{g}_{0\mu}^2 \tilde{v}_z)_z + \alpha \tilde{v} = -f(\tilde{v}) \text{ in } (0, 1), \\ \text{(ii)} \quad \tilde{v}_z = 0 \text{ in } z = 0, 1 . \end{cases}$$

If $\tilde{\varphi}$ is an equilibrium point of $(3.59)_\mu$, we denote by $\lambda_1(\tilde{\varphi}, \tilde{g}_{0\mu}, f)$ the first eigenvalue of the differential operator

$$-\frac{\partial}{\partial z}(\tilde{g}_{0\mu}^2 \frac{\partial}{\partial z}) + \alpha + Df(\tilde{\varphi})$$

with the boundary conditions (3.59)(ii). Let us recall the following interesting result of (Yanagida (1982)).

THEOREM 3.10. — (i) If $\tilde{g}''_{0\mu}(z) \leq 0$, $0 \leq z \leq 1$, and $\tilde{\varphi}$ is a nonconstant equilibrium point of $(3.59)_\mu$, then $\lambda_1(\tilde{\varphi}, \tilde{g}_{0\mu}, f) < 0$ for any function f. In particular, for every function f, each nonconstant equilibrium solution is unstable.

(ii) If there is a $z_0 \in (0,1)$ such that $g''_{0\mu}(z) > 0$, then there is a C^2-function f such that $(3.59)_\mu$ has a non-constant equilibrium solution $\tilde{\varphi}$ with $\lambda_1(\tilde{\varphi}, \tilde{g}_{0\mu}, f) > 0$; this means that, for this f, there is a non-constant equilibrium solution which is stable.

A few calculations give us that

$$\tilde{g}''_{0\mu}(z) = (\tilde{g}_{0\mu}(z))^{-2} H_\mu(x)$$

where

$$H_\mu(x) = -\frac{(g'_{0\mu})^2}{g_{0\mu}} + g''_{0\mu} .$$

Using the above result of Yanagida as well as Theorem 3.9, we can prove the following properties.

THEOREM 3.11. — The following statements hold:

(i) If $H_\mu(x) \leq 0$, $0 \leq x \leq 1$, and f is a given C^2-function such that the zeros of $f + \alpha I$ are simple, then there exists $\epsilon_1 > 0$ such that, for $0 < \epsilon \leq \epsilon_1$, any non-constant equilibrium solution of $(3.11)_\epsilon$ is unstable.

(ii) If there is an $x_0 \in (0,1)$ such that $H_\mu(x_0) > 0$, then there exists a C^2-function f and $\epsilon_1 > 0$ such that, for $0 < \epsilon \leq \epsilon_1$, $(3.11)_\epsilon$ has a stable equilibrium solution.

For the proof of this theorem, see (Hale, Raugel (1992d)). The statement (ii) has also been proved by (Yanagida (1990)), by using upper and lower solutions.

REMARK 3.5. — If Q_ϵ is a convex region, that is, $g''_{0\mu} \leq 0$, $0 \leq x \leq 1$, then it is known (see (Casten and Holland (1978))) that any nonconstant equilibrium point of $(3.6)_0$ is unstable for any function f. Since the condition $H_\mu(x) \leq 0$, $x \in [0,1]$ is weaker than convexity, we can use theorem 3.11 to assert that there is a family of nonconvex regions Q_ϵ, $0 < \epsilon \leq \epsilon_1 = \epsilon_1(f)$, for which every nonconstant equilibrium solution of $(3.6)_\epsilon$ is unstable.

3.1.2 A reaction–diffusion equation on a thin L–shaped domain.

We will not give any details and refer to (Hale, Raugel (1993a)).

Let $g_i \in C^2([0,1], \mathbb{R}^{+*})$, $i = 1,2$. For $\epsilon > 0$, we define a general L-shaped domain Q_ϵ by the relations

$$Q_\epsilon = Q_\epsilon^1 \cup Q_\epsilon^2 ,$$
$$Q_\epsilon^1 = \{(\tilde{x}_1, \tilde{x}_2) \in \mathbb{R}^2 : 0 < \tilde{x}_2 < \epsilon g_1(\tilde{x}_1), 0 < \tilde{x}_1 < 1\},$$
$$Q_\epsilon^2 = \{(\tilde{x}_1, \tilde{x}_2) \in \mathbb{R}^2 : 0 < \tilde{x}_1 < \epsilon g_2(\tilde{x}_2), 0 < \tilde{x}_2 < 1\} .$$

The set $\bar{J}_\epsilon = \overline{Q_\epsilon^1 \cap Q_\epsilon^2}$ is called the junction set of Q_ϵ and is the closure of the set

$$J_\epsilon = \{(\tilde{x}_1, \tilde{x}_2) : 0 < \tilde{x}_2 < \epsilon g_1(\tilde{x}_1), 0 < \tilde{x}_1 < \epsilon g_2(\tilde{x}_2), \tilde{x}_1 \in (0,1), \tilde{x}_2 \in (0,1)\} .$$

To simplify, we suppose that $G \in W^{1,\infty}(\widetilde{Q})$ where $\widetilde{Q} \supset \bigcup_{0 < \epsilon \leq \epsilon_0} \overline{Q}_\epsilon$ and that $f \in C^2(\mathbb{R}, \mathbb{R})$ satisfies the conditions (3.3) with $\alpha_0 = 0$ and (3.4) with $0 \leq \gamma < +\infty$. We set:

$$\Gamma_\epsilon = \Gamma_\epsilon^1 \cup \Gamma_\epsilon^2, \ \Gamma_\epsilon^1 = \partial Q_\epsilon^1 \cup \{\widetilde{x}_1 = 1\}, \ \Gamma_\epsilon^2 = \partial Q_\epsilon^2 \cup \{\widetilde{x}_2 = 1\}.$$

We consider the equation

(3.60)
$$\begin{cases} u_t - \Delta u = -f(u) - G & \text{in} \quad Q_\epsilon, \\ u = 0 & \text{in} \quad \Gamma_\epsilon, \\ \dfrac{\partial u}{\partial \nu_\epsilon} = 0 & \text{in} \quad \partial Q_\epsilon \backslash \Gamma_\epsilon, \end{cases}$$

where the initial data is chosen from the space $H^1_{\Gamma_\epsilon}(Q_\epsilon) = \{u \in H^1(Q_\epsilon) : u = 0 \text{ in } \Gamma_\epsilon\}$. To study the problem (3.60), it is more convenient to write it in variational form. To this end, we write the inner product in $L^2(Q_\epsilon)$ in the form (see (Le Dret (1991)),

$$(u, v)_{L^2(Q_\epsilon)} = \int_{Q_\epsilon^1 \backslash J_\epsilon} uv d\widetilde{x}_1 d\widetilde{x}_2 + \frac{1}{2} \int_{J_\epsilon} uv d\widetilde{x}_1 d\widetilde{x}_2$$
$$+ \int_{Q_\epsilon^2 \backslash J_\epsilon} uv d\widetilde{x}_1 d\widetilde{x}_2 + \frac{1}{2} \int_{J_\epsilon} uv d\widetilde{x}_1 d\widetilde{x}_2.$$

With this notation, u is a solution of (3.60) if and only if, for all $\widetilde{u} \in H^1_{\Gamma_\epsilon}(Q_\epsilon)$, we have

(3.61)
$$(u_t, \widetilde{u})_{L^2(Q_\epsilon)} + (\nabla u, \nabla \widetilde{u})_{L^2(Q_\epsilon)} = -(f(u) + G, \widetilde{u})_{L^2(Q_\epsilon)}.$$

We now make a change of variables. We set $Q = Q^1 \cup Q^2$, where $Q^1 = Q^2 = (0,1) \times (0,1)$ and consider the change of variables

$$\varphi_\epsilon : \overline{Q}^1 \cup \overline{Q}^2 \to \overline{Q}_\epsilon$$

with

$$(\varphi_\epsilon)|\overline{Q}^j = \varphi_\epsilon^j, \quad j = 1, 2,$$

where

$$\varphi_\epsilon^1 : (x_1, x_2) \in \overline{Q}^1 \mapsto (x_1, \epsilon g_1(x_1)x_2) \in \overline{Q}_\epsilon^1,$$
$$\varphi_\epsilon^2 : (x_1, x_2) \in \overline{Q}^2 \mapsto (\epsilon g_2(x_2)x_1, x_2) \in \overline{Q}_\epsilon^2.$$

We set:

$$\Gamma^1 \equiv (\varphi_\epsilon^1)^{-1}\Gamma_\epsilon^1 \equiv \{(1, x_2), \ 0 < x_2 < 1\},$$
$$\Gamma^2 \equiv (\varphi_\epsilon^2)^{-1}\Gamma_\epsilon^2 \equiv \{(x_1, 1), \ 0 < x_1 < 1\},$$
$$\varphi_\epsilon^{-1} J_\epsilon = J_\epsilon^1 \cup J_\epsilon^2 \equiv (\varphi_\epsilon^1)^{-1} J_\epsilon \cup (\varphi_\epsilon^2)^{-1} J_\epsilon.$$

We now introduce some spaces which arise naturally from the transformation $\varphi_\epsilon^{-1} : Q_\epsilon \to Q^1 \cup Q^2$. We introduce the space $\mathcal{H}_\epsilon \cong L^2(Q) \equiv L^2(Q^1) \times L^2(Q^2)$ endowed with the norm $\|\cdot\|_{\mathcal{H}_\epsilon}$ induced by the inner product

$$(\xi, \zeta)_{\mathcal{H}_\epsilon} = \sum_{j=1}^{2} \left(\int_{Q^j \setminus J_\epsilon^j} g_j \xi_j \zeta_j dx_1 dx_2 + \frac{1}{2} \int_{J_\epsilon^j} g_j \xi_j \zeta_j dx_1 dx_2 \right)$$

where $\xi = (\xi_1, \xi_2)$, $\zeta = (\zeta_1, \zeta_2)$. We denote by $H_\epsilon^1(Q)$ the space $H^1(Q) \equiv H^1(Q^1) \times H^1(Q^2)$ equipped with the norm $\|\cdot\|_{H_\epsilon^1(Q)}$ defined by

$$\|\xi\|_{H_\epsilon^1(Q)} = \left(\|\xi\|_{1,Q}^2 + \frac{1}{\epsilon^2} \|\frac{\partial \xi_1}{\partial x_2}\|_{0,Q^1}^2 + \frac{1}{\epsilon^2} \|\frac{\partial \xi_2}{\partial x_1}\|_{0,Q^2}^2 \right)^{1/2},$$

for $\xi = (\xi_1, \xi_2)$. Let $H_{\Gamma^j}^1(Q^j) = \{u \in H^1(Q^j) \; ; \; u_{|\Gamma^j} = 0\}$, $j = 1, 2$. We let $\mathcal{X}_\epsilon$ be the space $H_{\Gamma^1}^1(Q^1) \times H_{\Gamma^2}^1(Q^2)$ equipped with the norm $\|\cdot\|_{\mathcal{X}_\epsilon} = \|\cdot\|_{H_\epsilon^1(Q)}$. Finally we introduce the spaces H_ϵ and X_ϵ,

$$H_\epsilon = \{(\xi_1, \xi_2) \in \mathcal{H}_\epsilon \; ; \; (\xi_1 \circ (\varphi_\epsilon^1)^{-1})_{|J_\epsilon} = (\xi_2 \circ (\varphi_\epsilon^2)^{-1})_{|J_\epsilon} \text{ a.e.}\}$$

$$X_\epsilon = \{(\xi_1, \xi_2) \in \mathcal{X}_\epsilon \; ; \; (\xi_1 \circ (\varphi_\epsilon^1)^{-1})_{|J_\epsilon} = (\xi_2 \circ (\varphi_\epsilon^2)^{-1})_{|J_\epsilon} \text{ a.e.}\}$$

endowed respectively with the norms $\|\cdot\|_{\mathcal{H}_\epsilon}$ and $\|\cdot\|_{\mathcal{X}_\epsilon}$.

If we make the above change of variables φ_ϵ and let

$$G_{1\epsilon}(x_1, x_2) = G(x_1, \epsilon g_1(x_1) x_2), \quad G_{2\epsilon}(x_1, x_2) = G(\epsilon g_2(x_2) x_1, x_2),$$

then the variational problem (3.61) is equivalent to finding a function $u^\epsilon \in X_\epsilon$ such that, for every $w \in X_\epsilon$, we have

$(3.62)_\epsilon$ $\qquad (u_t^\epsilon, w)_{\mathcal{H}_\epsilon} + a_\epsilon(u^\epsilon, w) = -(\widehat{f}(u^\epsilon) + \widehat{G}_\epsilon, w)_{\mathcal{H}_\epsilon},$

where

$$\widehat{f}(u) = (f(u_1), f(u_2)), \quad \widehat{G}_\epsilon = (G_{1\epsilon}, G_{2\epsilon}),$$

belong to H_ϵ, where $a_\epsilon(\xi, \zeta) = (\mathcal{L}_\epsilon^{1/2} \xi, \mathcal{L}_\epsilon^{1/2} \zeta)_{\mathcal{H}_\epsilon}$ with, for $\xi = (\xi_1, \xi_2)$,

$$\mathcal{L}_\epsilon^{1/2} \xi = (\mathcal{L}_{1\epsilon}^{1/2} \xi_1, \mathcal{L}_{2\epsilon}^{1/2} \xi_2) = \begin{pmatrix} \xi_{1x_1} - \frac{g_{1x_1}}{g_1} x_2 \xi_{1x_2} & \frac{1}{\epsilon g_2} \xi_{2x_1} \\ \frac{1}{\epsilon g_1} \xi_{1x_2} & \xi_{2x_2} - \frac{g_{2x_2}}{g_2} x_1 \xi_{2x_1} \end{pmatrix}$$

and, for $\underline{\xi} = \begin{pmatrix} \xi^1 \\ \xi^2 \end{pmatrix} = \begin{pmatrix} \xi_1^1 & \xi_2^1 \\ \xi_1^2 & \xi_2^2 \end{pmatrix}$ and $\underline{\zeta} = \begin{pmatrix} \zeta^1 \\ \zeta^2 \end{pmatrix}$,

$$(\underline{\xi}, \underline{\zeta})_{H_\epsilon} = \sum_{j=1}^{2} \left(\int_{Q^1 \setminus J_\epsilon^1} g_1 \xi_1^j \zeta_1^j dx_1 dx_2 + \frac{1}{2} \int_{J_\epsilon^1} g_1 \xi_1^j \zeta_1^j dx_1 dx_2 \right.$$
$$\left. + \int_{Q^2 \setminus J_\epsilon^2} g_2 \xi_2^j \zeta_2^j dx_1 dx_2 + \frac{1}{2} \int_{J_\epsilon^2} g_2 \xi_2^j \zeta_2^j dx_1 dx_2 \right).$$

In the classical way, the triple $\{X_\epsilon, H_\epsilon, a_\epsilon(\cdot, \cdot)\}$ defines a unique unbounded operator A_ϵ on X_ϵ with domain $D(A_\epsilon)$ in the following way: an element $u \in X_\epsilon$ belongs to $D(A_\epsilon)$ if the form

$$w \mapsto a_\epsilon(u, w)$$

is continuous on X_ϵ for the topology induced by H_ϵ. Thus the equation $(3.62)_\epsilon$ can be written in a shorter way as

$(3.63)_\epsilon$ $$u_t^\epsilon + A_\epsilon u^\epsilon = -\widehat{f}(u^\epsilon) - \widehat{G}_\epsilon .$$

Arguing as in Section 3.1.1, one shows that $(3.62)_\epsilon$ or $(3.63)_\epsilon$ generates a C^0–semigroup $T_\epsilon(t)$ on X_ϵ, which is a C^1-gradient system and has a connected global attractor $\mathcal{A}_\epsilon$ in X_ϵ.

Now we define the limit problem for $\epsilon = 0$. To this end, we let $H_0 = L^2((0,1)) \times L^2((0,1))$ with the inner product

$$(\xi, \zeta)_{H_0} = \int_0^1 g_1 \xi_1 \zeta_1 dx_1 + \int_0^1 g_2 \xi_2 \zeta_2 dx_2$$

and let

$$X_0 = \{\xi = (\xi_1, \xi_2) \in H^1((0,1)) \times H^1((0,1)) \;;\; \xi_1(1) = \xi_2(1) = 0, \; \xi_1(0) = \xi_2(0)\}$$

endowed with the norm $\|\cdot\|_{X_0}$ induced by the inner product

$$a_0(v, \xi) = \int_0^1 g_1 v_{1x_1} \xi_{1x_1} dx_1 + \int_0^1 g_2 v_{2x_2} \xi_{2x_2} dx_2 ,$$

for $v = (v_1, v_2)$, $\xi = (\xi_1, \xi_2)$ in X_0. Finally, we set

$$\widehat{G}_0 = (G_{10}, G_{20}), \text{ with } G_{10}(x_1) = G(x_1, 0), \; G_{20}(x_2) = G(0, x_2) .$$

Then the limit variational problem is to find a function v in X_0 such that, for each $\xi \in X_0$, we have

$(3.62)_0$ $$(v_t, \xi)_{H_0} + a_0(v, \xi) = -(\widehat{f}(v) + \widehat{G}_0, \xi)_{H_0} .$$

As above the triple $\{X_0, H_0, a_0(\cdot, \cdot)\}$ defines a unique unbounded operator A_0 on X_0 and thus we can also write $(3.62)_0$ in the abstract form

$(3.63)_0$ $$v_t + A_0 v = -\widehat{f}(v) - \widehat{G}_0 .$$

As above, the equations $(3.62)_0$ and $(3.63)_0$ generate a C^0–semigroup $T_0(t)$ on X_0, which is a C^1-gradient system and has a connected attractor $\mathcal{A}_0$ in X_0. Since $\mathcal{A}_0 \not\subset X_\epsilon$, the analysis will be more complicated as in the simple thin domain case. But we can consider that $\mathcal{A}_0 \subset \mathcal{X}_\epsilon$.

We remark that the problem $(3.63)_0$ is equivalent to the problem: find $v = (v_1, v_2) \in X_0$ such that

$(3.64)_0$
$$\begin{cases} v_{1t} - \frac{1}{g_1}(g_1 v_{1x_1})_{x_1} = -f(v_1) - G_{10} & \text{in} \quad \Omega^1 = (0,1), \\ v_1(1,t) = 0 \\ v_{2t} - \frac{1}{g_2}(g_2 v_{2x_2})_{x_2} = -f(v_2) - G_{20} & \text{in} \quad \Omega^2 = (0,1), \\ v_2(1,t) = 0 \\ v_1(0,t) = v_2(0,t) \\ g_1(0)v_{1x_1}(0,t) = -g_2(0)v_{2x_2}(0,t) \end{cases}$$

REMARK 3.6. — In the special case where $g_1(0) = g_2(0)$, $g_{1x_1}(0) = g_{2x_2}(0)$, then $(3.64)_0$ is equivalent to an equation on $(-1,1)$. Indeed, if $v = (v_1, v_2)$ is a solution of $(3.64)_0$ and if we set

$$v^*(s,t) = \begin{cases} v_1(s,t) & (0 \le s \le 1) \\ v_2(-s,t) & (-1 \le s \le 0) \end{cases}$$

$$g^*(s) = \begin{cases} g_1(s) & (0 \le s \le 1) \\ g_2(-s) & (-1 \le s \le 0) \end{cases}$$

$$G^*(s) = \begin{cases} G_{10}(s) & (0 \le s \le 1) \\ G_{20}(-s) & (-1 \le s \le 0) \end{cases}$$

then v^* satisfies the equation

$(3.64)^*$
$$\begin{cases} v_t^* - \frac{1}{g^*}(g^* v_s^*)_s = -f(v^*) - G^* & \text{in} \quad (-1,1) \\ v^*(-1,t) = v^*(1,t) = 0 . \end{cases}$$

We remark that we have already encountered this equation in Section 3.1.1 as a limit equation on a simple thin domain.

As we already have explained it, we introduce an operator $M_\epsilon \in \mathcal{L}(X_\epsilon; X_0)$ which is approximately a partial mean value operator and an operator $I_\epsilon \in \mathcal{L}(X_0; X_\epsilon)$ which is approximately the identity. We construct the operators M_ϵ and I_ϵ so that all the requirements of Chapter 2 are satisfied and also so that $I_\epsilon \in \mathcal{L}(X_0; X_\epsilon \cap X_0)$. In particular we have:

(3.65)
$$\begin{cases} \text{(i)} & \|M_\epsilon\|_{\mathcal{L}(X_\epsilon; X_0)} \le C \\ \text{(ii)} & \|I - M_\epsilon\|_{\mathcal{L}(X_\epsilon; \mathcal{H}_\epsilon)} \le C\epsilon, \end{cases}$$

(3.66)
$$\begin{cases} \text{(i)} & \|I_\epsilon\|_{\mathcal{L}(X_0; X_\epsilon)} \le C \\ \text{(ii)} & \|I - I_\epsilon\|_{\mathcal{L}(X_0; \mathcal{H}_\epsilon)} \le C\epsilon \\ \text{(iii)} & \|I - I_\epsilon\|_{\mathcal{L}(X_0 \cap (H^2(0,1) \times H^2(0,1)); X_\epsilon)} \le C\epsilon^{1/2}, \end{cases}$$

and

(3.67)
$$\|I - M_\epsilon I_\epsilon\|_{\mathcal{L}(X_0 \cap (H^2(0,1) \times H^2(0,1)); X_0)} \le C\epsilon^{1/2} .$$

In (Hale and Raugel (1993a)), we show the following results.

PROPOSITION 3.12. — (i) *There is a positive constant C_0 such that, for* $0 \leq \epsilon \leq \epsilon_0$,

$$\|\varphi\|_{X_\epsilon} \leq C_0, \quad \forall \varphi \in \mathcal{A}_\epsilon. \tag{3.68}$$

(ii) *There exists a positive, non-decreasing function $K_0(\cdot)$ such that, for any positive number r_0, for $0 \leq \epsilon \leq \epsilon_0$, for any u_0 in $B_{X_\epsilon}(0, r_0)$, we have*

$$\|t(T_\epsilon(t)u_0 - T_0(t)M_\epsilon u_0)\|_{X_\epsilon} \leq \epsilon^{1/2} K_0(r_0) \exp(K_0(r_0)t). \tag{3.69}$$

From the propositions 2.2 and 3.12, we at once deduce the following result.

COROLLARY 3.13. — *Under the above hypotheses, the attractors $\mathcal{A}_\epsilon$ are u.s.c. at $\epsilon = 0$, i.e. $\delta_{X_\epsilon}(\mathcal{A}_\epsilon, \mathcal{A}_0) \to 0$ as $\epsilon \to 0$.*

Assume now that all the equilibrium points of $(3.62)_0$ are hyperbolic. Then, by the results of Section 2.3, the sets of equilibrium points E_0 and E_ϵ of $(3.62)_0$ and $(3.62)_\epsilon$ are finite sets of N_0 elements u_ℓ^0, $1 \leq \ell \leq N_0$, and u_ℓ^ϵ, $1 \leq \ell \leq N_0$, respectively and

$$\|u_\ell^0 - u_\ell^\epsilon\|_{X_\epsilon} \leq C\epsilon^{1/2}. \tag{3.70}$$

We can prove (but the proof is a little long and technical) that, for $0 \leq \epsilon \leq \epsilon_0$, there exist neighbourhoods U_ℓ^ϵ and $\widetilde{U}_\ell^\epsilon$ of u_ℓ^ϵ, $1 \leq \ell \leq N_0$, such that

$$\max(\delta_{X_\epsilon}(W_0^u(u_\ell^0, U_\ell^0), W_\epsilon^u(u_\ell^\epsilon, \widetilde{U}_\ell^\epsilon)), \delta_{X_\epsilon}(W_\epsilon^u(u_\ell^\epsilon, U_\ell^\epsilon), W_0^u(u_\ell^0, \widetilde{U}_\ell^0)) \leq C\epsilon^{1/2}. \tag{3.71}$$

(For a proof, see Hale and Raugel (1992d)). Applying now the results of Section 2.2 and using (3.69), (3.70) and (3.71), we deduce the following l.s.c. result.

THEOREM 3.14. — *Assume that all the equilibrium points of $T_0(t)$ are hyperbolic, then the attractors $\mathcal{A}_\epsilon$ are l.s.c. at $\epsilon = 0$ and there exist positive numbers ϵ_2, C and $\widetilde{p}$, $0 < \widetilde{p} \leq \frac{1}{2}$, such that, for $0 < \epsilon \leq \epsilon_2$,*

$$\delta_{X_\epsilon}(\mathcal{A}_0, \mathcal{A}_\epsilon) + \delta_{X_\epsilon}(\mathcal{A}_\epsilon, \mathcal{A}_0) \leq C\epsilon^{\widetilde{p}}. \tag{3.72}$$

Arguing as in (Hale, Raugel (1992a)), we can show that the statement (i) of the theorem 3.9 still holds. The statement (ii) is no longer true even if $g_1(x_1) \equiv g_2(x_2) \equiv 1$.

3.2 The complex Ginzburg–Landau equation on a thin domain.

Now we consider the complex Ginzburg–Landau equation on a thin domain Q_ϵ,

$$(3.73) \quad \begin{cases} (\text{i})u_t - (\nu + ia)\Delta u + (\kappa + ib)|u|^2 u - du = 0 \text{ in } Q_\epsilon \\ (\text{ii})\dfrac{\partial u}{\partial \nu_\epsilon} = 0 \text{ in } \partial Q_\epsilon, \end{cases}$$

where the parameters ν, a, κ, b and d are real numbers and

$$\nu > 0, \qquad \kappa > 0.$$

We keep the model simple thin domain Q_ϵ described in Section 3.1.1. But, in order to avoid problems of global existence of strong solutions of the Ginzburg–Landau equations in three–dimensional domains, we restrict our study to the case $n = 1$, that is, Q_ϵ is a thin two–dimensional domain. Of course, using techniques similar to those developped in Sections 5.1 and 5.2, we could show global existence of strong solutions of (3.73) for large initial data.

As in Section 3.1.1, we transform coordinates to the canonical domain $Q = \Omega \times (0,1)$ and obtain the system

$$(3.74)_\epsilon \qquad \begin{cases} \text{(i)} u_t - (\nu + ia)L_\epsilon u + (\kappa + ib)|u|^2 u - du = 0 \text{ in } Q \\[2mm] \text{(ii)} \dfrac{\partial u}{\partial \nu_{B_\epsilon}} = 0. \end{cases}$$

We also want to write the equation $(3.74)_\epsilon$ as an abstract evolutionary equation. We keep the notation of Section 3.1.1 for the spaces $L^2(Q)$, $H^m(Q)$ and their corresponding inner products and norms. Here, of course, these spaces concern (classes of) complex functions and the norms and inner products must be modified accordingly. We make the corresponding needed modifications in the definitions of the spaces H_ϵ and H_0. Here we define the unbounded linear operator A_ϵ on $H^1(Q)$ by $A_\epsilon = L_\epsilon$ with homogeneous Neumann boundary conditions. Thus we have

$$a_\epsilon(u,u)^{1/2} = \|(A_\epsilon + \alpha \mathrm{Id})^{1/2} u\|_{H_\epsilon},$$

where $\alpha > 0$ has been given in (3.2). And we let $X_\epsilon = D((A_\epsilon + \alpha \mathrm{Id})^{1/2})$ endowed with the norm $\|u\|_{X_\epsilon} = \|(A_\epsilon + \alpha \mathrm{Id})^{1/2} u\|_{H_\epsilon}$. Accordingly we define the space X_ϵ^2. With these changes, the equation $(3.74)_\epsilon$ is equivalent to the abstract evolutionary equation

$$(3.75)_\epsilon \qquad u_t + (\nu + ia)A_\epsilon u + (\kappa + ib)|u|^2 u - du = 0.$$

The limit equation is given by

$$(3.74)_0 \qquad \begin{cases} \text{(i)} v_t - (\nu + ia)\dfrac{1}{g_0}(g_0 v_x)_x + (\kappa + ib)|v|^2 v - dv = 0 \text{ in } \Omega \\[2mm] \text{(ii)} \dfrac{\partial v}{\partial n} = 0 \text{ in } \partial\Omega. \end{cases}$$

If we do the corresponding modifications in the definitions of A_0, X_0 (and X_0^2), we see that the abstract evolutionary equation corresponding to $(3.74)_0$ is

$$(3.75)_0 \qquad v_t + (\nu + ia)A_0 v + (\kappa + ib)|v|^2 v - dv = 0.$$

Since the mapping $u \in X_\epsilon \mapsto |u|^2 u \in H_\epsilon$ is locally Lipschitz–continuous, $(3.75)_\epsilon$ defines a (local) C^0 semigroup $T_\epsilon(t)$ on X_ϵ given by $u_\epsilon(t) = T_\epsilon(t)u^0$ where $u_\epsilon(t)$ is the solution of $(3.75)_\epsilon$ satisfying $u_\epsilon(0) = u^0$ (see Henry (1981)). Likewise, $(3.75)_0$

defines a (local) C^0 semigroup $T_0(t)$ on X_0. In (Ghidaglia and Héron (1987)), the semigroup $T_\epsilon(t)$ is defined on H_ϵ. Repeating exactly the proofs of (Ghidaglia and Héron (1987)) in our framework, one shows that there exist a positive non–decreasing function of a real variable $K(\cdot)$ and a positive constant r_1 such that, for any positive number r, for any $0 \leq \epsilon \leq \epsilon_0$, there is a positive time $T_1(\epsilon, r)$ so that, if $\|u_0\|_{X_\epsilon} \leq r$, then

$$(3.76) \qquad \|T_\epsilon(t)u_0\|_{X_\epsilon} \leq K(r) \qquad \text{for } t \geq 0,$$

and

$$(3.77) \qquad \|T_\epsilon(t)u_0\|_{X_\epsilon} \leq r_1 \qquad \text{for } t \geq T_1(\epsilon, r).$$

On the other hand, it is easy to show that $T_\epsilon(t)$, for $\epsilon \geq 0$, is a compact mapping in X_ϵ, for $t > 0$. If $u_0 \in X_\epsilon$, one shows that $T_\epsilon(t)u_0$ belongs to $L^2((0,T); D(A_\epsilon))$, for any positive number T. Since the semigroup $T_\epsilon(t)$ is bounded dissipative in X_ϵ and is a compact mapping in X_ϵ, for $t > 0$, it admits a global attractor $\mathcal{A}_\epsilon$ in X_ϵ. Thanks to (3.77), we obtain the uniform bound, for $0 \leq \epsilon \leq \epsilon_0$,

$$(3.78) \qquad \|\varphi_\epsilon\|_{X_\epsilon} \leq r_1, \qquad \forall \varphi_\epsilon \in \mathcal{A}_\epsilon.$$

REMARK 3.7. — In the case $d \leq 0$, it is easily seen that all the solutions of $(3.75)_\epsilon$, $0 \leq \epsilon \leq \epsilon_0$, decay uniformly to 0 and that thus $\mathcal{A}_\epsilon = \{0\}$. This decay is exponential if $d < 0$, whereas it is polynomial if $d = 0$.

In (Hale and Raugel (1992d)) (see also (Mischaikow and Raugel (1994), Raugel (1993)), the following result is proved.

LEMMA 3.15. — Let $0 \leq \epsilon \leq \epsilon_0$. There exists a positive, non–decreasing function $K_0(\cdot)$ such that, for any positive number r_0, for any u_0 in $B_{X_\epsilon}(0, r_0)$, we have, for $t \geq 0$,

$$(3.79) \quad \|t(T_\epsilon(t)u_0 - T_0(t)Mu_0)\|_{X_\epsilon}^2 + \int_0^t \|\frac{d}{ds}(s(T_\epsilon(s)u_0 - T_0(s)Mu_0))\|_{H_\epsilon}^2 ds$$
$$\leq \epsilon^2 K_0(r_0) \exp(K_0(r_0)t).$$

Thanks to (3.78) and (3.79), the hypotheses (H.1) and (H.2) of Proposition 2.1 are satisfied, which gives us the first part of the following theorem. The second statement is proved like in Proposition 3.4.

THEOREM 3.16. — (i) Under the above hypotheses, the attractors $\mathcal{A}_\epsilon$ are u.s.c. at $\epsilon = 0$, i.e., $\delta_{X_\epsilon}(\mathcal{A}_\epsilon, \mathcal{A}_0) \to 0$ as $\epsilon \to 0$.

(ii) *If $Q_\epsilon = \Omega \times (0, \epsilon)$, there exists a positive constant ϵ_1 such that, for* $0 \leq \epsilon \leq \epsilon_1$,

$$A_\epsilon = A_0 .$$

The Ginzburg–Landau equation is not a gradient system. Therefore, in the general case, no lower semicontinuity result is known. However, in the case where

$$d = 1, \quad \kappa = 1, \quad b = \frac{a}{\nu},$$

Mischaikow and Morita (1994) have shown that the Ginzburg–Landau equation can be "transformed" into a gradient system. In this case, when the periodic orbits and equilibrium points of $(3.75)_0$ are all hyperbolic, one shows that the attractors A_ϵ are lower–semicontinuous in X_ϵ at $\epsilon = 0$ and one obtains an estimate of the Hausdorff distance in X_ϵ between A_ϵ and A_0 (for more details, see (Hale and Raugel (1992d) and Raugel (1993)). In the case where $g_0 \equiv 1$ in $(3.74)_0$, Mischaikow and Morita gave a Morse decomposition of the attractor A_0 and computed the cohomology Conley indices of the Morse sets. In (Mischaikow and Raugel (1994)), we show that these properties persist for the semigroups $T_\epsilon(t)$, $0 \leq \epsilon \leq \epsilon_0$.

Since the operator A_ϵ has still the gap property, we can construct inertial manifolds $\mathcal{M}^\epsilon$ and $\mathcal{M}^0$ and compare them. A general result of (Hale and Raugel (1992d)) implies the following theorem.

THEOREM 3.17. — *Under the above hypotheses, the statements* (i) *and* (ii) *of Theorem 3.9 hold for the equations* $(3.75)_\epsilon$ *and* $(3.75)_0$.

We end this section by two remarks.

REMARK 3.8. — If one considers the Ginzburg–Landau equation (3.73) with homogeneous Dirichlet boundary conditions on Q_ϵ, one can show, like for the reaction–diffusion equation, that $A_\epsilon \to 0$ in X_ϵ, as $\epsilon \to 0$.

REMARK 3.9. — We can also consider the Ginzburg–Landau equation on a thin L–shaped domain Q_ϵ. In this case, Lemma 3.15, Theorem 3.16 and the first statement of Theorem 3.17 still hold. We let the proofs to the reader.

3.3 Damped wave equations on a simple thin domain.

Since, in our estimates below, we use H^2–regularity results, we will study the damped wave equations only on simple thin domains, which are regular or which have convex angles. Note that this restriction in the hypothesis is not needed in the Chapter 4.

We keep *all the notation* of Section 3.1. In particular, we consider the same model simple thin domain Q_ϵ, where $\Omega \subset \mathbb{R}^n$, $n \leq 2$, is a bounded domain, which now is regular or a $\mathcal{C}^1$–polygonal domain with convex angles. For α and β positive constants, we consider the damped wave equation

$$(3.80) \quad \begin{cases} \text{(i)} \quad u_{tt} + \beta u_t - \Delta u + \alpha u = -f(u) - G \quad & \text{in} \quad Q_\epsilon, \\ \text{(ii)} \quad \dfrac{\partial u}{\partial \nu_\epsilon} = 0 \quad & \text{in} \quad \partial Q_\epsilon, \end{cases}$$

where f and G are the functions introduced in Section 3.1. As in Section 3.1.1, we transform coordinates to the canonical domain $Q = \Omega \times (0,1)$ and we obtain the system

$$(3.81)_\epsilon \qquad \begin{cases} \text{(i)} & u_{tt} + \beta u_t + L_\epsilon u + \alpha u = -f(u) - G_\epsilon \quad \text{in} \quad Q, \\[2mm] \text{(ii)} & \dfrac{\partial u}{\partial \nu_{B_\epsilon}} = 0 \qquad\qquad\qquad\qquad\quad \text{in} \quad \partial Q. \end{cases}$$

As in section 3.1.1, we introduce the operator A_ϵ and the spaces X_ϵ^s, $s = 0, 1, 2$, where $X_\epsilon^0 = H_\epsilon$, $X_\epsilon^1 = X_\epsilon$. For $s = 1, 2$, we also define the space $Y_\epsilon^s = D(A_\epsilon^{s/2}) \times D(A_\epsilon^{(s-1)/2}) = X_\epsilon^s \times X_\epsilon^{s-1}$ endowed with the norm

$$\|(\varphi, \psi)\|_{Y_\epsilon^s} = (\|\varphi\|_{X_\epsilon^s}^2 + \|\psi\|_{X_\epsilon^{s-1}}^2)^{1/2}.$$

Clearly, Y_ϵ^1 is isomorphic to $H^1(Q) \times L^2(Q)$ and Y_ϵ^2 is isomorphic to

$$\{\varphi \in H^2(Q) \ : \ \frac{\partial \varphi}{\partial \nu_{B_\epsilon}} = 0 \text{ in } \partial Q\} \times H^1(Q).$$

With this notation, the equation $(3.81)_\epsilon$, with initial data in Y_ϵ^1, is equivalent to the abstract evolutionary equation

$$(3.82)_\epsilon \qquad u_{tt} + \beta u_t + A_\epsilon u = -f(u) - G_\epsilon.$$

We compare it with the following limit problem on Ω:

$$(3.81)_0 \qquad \begin{cases} \text{(i)} & v_{tt} + \beta v_t - \frac{1}{g_0}\sum_{i=1}^n (g_0 v_{x_i})_{x_i} + \alpha v = -f(v) - G_0 \quad \text{in} \quad \Omega, \\[2mm] \text{(ii)} & \dfrac{\partial v}{\partial n} = 0 \qquad\qquad\qquad\qquad\qquad\qquad\qquad\quad \text{in} \quad \partial \Omega. \end{cases}$$

As in section 3.1.1, we introduce the operator A_0 and the spaces X_0^s, $s = 0, 1, 2$, where $X_0^0 = H_0$, $X_0^1 = X_0$. For $s = 1, 2$, we also define the space $Y_0^s = X_0^s \times X_0^{s-1}$ endowed with the norm

$$\|(\varphi, \psi)\|_{Y_0^s} = (\|\varphi\|_{X_0^s}^2 + \|\psi\|_{X_0^{s-1}}^2)^{1/2}.$$

With this notation, the equation $(3.81)_0$, with initial data in Y_0^1, is equivalent to the abstract evolutionary equation

$$(3.82)_0 \qquad v_{tt} + \beta v_t + A_0 v = -f(v) - G_0.$$

For $\epsilon > 0$ we introduce the operator $T_\epsilon(t) : (u_0, u_1) \in Y_\epsilon^1 \mapsto (u^\epsilon(t), u_t^\epsilon(t)) \in Y_\epsilon^1$, where $u^\epsilon(t)$ is the solution of $(3.82)_\epsilon$ with initial data $(u^\epsilon(0), u_t^\epsilon(0)) = (u_0, u_1)$. Likewise, we introduce the operator $T_0(t)$ on Y_0^1, generated by the equation $(3.82)_0$.

Let us quickly recall the properties of $T_\epsilon(t)$ and $T_0(t)$ that we will use here. To this end, we introduce a general smooth enough bounded domain U in $\mathbb{R}^N$, $N \leq 3$, we let $Y^1 = H^1(U) \times L^2(U)$, $Y^2 = \{\varphi \in H^2(U) \ : \ \frac{\partial \varphi}{\partial \nu_U} = 0 \text{ in } \partial U\} \times H^1(U)$ and define the operator $T(t)$ generated on Y^1 by the damped wave equation (3.80) on

U. Under the hypothesis (3.4), $T(t)$ is a C^0-group on Y^1 and also on Y^2 ; and under the hypotheses (3.3) and (3.4), the positive orbits of bounded sets in Y^1 (resp. Y^2) are bounded in Y^1 (resp. Y^2) (for all these properties, see, for example, Segal (1963), Lions (1969), Babin and Vishik (1983)). Moreover, the group $T(t)$ is bounded dissipative in Y^1 (resp. Y^2) (see, for instance, Hale (1985,1988), Haraux (1985), Ghidaglia and Temam (1987)). Furthermore, $T(t)$ has a global attractor $\mathcal{A}$ in Y^1. Hale (1985) and Haraux (1985) independently proved it for $N \leq 2$ and $N = 3, \gamma < 1$. Hale used the variation of constants formula to show that $T(t)$ can be written as a sum of a linear exponentially contracting group and a compact group. This argument does not use the smoothness property of the domain U. Using also the variation of constants formula together with a "bootstrap" argument, Haraux showed that all the orbits, that are globally bounded in Y^1, are actually bounded in Y^2. This allowed him to show that $T(t)$ is asymptotically smooth and that $T(t)$ has a global attractor $\mathcal{A}$ in Y^1, which is bounded in Y^2. Actually the same method of proof of Hale shows that the group $T(t)$, acting on Y^2, has a global attractor $\mathcal{A}_2$ in Y^2, for $N \leq 3$. Since $\mathcal{A}$ is bounded in Y^2, we conclude that $\mathcal{A}_2$ and $\mathcal{A}$ coincide in the cases $N \leq 2$ or $N = 3, \gamma < 1$. In the case $N = 3, \gamma = 1$, Hale and Raugel (1992c) proved the existence of a global attractor $\mathcal{A}$ in Y^1, when U is a thin domain over a domain $\Omega \in \mathbb{R}^2$, whereas Babin and Vishik (1989a) gave a proof for general domains, but special functions satisfying (3.4). Finally, Arrieta, Carvalho and Hale (1992) proved it for general domains and general functions satisfying (3.4). In the case $N = 3, \gamma = 1$, for general domains, it is not yet known that the attractor $\mathcal{A}$ in Y^1 is also bounded in Y^2. However, when U is a thin domain over a domain $\Omega \in \mathbb{R}^2$, Hale and Raugel (1992d) have shown that $\mathcal{A}$ is also bounded in Y^2, which implies, like above, that $\mathcal{A}$ is also the global attractor of $T(t)$ in Y^2. Finally, let us also remark that $T(t)$ is a gradient system on Y^1. In the case of our thin domain Q_ϵ, the Lyapunov functional $V_\epsilon(\varphi, \psi)$ for $T_\epsilon(t)$ is given by

$$(3.83) \qquad V_\epsilon(\varphi, \psi) = \frac{1}{2}\|(\varphi,\psi)\|^2_{Y^1_\epsilon} + (F(\varphi), 1)_{H_\epsilon} + (G_\epsilon, \varphi)_{H_\epsilon} ,$$

where $F(s) = \int_0^s f(\sigma)d\sigma$. Let us also remark that the set of equilibrium points $\tilde{E}_\epsilon$ of $(3.82)_\epsilon$ is given by

$$(3.84) \qquad \tilde{E}_\epsilon = E_\epsilon \times \{0\} ,$$

where E_ϵ denotes the set of equilibrium points of the reaction–diffusion equation $(3.11)_\epsilon$. Like in this case, the attractor $\mathcal{A}_\epsilon$ of $T_\epsilon(t)$ is the unstable set of the set $\tilde{E}_\epsilon$. Using this property as well as the uniform bound (3.21), we obtain the following result.

LEMMA 3.18. — *There is a positive constant* C_1 *such that, for* $0 \leq \epsilon \leq \epsilon_0$,

$$(3.85) \qquad \|(\varphi,\psi)\|_{Y^1_\epsilon} \leq C_1, \qquad \forall(\varphi,\psi) \in \mathcal{A}_\epsilon .$$

In (Hale and Raugel (1992c), Theorem 3.4), we have proved the following result:

LEMMA 3.19. — *There exist a positive constant K and, for any $r_1 > 0$, $r_2 > 0$, two positive constants $K_1(r_1)$, $K_2(r_1, r_2)$ such that, for $0 \le \epsilon \le \epsilon_0$, any solution $U_\epsilon(t) = (u_\epsilon(t), u_{\epsilon t}(t))$ of $(3.82)_\epsilon$ with $\|U_\epsilon(0)\|_{Y^i_\epsilon} \le r_i$, $i = 1, 2$, satisfies the following estimate for $t \ge 0$,*

$$(3.86) \qquad \|u_{\epsilon t t}\|^2_{X^1_\epsilon} + \|U_\epsilon(t)\|^2_{Y^2_\epsilon} \le K_1(r_1) + K_2(r_1, r_2) e^{-Kt} .$$

Since $\mathcal{A}_\epsilon$ is bounded in Y^2_ϵ (for the proof of this property in the case $N = 2$, $\gamma = 1$, see (Hale and Raugel (1992c)), using the invariance property of $\mathcal{A}_\epsilon$, we at once deduce from the lemmas 3.18 and 3.19 the following property.

PROPOSITION 3.20. — *There is a positive constant C_2 such that, for $0 \le \epsilon \le \epsilon_0$,*

$$(3.87) \qquad \|(\varphi, \psi)\|_{Y^2_\epsilon} \le C_2, \qquad \forall (\varphi, \psi) \in \mathcal{A}_\epsilon .$$

In (Hale and Raugel (1992c), Proposition 5.1), we have compared the orbits of the equations $(3.82)_\epsilon$ and $(3.82)_0$.

PROPOSITION 3.21. — *There exists a positive non-decreasing function $K(\cdot)$ such that, for $0 < \epsilon \le \epsilon_0$, for any solution $U_\epsilon(t) = (u_\epsilon(t), u_{\epsilon t}(t))$ of $(3.82)_\epsilon$ with $\|U_\epsilon(0)\|_{Y^2_\epsilon} \le r$, we have, for $t \ge 0$,*

$$(3.88) \qquad \|U_\epsilon(t) - T_0(t) M U_\epsilon(0)\|_{Y^1_\epsilon} \le \epsilon^{1/2} K(r) \exp K(r) t .$$

The first statement of the next theorem is then a direct consequence of the propositions 2.1, 3.20 and 3.21. The second statement is proved in (Hale and Raugel (1992c)) and uses a decomposition like $(3.42)_\epsilon$.

THEOREM 3.22. — (i) *The attractors $\mathcal{A}_\epsilon$ are u.s.c. at $\epsilon = 0$, i.e.,*

$$\delta_{Y^1_\epsilon}(\mathcal{A}_\epsilon, \mathcal{A}_0) \to 0 \quad as \quad \epsilon \to 0 .$$

(ii) *If $Q_\epsilon = \Omega \times (0, \epsilon)$ and $G(\tilde{x}, \tilde{y}) = G(\tilde{x})$, there exists a positive constant ϵ_1 such that, for $0 \le \epsilon \le \epsilon_1$,*

$$\mathcal{A}_\epsilon = \mathcal{A}_0 .$$

Since, for $\epsilon \ge 0$, $T_\epsilon(t)$ is a gradient system, we can also give the following lower–semicontinuity result.

THEOREM 3.23. — *If the equilibrium points of $(3.82)_0$ are all hyperbolic, then the attractors $\mathcal{A}_\epsilon$ are lower–semicontinuous at $\epsilon = 0$ and there exist positive constants ϵ_1, C and p, $0 < p \le 1/2$, such that*

$$(3.89) \qquad \delta_{Y^1_\epsilon}(\mathcal{A}_0, \mathcal{A}_\epsilon) + \delta_{Y^1_\epsilon}(\mathcal{A}_\epsilon, \mathcal{A}_0) \le C \epsilon^p .$$

Using the propositions 3.20 and 3.21, the results of Sections 2.3 and 2.4 as well as the comparison of the local unstable manifolds of the equilibria given in (Hale and Raugel (1992d)), we see that the hypotheses of the theorems 2.11 and 2.12 are satisfied, which at once implies the above result (see also Raugel (1993)).

REMARK 3.10. — We could also replace the constant damping term βu_t by a local nonlinear damping term $\beta_\epsilon(\tilde{x}, \tilde{y}) h(u_t)$. For some results in this case, see (Hale and Raugel (1992d)).

Chapter 4

Convergence to singletons in problems on thin domains

In this section we present some examples of gradient-like systems on thin domains for which it is possible to prove that the ω-limit set of any precompact orbit is a single equilibrium point. If this property is satisfied, we will say that we have convergence of solutions. In asymptotically smooth gradient systems, the equilibria of which are all isolated, the convergence of all bounded solutions is an immediate consequence of the connectedness of the ω-limit sets. But what happens if some equilibria are no longer isolated? For a scalar autonomous first order ordinary differential equation, we always have convergence of solutions. On the other hand, there are gradient systems in $\mathbb{R}^2$ for which convergence of solutions does not hold ; Palis and de Melo (1982) constructed a gradient planar system where the ω-limit set of some trajectories is a circle of equilibrium points.

For ordinary differential equations, Malkin (1952) proved convergence of solutions in the neighbourhood of an equilibrium point x_0 under the conditions that the linear variational equation about x_0 has k zero eigenvalues with the remaining ones having negative real parts and x_0 belongs to a smooth manifold of equilibrium points of dimension k. Henry (1981) gave an extension of the result of Malkin to a scalar parabolic equation in one space dimension and gave an application to the stability of travelling waves. In the general case of scalar semilinear parabolic equations on $(0, 1)$, Zelenyak (1968) and, independently, Matano (1978) proved that there is convergence of solutions. Zelenyak used a special Lyapunov function to obtain this result whereas Matano used a nontrivial application of the maximun principle. In several space dimensions and for some types of analytic nonlinearities, Simon (1983) has shown that the convergence property still holds for semilinear parabolic equations. Hale and Massat (1982) gave an indication of how geometric properties of the flow defined by the equation could be used to obtain the result of Matano and Zelenyak. This approach makes it possible to consider the convergence problem in more general gradient–like systems in several space dimensions. In finite dimensions, Aulbach (1983) obtained a convergence result which includes part of the statement of Hale and Massat. In Section 4.1, we give an extension of the result of Hale and Massat which is more convenient for applications to several space dimensions. In the remaining part of Chapter 4, we apply this general convergence result to semilinear parabolic equations and semilinear damped wave equations on thin domains around an arc of a curve or on some thin L–shaped domains. Haraux and Poláčik (1992) used this general convergence result in order to prove convergence of all nonnegative bounded solutions of spatially homogeneous semilinear parabolic equations on a ball in $\mathbb{R}^N$. In view of these convergence results, the problem of existence of a semilinear parabolic equation of gradient type, for which the convergence property does not hold, becomes very interesting. Very recently, by modifying the above mentioned construction of Palis and de Melo and using center–manifold theory, Poláčik and Rybakowski constructed a family of semilinear parabolic equations of gradient type defined on the unit disk in $\mathbb{R}^2$, for which some bounded solutions have an ω–limit set diffeomorphic to the unit circle S^1.

4.1 A general convergence result.

Let Z be a Banach space normed by a C^1-norm $||\cdot||_Z$. We consider the abstract evolutionary equation

$$(4.1) \qquad \frac{dz}{dt} = C_1 z + H_1(z)$$

where C_1 is the generator of a strongly continuous linear semigroup. We assume that C_1 and H_1 satisfy the following property:
(H1) *either*

(i) H_1 *is a C^1-mapping from Z into Z, or*

(ii) $-C_1$ *is a sectorial operator and there exists a real number b, $0 \le b < 1$ such that H_1 is a C^1-mapping from Z^b into Z.*(Without loss of generality, we can and do assume that $\mathrm{Re}\,\sigma(-C_1) > 0$).

If z_0 is an equilibrium point of (4.1), then, replacing z by $z + z_0$, we can rewrite the equation (4.1) as

$$(4.2) \qquad \frac{dz}{dt} = Cz + H(z),$$

where

$$(4.3) \qquad Cz = C_1 z + DH_1(z_0)z, \quad H(z) = H_1(z_0 + z) - H_1(z_0) - DH_1(z_0)z .$$

We remark that $C \equiv C_{z_0}$ is the generator of a strongly continuous linear semigroup $S(t) \equiv S_{z_0}(t) \equiv e^{Ct}$ and, in the case of the hypothesis (H.1)(ii), $-C$ is a sectorial operator and we assume also that:

$$(H.1\mathrm{bis}) \qquad D(C_1) = D(C) .$$

Below, we set $b = 0$ if we are in the case (H.1)(i). We further assume that:
(H.2) *There is a decomposition $Z = Y_{1z_0} \oplus X_{z_0} \oplus Y_{2z_0} \equiv Y_1 \oplus X \oplus Y_2$ with associated continuous projection operators $P_1 \equiv P_{1z_0}$, $P_0 \equiv P_{0z_0}$, $P_2 \equiv P_{2z_0}$ which commute with $S_{z_0}(t)$.*
(H.3) *The ranges X_{z_0} and Y_{2z_0} of P_{0z_0} and P_{2z_0} are finite dimensional ($\dim X_{z_0} = m_{0z_0} \equiv m_0$, $\dim Y_{2z_0} = m_{2z_0} \equiv m_2$).*
(H.4) *The spectrum $\sigma(S_{z_0}(1))$ of $S_{z_0}(1)$ can be written as $\sigma(S_{z_0}(1)) = \sigma_{-z_0} \cup \sigma_{0z_0} \cup \sigma_{+z_0} \equiv \sigma_- \cup \sigma_0 \cup \sigma_+$ where $\sigma_{-z_0} = \sigma(S_{z_0}(1)P_{1z_0})$, $\sigma_{0z_0} = \sigma(S_{z_0}(1)P_{0z_0})$, $\sigma_{+z_0} = \sigma(S_{z_0}(1)P_{2z_0})$ lies respectively inside, on, outside the unit circle with centre 0 in $\mathbb{C}$. Moreover, the distance of σ_- to the unit circle is positive.*

If we set: $A = P_0 C$, $B = (I - P_0)C$, $B_1 = P_1 C$, $B_2 = P_2 C$, then the equation (4.2) also can be written as

$$(4.4) \qquad \begin{cases} x_t = Ax + f(x, y) \\ y_t = By + g(x, y) \end{cases}$$

or

$$(4.4\text{bis}) \qquad \begin{cases} x_t = Ax + f(x,y) \\ y_{1t} = B_1 y_1 + g_1(x, y_1, y_2) \\ y_{2t} = B_2 y_2 + g_2(x, y_1, y_2) \end{cases}$$

where

$$y = (y_1, y_2), \ f = P_0 H, \ g = (I - P_0)H, \ g_1 = P_1 H, \ g_2 = P_2 H.$$

Let E be the set of equilibrium points of the equation (4.1). The following convergence result is given in (Hale, Raugel (1992b)) (see also (Hale, Raugel (1992d)).

THEOREM 4.1. — *Assume that the hypotheses* (H.1), (H.1bis) *hold, that the positive orbit of* $z_1 \in Z^b$ *is precompact and that the* ω-*limit set* $\omega(z_1)$ *of* z_1 *belongs to* E. *Suppose that, for any* $z_0 \in \omega(z_1)$, *the semigroup* $S(t) = S_{z_0}(t)$ *generated by the linear operator* $C_1 + DH_1(z_0)$ *satisfies the hypotheses* (H.2) *to* (H.4) *and moreover there exists* $\widetilde{z}_0 \in \omega(z_1)$ *such that* $\widetilde{\sigma}_0 = \sigma(S_{\widetilde{z}_0}(1)P_{0\widetilde{z}_0})$ *is either the empty set or contains only the point* 1 *and it is a simple eigenvalue of* $S_{\widetilde{z}_0}(1)$. *Then there is a unique point* $\varphi = \varphi_{z_1}$ *such that* $\omega(z_1) = \varphi_{z_1}$.

Comments and outline of the proof of Theorem 4.1 : In (Hale, Raugel (1992b)), this theorem was stated in a weaker way ; namely, the expression "there exists $\widetilde{z} \in \omega(z_1)$ such that" was replaced by "for any $\widetilde{z} \in \omega(z_1)$". However, the proof given there needs no changes in order to obtain the above mentioned result. P. Poláčik made this observation to the authors. This observation has been used by (Haraux, Poláčik (1992)) in the aboved mentioned application about convergence to nonnegative equilibrium points.

Outline of the proof of Theorem 4.1 *as given in* (Hale, Raugel (1992b)) : Since $\omega(z_1)$ is connected, it is only necessary to consider the case where $\omega(z_1)$ is not an isolated equilibrium point. In a neighbourhood of the point $\widetilde{z} \in \omega(z_1)$, the center manifold theory implies that $\omega(z_1)$ is a *local* C^1-manifold of dimension one. Using the properties of the center, center-stable and center-unstable manifolds of the equilibrium point $\widetilde{z}$, writing the system (4.2) in an adequate coordinates system and using several technical estimates, that generalize those of Aulbach (1983), we showed in (Hale, Raugel (1992b)) that the orbit through z_1 must intersect the strongly stable manifold of $\widetilde{z}$. Therefore, the solution converges to $\widetilde{z}$. This contradicts the supposition that $\omega(z_1)$ is not an isolated equilibrium point. P. Poláčik has pointed out to us that another proof, which uses more of the theory of invariant manifolds, is possible. ☐

4.2 Convergence in a reaction-diffusion equation on a thin domain over an arc of a curve.

4.2.1 The case of a simple thin domain over an arc of a curve.

We could state the convergence results given below on general thin domains over an arc of a curve (see (Hale, Raugel (1992b), Section 4.6)). The *same proof* (except for some additional notational complexities) *would hold*. Thus, for the sake of clarity, we consider the reaction-diffusion equation (3.2) on the model simple thin domain Q_ϵ given in Section 3.1.1 with *the restriction $n = 1$* ; that is, we assume, for simplicity, that $\Omega = (0, 1)$. Here the function f needs only to satisfy the condition (3.4). The limit problem $(3.6)_0$ is then reduced to the equation

$$(4.5) \quad \begin{cases} \text{(i)} & v_t - \frac{1}{g_0}(g_0 v_x)_x + \alpha v = -f(v) - G_0, \quad x \in (0, 1) \\ \text{(ii)} & v_x(0) = v_x(1) = 0 . \end{cases}$$

For any $v_0 \in H^1(\Omega)$, we consider the eigenvalue problem

$$(4.6) \quad \begin{cases} \text{(i)} & -\frac{1}{g_0}(g_0 v_x)_x + (\alpha + Df(v_0))v = \lambda v, \quad x \in (0, 1) \\ \text{(ii)} & v_x(0) = v_x(1) = 0 . \end{cases}$$

By the Sturm-Liouville theory, we know that the eigenvalues of (4.6) form a denumerable sequence of real numbers

$$(4.7) \quad \tilde{\lambda}_1(v_0) < \tilde{\lambda}_2(v_0) < \cdots < \tilde{\lambda}_n(v_0) \to +\infty$$

with each eigenvalue being simple. Thus, we can find a positive integer $N_0 \equiv N_0(v_0)$ and positive constants $B_1 \equiv B_1(v_0)$, $B_2 \equiv B_2(v_0)$, $\delta_0 \equiv \delta_0(v_0)$ such that

$$(4.8) \quad \begin{cases} \text{(i)} & \tilde{\lambda}_{N_0-1}(v_0) > \delta_0 , \\ \text{(ii)} & -B_1 < \tilde{\lambda}_1(v_0) < \tilde{\lambda}_2(v_0) < \cdots < \tilde{\lambda}_{N_0}(v_0) < B_2 , \\ \text{(iii)} & \min\{|\tilde{\lambda}_{j+1}(v_0) - \tilde{\lambda}_j(v_0)| : 1 \le j \le N_0 - 1\} > \delta_0 . \end{cases}$$

If u_0 is an element of $H^1(Q)$ or of X_ϵ, we consider now the eigenvalue problem

$$(4.9)_\epsilon \quad (A_\epsilon + Df(u_0))u_\epsilon = \lambda_\epsilon u_\epsilon .$$

The linear operator $A_\epsilon + Df(u_0)$ is selfadjoint and its inverse is compact. Therefore the spectrum $\sigma(A_\epsilon + Df(u_0))$ of $A_\epsilon + Df(u_0)$ is composed only of eigenvalues and the eigenvalues of $(4.9)_\epsilon$ form a denumerable sequence of real numbers

$$(4.10)_\epsilon \quad \tilde{\lambda}_{1,\epsilon}(u_0) < \tilde{\lambda}_{2,\epsilon}(u_0) \le \cdots \le \tilde{\lambda}_{n,\epsilon}(u_0) \to +\infty .$$

In particular, $\sigma(A_\epsilon + Df(u_0))$ or $\sigma(A_\epsilon + Df(u_0))\backslash\{0\}$ if zero is an eigenvalue is always bounded away from the imaginary axis.

Applying the results of Section 2.4 (see, in particular, the last remark of Section 2.4), one proves the following result.

PROPOSITION 4.2. — *For any $v_0 \in H^1(\Omega)$, for any positive integer n_0 and any positive constant δ, there are positive numbers $\eta_0 = \eta(v_0, n_0, \delta)$, $\epsilon_0 = \epsilon(v_0, n_0, \delta)$ such that, for $0 \leq \epsilon \leq \epsilon(v_0, n_0, \delta)$, for $u_0 \in B_{H^1(Q)}(v_0; \eta(v_0, n_0, \delta))$, the first n_0 eigenvalues of $A_\epsilon + Df(u_0)$ are simple, contained in $[-B_1(v_0), \infty)$ and satisfy*

$$(4.11) \qquad \max_{1 \leq j \leq n_0} |\widetilde{\lambda}_j(v_0) - \widetilde{\lambda}_{j,\epsilon}(u_0)| \leq c(v_0)(\epsilon + \eta(v_0, n_0, \delta)) \leq \delta,$$

where $c(v_0)$ is a positive constant depending only on v_0.

Proof: Let $v_0 \in H^1(\Omega)$, we choose a positive constant $r_0 > 1$, such that

$$(4.12) \qquad \max(||v_0||_{1,\Omega}, ||v_0||_{X_0}) \leq r_0 .$$

At first, one observes that there exist positive constants ℓ_0, b_0 such that, for $0 < \epsilon \leq \epsilon_0$, $u_0 \in B_{H^1(Q)}(0; 2r_0)$, $v_0 \in B_{X_0}(0; 2r_0)$, $u \in X_\epsilon$, $v \in X_0$,

$$(4.13) \qquad \begin{cases} (C_\epsilon(u_0)u, u)_{H_\epsilon} \geq b_0 ||u||^2_{X_\epsilon} , \\ (C_0(v_0)v, v)_{H_0} \geq b_0 ||v||^2_{X_0} , \end{cases}$$

where

$$(4.14) \qquad \begin{cases} C_\epsilon(u_0) = A_\epsilon + Df(u_0) + \ell_0 I \\ C_0(v_0) = A_0 + Df(v_0) + \ell_0 I \end{cases}$$

One also needs the following auxiliary result, proved in (Hale, Raugel (1992b)).

LEMMA 4.3. — *There are positive constants c_1 and ϵ_1 such that, for $0 < \epsilon \leq \epsilon_1$ and any $u_0 \in B_{H^1(Q)}(0; 2r_0)$, $v_0 \in B_{H^1(\Omega)}(0; 2r_0)$ and $h \in H_0$, we have, for $0 < \epsilon \leq \epsilon_1$,*

$$(4.15) \qquad ||C_\epsilon(u_0)^{-1}||_{\mathcal{L}(H_\epsilon; X_\epsilon)} + ||C_0(v_0)^{-1}||_{\mathcal{L}(H_0; X_0)} \leq c_1 ,$$

$$(4.16) \qquad ||(C_\epsilon(u_0)^{-1} - C_0(v_0)^{-1})h||_{X_\epsilon} \leq c_1(\epsilon + ||u_0 - v_0||_{H^1(Q)})||h||_{H_0} .$$

The proposition 4.2 is now a direct consequence of the last remark of Section 2.4, with the spaces H_ϵ, H_0, X_ϵ, X_0 as described in Section 3.1.1 and with $B_0(\mu_0) = C_0(v_0)^{-1}$, $B_\epsilon(\mu) = C_\epsilon(u_0)^{-1}$, $\mu = u_0$, $Z = H^1(Q)$.

From Proposition 4.2, we at once deduce the following result.

PROPOSITION 4.4. — *For any bounded set $B \subset H^1(Q)$, there exists a real number $\epsilon_0(B) > 0$ such that, for $0 < \epsilon \leq \epsilon_0(B)$, the null space of $A_\epsilon + Df(u_0)$ has dimension no more than one for any equilibrium point $u_0 \in B$ of $(3.6)_\epsilon$ or $(3.11)_\epsilon$.*

Proof: Let E_ϵ be the set of equilibrium points of $(3.6)_\epsilon$. Since $E_\epsilon \cap B$ is bounded in $H^1(Q)$ and that the function f satisfies (3.4), there exist positive constants $K_0(B)$, $K_1(B)$ such that, for $0 \le \epsilon \le \epsilon_0$,

$$(4.17) \qquad \begin{cases} \text{(i)} \quad \| -f(u_0^\epsilon) - G_\epsilon \|_{H_\epsilon} \le K_0(B) \\ \text{(ii)} \quad \| u_0^\epsilon \|_{X_\epsilon} \le K_1(B) \end{cases}, \quad \forall u_0^\epsilon \in E_\epsilon \cap B \ .$$

From the regularity results in the Appendix of (Hale, Raugel (1992c)) and the inequalities (4.17), we deduce that there exists a positive constant $K_2(B)$ such that, for $0 \le \epsilon \le \epsilon_0$,

$$(4.18) \qquad \| u_0^\epsilon \|_{2,Q} \le c \| u_0^\epsilon \|_{X_\epsilon^2} \le K_2(B), \quad \forall u_0^\epsilon \in E_\epsilon \cap B \ .$$

We consider the set $S = \bigcup_{0 \le \epsilon \le \epsilon_0} (E_\epsilon \cap B)$. The set S is bounded in $H^2(Q)$ and thus the set $S_0 = \{ Mu \ : \ u \in S \}$ is compact in $H^1(\Omega)$. We now apply the proposition 4.2 with $v_0 \in S_0$, and use the property that

$$(4.19) \qquad \| u_0^\epsilon - M u_0^\epsilon \|_{1,Q} \le C\epsilon \| u_0^\epsilon \|_{X_\epsilon^2} \le C\epsilon K_2(B), \quad \forall u_0^\epsilon \in S \ .$$

(See Lemma 3.1).

From Theorem 4.1 and Proposition 4.4, we deduce the following two results.

THEOREM 4.5. — *Assume that $n = 1$ and the function f satisfies the condition* (3.4). *For any bounded set B in $H^1(Q)$, there exists a positive constant $\epsilon_0 = \epsilon_0(B)$ such that, for $0 < \epsilon \le \epsilon_0$, any bounded orbit of $(3.6)_\epsilon$, whose ω-limit set belongs to B, converges to a single equilibrium point.*

If we assume that the dissipative condition (3.3) also holds, then each positive orbit is bounded and the bound is "uniform in ϵ". Thus we can state the result in a simpler way.

THEOREM 4.6. — *Assume that $n = 1$ and the function f satisfies the conditions* (3.3), (3.4). *Then there exists a positive constant ϵ_1 such that, for $0 < \epsilon \le \epsilon_1$, the ω-limit set of any positive orbit of $(3.6)_\epsilon$ is a single equilibrium point.*

REMARK 4.1. — If the function f satisfies the conditions (3.3), (3.4), then, as we saw in Section 3.1.1, the system $(3.6)_\epsilon$ has a global attractor $\mathcal{A}_\epsilon$ in X_ϵ and the attractors $\mathcal{A}_\epsilon$, $\epsilon \ge 0$, are upper-semicontinuous at $\epsilon = 0$. Using the compactness of $\mathcal{A}_0$, the upper-semicontinuity of the attractors $\mathcal{A}_\epsilon$ as well as Proposition 4.2, we can show that there exists a positive number ϵ_1 such that, for $0 < \epsilon \le \epsilon_1$, the null space of $A_\epsilon + Df(u_0)$ has dimension no more than one for any u_0 in E_ϵ. This type of proof, which does not use any regularity result in the space $H^2(Q)$, allows us to generalize the theorem 4.6 to the case of the mixed boundary conditions (3.2)(iibis) or to the case of non-simple thin domains.

4.2.2 A reaction-diffusion equation on the thin L-shaped domain described in Section 3.1.2.

We now consider the equation (3.61) on the thin L-shaped domain $Q_\epsilon = Q_\epsilon^1 \cup Q_\epsilon^2$ described in Section 3.1.2. The first step is to show the simplicity of the eigenvalues of the limit equation. We remark, at first, that, for any $v_0 \in X_0$, the spectrum of

$A_0 + D\widehat{f}(v_0)$ consists only of simple real eigenvalues. Indeed, since $A_0 + D\widehat{f}(v_0)$ is a self-adjoint operator with compact resolvent, the spectrum $\sigma(A_0 + D\widehat{f}(v_0))$ consists only of real eigenvalues. Let us show that they are all simple. We set $v_0 = (v_{01}, v_{02})$. Let λ_0 be an eigenvalue of $A_0 + D\widehat{f}(v_0)$. For $i = 1, 2$, the space of solutions of the equation

$$(4.20) \qquad -\frac{1}{g_i}(g_i v_{ix_i})_{x_i} + Df(v_{0i})v_i = \lambda_0 v_i, \quad v_i(1) = 0$$

is of dimension 1, given say, by $\mathbb{R}\varphi_i$, where $\varphi_i \in H^2((0,1))$ and $\varphi_i^2 + \varphi_{ix_i}^2 \neq 0$. We note that λ_0 is an eigenvalue of $A_0 + D\widehat{f}(v_0)$ if and only if the system

$$(4.21) \qquad \begin{cases} \mu_1\varphi_1(0) - \mu_2\varphi_2(0) = 0 \\ \mu_1 g_1(0)\varphi_{1x_1}(0) + \mu_2 g_2(0)\varphi_{2x_2}(0) = 0 \end{cases}$$

has a solution $(\mu_1, \mu_2) \neq (0,0)$. Since $g_i(0) \neq 0$ and $\varphi_i^2 + \varphi_{ix_i}^2 \neq 0$, the space of the solutions of (4.21) is at most one dimensional. Since λ_0 is an eigenvalue, the dimension is 1.

We now argue as in Section 4.2.1 to show that Proposition 4.2 still holds for any v_0 in X_0. Then, using the upper-semicontinuity of the attractors $\mathcal{A}_\epsilon$, like in Remark 4.1 or in (Hale, Raugel (1992b)), we show that, there exists a positive number ϵ^* such that, for $0 < \epsilon \leq \epsilon^*$, if u_0 is an equilibrium point of $(3.63)_\epsilon$, then the null space of $A_\epsilon + D\widehat{f}(u_0)$ has dimension no more than one. Finally, applying the theorem 4.1, we obtain the convergence result below.

THEOREM 4.7. — *The ω-limit set of any orbit of $(3.63)_0$ is a single equilibrium point. Furthermore, there exists a positive number ϵ_1 such that, for $0 < \epsilon \leq \epsilon_1$, the ω-limit set of any orbit of $(3.63)_\epsilon$ is a single equilibrium point.*

4.3 A linearly damped hyperbolic equation on a thin domain.

As in the section 4.2, we could state the convergence results given below on general thin domains over an arc of a curve (see (Hale, Raugel (1992b)), Section 5.3.2)). But, again for sake of clarity, we consider the linearly damped wave equation (3.80) with $\beta > 0$ on the model simple thin domain Q_ϵ given in Section 3.1.1 with *the restriction $n = 1$*. In (Hale, Raugel (1992b)), we have proved the following results.

THEOREM 4.8. — *Assume that $n = 1$.*

(i) *If f satisfies the condition (3.4), then, for any bounded set B in $H^1(Q) \times L^2(Q)$, there exists a positive number $\epsilon_0(B)$ such that, for $0 \leq \epsilon \leq \epsilon_0(B)$, any orbit of $(3.82)_\epsilon$, whose ω-limit set is in B, converges to a single equilibrium point.*

(ii) *If f satisfies the conditions* (3.3) *and* (3.4), *then there exists a positive constant ϵ_1 such that, for $0 < \epsilon \leq \epsilon_1$, the ω-limit set of any positive orbit of* $(3.82)_\epsilon$ *is a single equilibrium point.*

Theorem 4.8 is a direct consequence of Theorem 4.1 and the Lemma 4.9 below, proved in (Hale, Raugel (1992b)). We recall that the set of equilibrium points of $(3.82)_\epsilon$ is given by $E_\epsilon \times \{0\}$ where E_ϵ is the set of solutions of

$$(4.22)_\epsilon \qquad\qquad A_\epsilon \varphi + f(\varphi) + G_\epsilon = 0 \,,$$

that is, the set of equilibrium points of $(3.11)_\epsilon$.

Let $(u_0^\epsilon, 0)$ be an equilibrium point of $(3.82)_\epsilon$ and let us consider the linearized equation around $(u_0^\epsilon, 0)$:

$$(4.23) \qquad\qquad u_{tt} + \beta u_t + A_\epsilon u + Df(u_0^\epsilon)u = 0 \,.$$

This equation generates a C^0-group $S_\epsilon(u_0^\epsilon, t)$ on Y_ϵ^1 (and also Y_ϵ^2). We denote by $\sigma(S_\epsilon(u_0^\epsilon, 1))$ the spectrum of $S_\epsilon(u_0^\epsilon, 1)$ and by $\sigma_0(S_\epsilon(u_0^\epsilon, 1))$ its intersection with the unit circle with center 0 in $\mathbb{C}$. Since $\beta > 0$, there exists a positive number ν such that the essential spectrum of $S_\epsilon(u_0^\epsilon, 1)$ is contained in the disk $\{\xi \in \mathbb{C} \, : \, |\xi| \leq e^{-\nu}\}$.

LEMMA 4.9. — *For any bounded set $B \subset H^1(Q) \times L^2(Q)$, there exists a positive number $\epsilon_0^* = \epsilon_0^*(B)$ such that, for $0 \leq \epsilon \leq \epsilon_0^*$, if $(u_0^\epsilon, 0)$ belongs to $(E_\epsilon \times \{0\}) \cap B$, then $\sigma_0(S_\epsilon(u_0^\epsilon, 1))$ is either the empty set or contains only the point 1 and, if 1 is contained in $\sigma_0(S_\epsilon(u_0^\epsilon, 1))$, it is a simple eigenvalue of $S_\epsilon(u_0^\epsilon, 1)$.*

REMARKS 4.2. — 1. We can also consider the damped wave equation

$$(4.24) \qquad \begin{cases} u_{tt} + \beta u_t - \Delta u = -f(u) - G & \text{in} \quad Q_\epsilon \\[2mm] u = 0 & \text{in} \quad \Gamma_\epsilon \\[2mm] \dfrac{\partial u}{\partial \nu_\epsilon} = 0 & \text{in} \quad \partial Q_\epsilon \backslash \Gamma_\epsilon \end{cases}$$

on the L-shaped domain Q_ϵ considered in Section 3.1.2. We keep the hypotheses on f, G made there. Then the statement (ii) of Theorem 4.8 about convergence still holds. For more details, see (Hale, Raugel (1993b)).

2. We can also consider strongly damped wave equations on thin domains. The above convergence results still hold (see (Hale, Raugel (1992b))).

4.4 A locally damped hyperbolic equation on a thin domain.

We keep the model simple thin domain Q_ϵ described in Section 3.1.1. And we consider the equation

$$(4.25) \qquad \begin{cases} u_{tt} + \widetilde{\beta}(\epsilon, \widetilde{x}, \widetilde{y}) u_t - \Delta u + \alpha u = -f(u) - G & \text{in} \quad Q_\epsilon \\[2mm] \dfrac{\partial u}{\partial \nu_\epsilon} = 0 & \text{in} \quad \partial Q_\epsilon \,, \end{cases}$$

where the functions f and G satisfy the conditions given in Section 3.1.1. We assume that $\tilde{\beta}_\epsilon(\tilde{x}, \tilde{y}) = \tilde{\beta}(\epsilon, \tilde{x}, \tilde{y})$ is a continuous function in $(\tilde{x}, \tilde{y}) \in Q_\epsilon$, $\epsilon \geq 0$ satisfying

$$(4.26) \quad \begin{cases} \text{(i)} \quad \tilde{\beta}_\epsilon(\tilde{x}, \tilde{y}) \geq 0, \ \forall (\tilde{x}, \tilde{y}) \in Q_\epsilon, \ \epsilon \geq 0 \\ \text{(ii)} \quad \tilde{\omega}_\epsilon \equiv \text{support of } \tilde{\beta}_\epsilon \text{ contains a neighbourhood of } \partial Q_\epsilon, \ \epsilon > 0 \ . \end{cases}$$

An example of such a function $\tilde{\beta}$ is the function

$$\tilde{\beta}(\epsilon, \tilde{x}, \tilde{y}) = \beta_0(\tilde{x}) + \epsilon^p \tilde{\beta}_1(\epsilon, \tilde{x}, \tilde{y})$$

where $p > 0$ is a constant, $\beta_0(\tilde{x}) \geq 0$ (and may actually be equal to zero everywhere), and the function $\tilde{\beta}_{1,\epsilon}$ has a support containing a neighbourhood of ∂Q_ϵ.

Let $S_\epsilon(t) : H^1(Q_\epsilon) \times L^2(Q_\epsilon) \rightarrow H^1(Q_\epsilon) \times L^2(Q_\epsilon)$ be the linear semigroup generated by the linear equation

$$(4.27) \quad \begin{cases} u_{tt} + \tilde{\beta}_\epsilon(\tilde{x}, \tilde{y})u_t - \Delta u + \alpha u = 0 & \text{in } Q_\epsilon, \\ \dfrac{\partial u}{\partial \nu_\epsilon} = 0 & \text{in } \partial Q_\epsilon \ . \end{cases}$$

By a result of Zuazua (1990), there exist two positive constants $\tilde{k}_\epsilon, \tilde{\gamma}_\epsilon$ such that

$$(4.28) \qquad \|S_\epsilon(t)\|_{\mathcal{L}(H^1(Q_\epsilon) \times L^2(Q_\epsilon); H^1(Q_\epsilon) \times L^2(Q_\epsilon))} \leq \tilde{k}_\epsilon e^{-\tilde{\gamma}_\epsilon t}, \ t \geq 0 \ .$$

If we make the usual change of variables to the canonical domain $Q = \Omega \times (0, 1)$, then the system (4.25) becomes

$$(4.29)_\epsilon \qquad u_{tt} + \beta_\epsilon u_t + A_\epsilon u = -f(u) - G_\epsilon$$

where $\beta_\epsilon(x, y) = \tilde{\beta}(\epsilon, x, g(x, \epsilon)y)$. We denote by ω_ϵ the support of β_ϵ. The equation $(4.29)_\epsilon$ can be written as the following system

$$(4.30)_\epsilon \qquad Z_t = C_\epsilon Z - \tilde{f}(Z) - \tilde{G}_\epsilon,$$

where

$$(4.31)_\epsilon \quad Z = \begin{pmatrix} u \\ u_t \end{pmatrix}, \ C_\epsilon = \begin{pmatrix} 0 & I \\ -A_\epsilon & -\beta_\epsilon I \end{pmatrix}, \ \tilde{f}(Z) = \begin{pmatrix} 0 \\ f(u) \end{pmatrix}, \ \tilde{G}_\epsilon = \begin{pmatrix} 0 \\ G_\epsilon \end{pmatrix} \ .$$

From (4.28), we deduce that

$$(4.32) \qquad \|e^{C_\epsilon(\cdot)t}\|_{\mathcal{L}(Y_\epsilon; Y_\epsilon)} \leq \tilde{k}_\epsilon e^{-\tilde{\gamma}_\epsilon t}, \ t \geq 0 \ .$$

Let $T_\epsilon(t)$ be the semigroup associated with $(4.29)_\epsilon$ or $(4.30)_\epsilon$. Using the property (4.32), one shows that $T_\epsilon(t)$ is an α-contraction. Therefore, if $T_\epsilon(t)$ is point dissipative, then $T_\epsilon(t)$ has a global attractor (see (Hale (1988))). Using a "unique continuation property" of Ruiz, which holds if $\overline{\omega}_\epsilon \supset \partial Q$, we show that $T_\epsilon(t)$ is a gradient system. (For the details, see (Hale, Raugel (1992d))).

THEOREM 4.10. — *If* (4.26) *holds and* f *satisfies the conditions* (3.3) *and* (3.4) *with* $\gamma < 1$ *if* $n = 2$, *then the semigroup* $T_\epsilon(t)$, *defined by* (4.29)$_\epsilon$, *is a gradient system and has a global attractor* $\mathcal{A}_\epsilon$.

REMARK 4.3. — Using another method of proof, Feireisl and Zuazua (1992) showed the existence of a global attractor $\mathcal{A}_\epsilon$, even in the case $\gamma = 1$ if $n = 2$. Using then a unique continuation property of Ruiz, we show as above that $T_\epsilon(t)$ is a gradient system.

In *the case* $n = 1$, we can prove convergence of solutions as stated below.

THEOREM 4.11. — *Assume that* $n = 1$ *and* (4.26) *holds.*

(i) *If* f *satisfies* (3.4), *then, for any bounded set* B *in* $H^1(Q) \times L^2(Q)$, *there exists* $\epsilon_0 = \epsilon_0(B) > 0$ *such that, for* $0 < \epsilon \le \epsilon_0(B)$, *any orbit of* (4.29)$_\epsilon$ *whose* ω-*limit set belongs to* B *must converge to a single equilibrium point.*

(ii) *If* f *satisfies* (3.3) *and* (3.4), *then there exists* $\epsilon_1 > 0$ *such that, for* $0 < \epsilon \le \epsilon_1$, *the* ω-*limit set of any solution of* (4.29)$_\epsilon$ *is a single equilibrium point.*

REMARK 4.4. — With the weak hypotheses that we made in Theorem 4.11, we cannot assert for $\epsilon = 0$ the convergence of an orbit to an equilibrium point since β_0 could be zero and the limit equation would be conservative.

Proof of Theorem 4.11 : We will show the statement (ii). We apply the abstract theorem 4.1. If $z_{0\epsilon} = (\varphi_{0\epsilon}, 0)$ is an equilibrium point of (4.29)$_\epsilon$, then the linear variational equation around $(\varphi_{0\epsilon}, 0)$ is

$$(4.33)_\epsilon \qquad u_{tt} + \beta_\epsilon u_t + A_\epsilon u + Df(\varphi_{0\epsilon})u = 0 .$$

Let $\Sigma_\epsilon(t)$ be the linear semigroup generated by (4.33)$_\epsilon$. As in the proof of Theorem 4.10, $\Sigma_\epsilon(t)$ is an α-contraction with $r_e(\Sigma_\epsilon(1)) < 1$. To complete the proof of statement (ii), we only need to show that the spectrum of $\Sigma_\epsilon(1)$ with modulus 1 is either empty or contains only the point 1 and then it is a simple eigenvalue. The eigenvalues of $\Sigma_\epsilon(1)$ with modulus 1 correspond to the eigenvalues of the differential operator

$$D_\epsilon = \begin{pmatrix} 0 & I \\ -A_\epsilon - Df(\varphi_{0\epsilon}) & -\beta_\epsilon I \end{pmatrix}$$

on the imaginary axis. The eigenvalues μ of the operator D_ϵ and the corresponding eigenfunctions (φ, ψ) satisfy

$$\begin{cases} \psi = \mu\varphi \\ (-A_\epsilon - Df(\varphi_{0\epsilon}))\varphi - \beta_\epsilon\psi = \mu\psi , \end{cases}$$

i.e.

$$(4.34) \qquad \begin{cases} \psi = \mu\varphi \\ (-A_\epsilon - Df(\varphi_{0\epsilon}))\varphi - \beta_\epsilon\mu\varphi = \mu^2\varphi . \end{cases}$$

If $\mu = 0$, then $(-A_\epsilon - Df(\varphi_{0\epsilon}))\varphi = 0$. By Proposition 4.4, there exists $\epsilon_1 > 0$ such that, for $0 < \epsilon \le \epsilon_1$, the kernel of $(-A_\epsilon - Df(\varphi_{0\epsilon}))$ has dimension ≤ 1. Let us show that we have no generalized eigenfunction of $\mu = 0$. If (θ_1, θ_2) is a generalized eigenfunction of 0, we have

$$\begin{cases} \theta_2 = \varphi \\ (-A_\epsilon - Df(\varphi_{0\epsilon}))\theta_1 - \beta_\epsilon \theta_2 = 0 \end{cases}$$

where $\varphi \in \mathrm{Ker}(-A_\epsilon - Df(\varphi_{0\epsilon}))$. Therefore

$$(-A_\epsilon - Df(\varphi_{0\epsilon}))\theta_1 - \beta_\epsilon \varphi = 0 \ .$$

This equation can have a solution only if

$$\int_0^1 \frac{g}{\epsilon} \beta_\epsilon(x,y) \varphi^2(x,y)\,dx\,dy = 0 \ .$$

This implies that $\beta_\epsilon(x,y)\varphi(x,y) \equiv 0$. Therefore, θ_1 would be a solution of

$$(-A_\epsilon - Df(\varphi_{0\epsilon}))\theta_1 = 0$$

that is, $\theta_1 = \lambda\varphi$, $\lambda \neq 0$. As a consequence, φ must be a solution of the equation

$$\begin{cases} (-A_\epsilon - f'(\varphi_{0\epsilon}))\varphi = 0 \\ \varphi = 0 \text{ on } \omega_\epsilon \\ \dfrac{\partial\varphi}{\partial\nu_{B_\epsilon}} = 0 \text{ on } \partial Q \ . \end{cases}$$

Since φ is a solution of an elliptic equation and $\varphi = 0$ on ω_ϵ, we conclude that $\varphi \equiv 0$ and thus $\mu = 0$ is a simple eigenvalue.

It remains to show that μ cannot be on the imaginary axis. If $(\varphi, \mu\varphi)$ is an eigenfunction corresponding to an eigenvalue μ, then

$$\mu^2 \|\varphi\|_{H_\epsilon}^2 + \mu m_0 + m_1 = 0 \ ,$$

where $m_0 = \int_Q \frac{g}{\epsilon} \beta_\epsilon \varphi\overline{\varphi}\,dx$, $m_1 = \int_Q \frac{g}{\epsilon}(A_\epsilon + Df(\varphi_{0\epsilon}))\varphi\overline{\varphi}\,dx$ are real numbers. If μ is purely imaginary, $m_0 = 0$, which implies that $\varphi = 0$ on the support ω_ϵ of β_ϵ. This implies that

(4.35) $$\begin{cases} (A_\epsilon + Df(\varphi_{0\epsilon}) + \mu^2)(\mathrm{Re}\varphi) = 0 \ , \\ \mathrm{Re}\varphi = 0 \text{ on } \omega_\epsilon \ , \ \dfrac{\partial(\mathrm{Re}\varphi)}{\partial\nu_{B_\epsilon}} = 0 \text{ on } \partial Q \ . \end{cases}$$

As above, this implies that $\mathrm{Re}\varphi = 0$. Since the function $\mathrm{Im}\varphi$ also satisfies (4.35), we have: $\mathrm{Im}\varphi = 0$. This completes the proof of the theorem.

$\Box$

Chapter 5

Global existence of solutions.

5.1 The Navier–Stokes equations: classical results and statements of the results on thin 3D domains.

The Navier–Stokes equations describe the time evolution of solutions of a mathematical model of viscous incompressible fluid flows. One of the main concerns has always been the existence of regular (or strong) global solutions. During the last few years, much attention has also been focused on the study of attractors for the Navier–Stokes equations. Let us recall some classical results.

The Navier–Stokes equations on a bounded region $\Omega \subset \mathbb{R}^n$, $n = 2, 3$, are given by

$$(5.1) \qquad \begin{cases} U_t - \nu \Delta U + (U.\nabla)U + \nabla P = F \\ \nabla.U = 0 \end{cases}$$

with various boundary conditions. Below we will treat the case where $\Omega = Q_\epsilon$ is the thin 3-dimensional domain $Q_\epsilon = Q_2 \times (0, \epsilon)$, where ϵ is a small positive parameter, and Q_2 is a smooth bounded region in $\mathbb{R}^2$ or $Q_2 = (0, \ell_1) \times (0, \ell_2)$, with $0 < \epsilon \le \ell_2 \le \ell_1$. When $Q_2 = (0, \ell_1) \times (0, \ell_2)$, (5.1) is supplemented with the periodic and zero mean conditions

$$(5.2) \qquad \begin{cases} \text{(i)} \ U(y + \ell_i e_i, t) = U(y, t), \ i = 1, 2, \ U(y + \epsilon e_3, t) = U(y, t) \\ \text{(ii)} \ \int_{Q_\epsilon} U dy = 0 \end{cases}$$

where $\{e_1, e_2, e_3\}$ is the natural basis of $\mathbb{R}^3$. In this case, one of course requires that F and the initial data satisfy the condition (5.2)(ii). It then follows that any solution U of (5.1) with $U(0) = U_0$ will satisfy $\int_{Q_\epsilon} U(t)dy = 0$ for $t > 0$. In the case where Q_2 is a *smooth* bounded domain in $\mathbb{R}^2$, we are interested in the following boundary conditions, here called *mixed conditions*,

$$(5.3) \qquad \begin{cases} \text{(i)} \ U(y + \epsilon e_3, t) = U(y, t) \\ \text{(ii)} \ U(y, t) = 0 \text{ on } \partial Q_2 \times (0, \epsilon) . \end{cases}$$

For the sake of simplicity, we assume here that F satisfies

$$(5.4) \qquad F(\cdot) \in L^\infty(0, \infty : L^2(\Omega))$$

and also satisfies (5.2)(ii) if $Q_2 = (0, \ell_1) \times (0, \ell_2)$. In all these cases, let V be the space of divergence-free vector fields U in $H^1(\Omega)^n$ satisfying the respective boundary conditions (or (5.2)), let H be the closure of V in $L^2(\Omega)^n$ and let $\mathbb{P}_n$

denote the orthogonal projection of $L^2(\Omega)^n$ onto H. The Navier–Stokes equations can be written in the abstract form

$$(5.5) \qquad\qquad U_t + \nu AU + B(U,U) = \mathbb{P}_n F,$$

where $AU = -\mathbb{P}_n \Delta U$, $B(U_1, U_2) = \mathbb{P}_n(U_1.\nabla)U_2$.

If U_0 belongs to H, then a weak (Leray) solution $U(t)$ on the time interval $[0,T]$ is a function in $L^2(0,T;V) \cap L^\infty(0,T;H) \cap C^0_w([0,T];H)$ satisfying

$$(5.6) \quad <U',w> + <\nu AU, w> + <B(U,U), w> = <\mathbb{P}_n F, w> \quad a.e.in\ t,\ \forall w \in V,$$

$$(5.7) \qquad\qquad\qquad U(0) = U_0$$

and the energy inequality

$$(5.8) \quad \frac{1}{2}\|U(t)\|_H^2 + \nu \int_{t_0}^t \|U(s)\|_V^2 ds \le \frac{1}{2}\|U(t_0)\|_H^2 + \int_{t_0}^t | <\mathbb{P}_n F, U(s)> |ds,$$

for $t_0 = 0$ and for almost all $0 \le t_0 \le t$ in $[0,T]$, where $< \cdot, \cdot >$ is the duality pairing between V and V'. (Recall that $C^0_w([0,T];H)$ is the subspace of $L^\infty(0,T;H)$ consisting of functions which are weakly continuous, that is, for each $h \in H$, the mapping $t \mapsto < U(t), h >$ is a continuous mapping. In particular, (5.7) is taken in this sense).

If $U_0 \in V$, by a *regular* or *strong* solution U on the time interval $[0,T]$, we mean a function $U(t) \in L^2(0,T;D(A)) \cap L^\infty(0,T;V) \cap C^0([0,T];V)$ satisfying (5.6) and (5.7). Remark that the strong solutions satisfy the energy inequality (5.8).

For the two-dimensional Navier–Stokes equations, it is known that, for any F satisfying (5.4) and $U_0 \in H$, there exists a unique weak solution $U(\cdot)$ on $[0,\infty)$ satisfying $U(0) = U_0$, $U(t) \in L^2(0,\tau;V) \cap L^\infty(0,\infty;H) \cap C^0_w([0,\infty);H)$, for every $\tau > 0$ (and $U_t \in L^2(0,\tau;V')$), see Lions and Prodi (1959). Furthermore, if $U_0 \in V$, there exists a unique global strong solution

$$U(t) \in L^2(0,\tau;D(A)) \cap L^\infty(0,\infty;V) \cap C^0([0,\infty);H), \text{ for every } \tau > 0$$

(and $U_t \in L^2(0,\tau;H)$) (that is, the weak solution becomes a strong solution). Moreover, there exists a positive constant L_1, independent of $U_0 \in V$ such that

$$(5.9) \qquad\qquad \limsup_{t \to +\infty} \|U(t)\|_V \le L_1 .$$

If, in addition, F belongs to $W^{1,\infty}([0,\infty);L^2(\Omega))$ or $L^\infty(0,\infty;V)$, then $U(t) \in C^0((0,\infty);D(A))$ and there is a positive constant L_2, independent of $U_0 \in V$ such that

$$(5.10) \qquad\qquad \limsup_{t \to +\infty} \|AU(t)\|_H \le L_2 .$$

(For all these classical results, see, for instance, Ladyzhenskaya (1969), Teman (1977, 1983, 1988) and Constantin and Foias (1988)). As a consequence of (5.9) and (5.10), one shows that, for instance, when F is time-independent, the nonlinear semigroup

$S(t)$ acting on H (or also on V) defined by the 2DNS has a global attractor $\mathcal{A}$ in H (or in V), bounded in $D(A)$ (see, for instance, Ladyzhenskaya (1972)). The attractor $\mathcal{A}$ is also the global attractor of the restricition of $S(t)$ to $H^2(\Omega) \cap V$. Furthermore, this global attractor $\mathcal{A}$ has finite dimension (see, for instance, Mallet-Paret (1976) and Foias and Teman (1979)). If F is time varying, but has some compactness property, one can still show that the 2DNS has a global attractor (see Raugel and Sell (1993a)). For the existence of inertial forms, see Kwak (1992a, 1992b).

For the three-dimensional Navier–Stokes equations, the situation is unfortunately not so good. For any F satisfying (5.4) and $U_0 \in H$, there exists *at least* a weak solution

$$U(t) \in L^2(0, \tau; V) \cap C_w^0([0, \infty); H) \cap L^\infty(0, \infty; H)$$

and $U_t \in L^{4/3}(0, \tau; V')$ for every $\tau > 0$. Furthermore, if U_0 belongs to V, there exists a time $T = T(F, U_0)$, $0 < T \leq \infty$, such that (5.5) has a unique strong or regular solution

$$U(t) \in L^2(0, \tau; D(A)) \cap C^0([0, T]; V) \cap L^\infty(0, T; V)$$

and $U_t \in L^2(0, \tau; H)$ for every finite time τ, $0 < \tau \leq T$. Moreover, if the data $U_0 \in V$ and F are small, then one has $T = \infty$, i.e. (5.5) has a global strong solution for small data. A first question arises: how large can be these data?

For all these results we again refer to Ladyzhenskaya (1969), Temam (1977, 1983, 1988) and Constantin and Foias (1988).

When F is time independent, by lack of uniqueness of the weak solutions of (5.5), one can no longer define a semigroup $S(t)$ on H. However, as in Foias and Temam (1987), one can construct a compact set $\mathcal{X}$ in H_{weak}, called weak global attractor of (5.5). Let $\mathcal{X}$ denote the set of those $U_0 \in H$ for which there exists a weak solution $U(t)$ of (5.5) on $(-\infty, \infty)$ that is bounded in H on $(-\infty, \infty)$, such that $U(0) = U_0$. This set $\mathcal{X}$ is not empty, is invariant in some sense and is compact in H_{weak}. Furthermore, for every weak solution $U(t)$ of (5.5) in $(0, \infty)$, we have

$$U(t) \to \mathcal{X} \text{ in } H_{\text{weak}} \text{ as } t \to +\infty .$$

A second question arises: Is $\mathcal{X}$ contained and bounded in V?

For the thin domains Q_ϵ above, one can show the following results (for more details see Raugel and Sell (1992, 1993a, 1993b)).

THEOREM 5.1. — *Consider the 3DNS equations* (5.5) *on the thin domain Q_ϵ with the boundary conditions* (5.2) *or* (5.3). *There exists a positive number $\epsilon_0 = \epsilon_0(\nu)$ such that, for every ϵ, $0 < \epsilon \leq \epsilon_0$, there are large sets $\mathcal{R}(\epsilon)$ and $\mathcal{S}(\epsilon)$ with*

$$\mathcal{R}(\epsilon) \subset V , \quad \mathcal{S}(\epsilon) \in L^\infty(0, \infty; L^2(Q_\epsilon))$$

such that, if $U_0 \in \mathcal{R}(\epsilon)$ and $F \in \mathcal{S}(\epsilon)$, then (5.5) *has a (unique) global strong solution $U(t)$ on $[0, \infty)$. Furthermore there exists a positive constant L_1 (which depends on F only) such that*

$$\limsup_{t \to \infty} \|U(t)\|_V \leq L_1 .$$

As ϵ goes to zero, the diameters of $\mathcal{R}(\epsilon)$ and $\mathcal{S}(\epsilon)$ go to infinity in some sense.

In the case of periodic boundary conditions in all the variables, one can choose the sets $\mathcal{R}(\epsilon)$ and $\mathcal{S}(\epsilon)$ much larger than in the other case.

Assume that F belongs to $\mathcal{S}(\epsilon)$ and is time-independent. Then, thanks to Theorem 5.1, one can define a C^0 semigroup $S_\epsilon(t)$ on $\mathcal{R}(\epsilon)$ and show that $S_\epsilon(t)$ has a local attractor $\mathcal{A}_\epsilon \equiv \mathcal{A}_\epsilon(F)$ attracting every bounded set contained in $\bigcup_{t \geq 0} S_\epsilon(t) \mathcal{R}(\epsilon)$. Of course, $\mathcal{A}_\epsilon$ is compact in $H^2(Q_\epsilon) \cap V$. And one proves the following result.

THEOREM 5.2. — *Under the hypotheses of the theorem 5.1, the local attractor $\mathcal{A}_\epsilon$ is actually the global attractor for all the weak solutions of* (5.5), *provided F belongs to $\mathcal{S}(\epsilon)$. In particular $\mathcal{A}_\epsilon$ coincides with the weak attractor $\mathcal{X}_\epsilon$ of* (5.5).

So we have answered the above two questions in the case of the thin 3D domain Q_ϵ. We can also compare the attractor $\mathcal{A}_\epsilon$ of (5.5) with the global attractor of a reduced equation (see Remarks 5.2 below).

Similar results seem possible for more general thin domains (other geometries ; see the remarks below about the Euler equations) and other boundary conditions (Neumann boundary conditions). These questions are still under investigation.

Because of the theorems 5.1 and 5.2, it is of interest to study more precisely the long time behaviour of the strong solutions of the above 3D Navier–Stokes equations. The existence of inertial forms has been studied very recently by Kwak, Sell and Shao (1994). See also (Pliss and Sell (1993)) for the other results about the 3D Navier–Stokes equations on thin domains.

In the next section, we give some details of the proofs in the purely periodic boundary condition case (5.2). We shall explain the main steps in the proof (short time and long time arguments). All the details of the proofs in the cases of the boundary conditions (5.2) or (5.3) have been given in Raugel and Sell (1992, 1993a). In the next section, we follow the lines of the proofs as written in Raugel and Sell (1992). Let us point out that, if we apply a reduction principle as explained in (Kirchgässner and Raugel (1994a, 1994b)), then we can write the proofs below in a much more elegant way.

5.2 The Navier–Stokes equations on Q_ϵ with periodic boundary conditions.

5.2.1 Notation. As earlier, we make the change of variables

$$(y_1, y_2, y_3) \mapsto (x_1, x_2, x_3) \text{ where } x_i = y_i, \ i = 1, 2, \ x_3 = \epsilon^{-1} y_3$$

mapping Q_ϵ onto $Q_3 = Q_2 \times (0, 1)$.

Setting $U = J_\epsilon u$ with

$$U(y_1, y_2, y_3) = u(y_1, y_2, \epsilon^{-1} y_3) = u(x_1, x_2, x_3),$$

one transforms the Navier–Stokes equations (5.1) on Q_ϵ into the system

(5.11)
$$\begin{cases} u_t - \nu \Delta_\epsilon u + (u \cdot \nabla_\epsilon) u + \nabla_\epsilon p = f \\ \nabla_\epsilon \cdot u = 0 \end{cases}$$

where $\nabla_\epsilon = (\partial_1, \partial_2, \epsilon^{-1}\partial_3)$, $\Delta_\epsilon = \partial_1^2 + \partial_2^2 + \epsilon^{-2}\partial_3^2$, $\partial_i = \frac{\partial}{\partial x_i}$, $u = J_\epsilon^{-1}U$, $f = J_\epsilon^{-1}F$, $p = J_\epsilon^{-1}P$. Of course, (5.11) is supplemented with periodic boundary conditions.

For $m = 0, 1, 2, \ldots$, we let $H_p^m(Q_3) \equiv H_p^m(Q_3; \mathbb{R}^3)$ be the closure in $H^m(Q_3; \mathbb{R}^3)$ of those smooth functions that are periodic in space. One then has $H_p^0(Q_3) = L^2(Q_3; \mathbb{R}^3) = L^2(Q_3)$. We denote by $H_\epsilon = H_\epsilon(Q_3)$ (resp. $V_\epsilon = V_\epsilon(Q_3)$) the closure in $L^2(Q_3)$ (resp. $H_p^1(Q_3)$) of those smooth functions that are periodic on Q_3 and satisfy

$$\int_{Q_3} u dx = 0 \text{ and } div_\epsilon u = \nabla_\epsilon \cdot u = \partial_1 u_1 + \partial_2 u_2 + \epsilon^{-1}\partial_3 u_3 = 0 .$$

We let $\mathbb{P}_\epsilon$ denote the orthogonal projection of $L^2(Q_3)$ onto H_ϵ, and introduce the operators

$$A_\epsilon u = -\mathbb{P}_\epsilon \Delta_\epsilon u, \quad B_\epsilon(u^1, u^2) = \mathbb{P}_\epsilon(u^1 \cdot \nabla_\epsilon)u^2 .$$

Applying the operator $\mathbb{P}_\epsilon$ to (5.11), we obtain the *dilated abstract* Navier–Stokes equations

$$(5.12) \qquad u_t + \nu A_\epsilon u + B_\epsilon(u, u) = \mathbb{P}_\epsilon f .$$

We set $V_\epsilon^m = V_\epsilon \cap H_p^m(Q_3)$. Then, $V_\epsilon = V_\epsilon^1 = D(A_\epsilon^{1/2})$, $V_\epsilon^2 = D(A_\epsilon)$. Here we have $A_\epsilon u = -\Delta_\epsilon u$, $u \in D(A_\epsilon)$.

Remark that V_ϵ^m, $A_\epsilon u$, etc... can be represented in terms of the Fourier series expansions (see Raugel and Sell (1993a)). One has the following inequalities

$$(5.13)(i) \qquad c_1(\|u\|_{H^1(Q_3)} + \epsilon^{-1}\|\partial_3 u\|) \leq \|A_\epsilon^{1/2}u\| \leq c_2(\|u\|_{H^1(Q_3)} + \epsilon^{-1}\|\partial_3 u\|)$$

and

$$(5.13)(ii) \qquad c_1(\|u\|_{H^2(Q_3)} + \epsilon^{-1}\|\partial_3 u\|_{H^1(Q_3)} + \epsilon^{-2}\|\partial_3^2 u\|) \leq \|A_\epsilon u\|$$
$$\leq c_2(\|u\|_{H^2(Q_3)} + \epsilon^{-1}\|\partial_3 u\|_{H^1(Q_3)} + \epsilon^{-2}\|\partial_3^2 u\|)$$

where $\|\cdot\|$ denotes the norm in $L^2(Q_3)$.

For $f \in L^\infty((0, \infty); L^2(Q_3))$, we define the norm $\|f\|_\infty$ by

$$\|f\|_\infty = \text{ess sup}_{0<t<\infty}\|f(t)\| .$$

5.2.2 The v and w equations. As before, we set

$$v(x_1, x_2) = (Mu)(x_1, x_2) = \int_0^1 u(x_1, x_2, s)ds$$

and

$$w = (I - M)u .$$

If we apply the projections M and $(I - M)$ to the equation (5.12), we obtain the system

$$(5.14) \quad \begin{cases} v_t + \nu A_\epsilon v + B_\epsilon(v, v) = M\mathbb{P}_\epsilon f - M(B_\epsilon(w, w) + B_\epsilon(w, v) + B_\epsilon(v, w)) \\ w_t + \nu A_\epsilon w = (I - M)\mathbb{P}_\epsilon f - (I - M)(B_\epsilon(w, w) + B_\epsilon(w, v) + B_\epsilon(v, w)) \end{cases}$$

where $v = Mu$, $w = (I - M)u$.

We are going to study the solutions $(v(t), w(t))$ of (5.14) that satisfy $v(0) = v_0 = Mu_0$, $w(0) = w_0 = (I - M)u_0$.

REMARK 5.1. — As we have already remarked it for other equations, the set $\{w = 0\}$ is positively invariant, that is, if $w_0 = 0$, then $w(t) = 0$ and $\bar{v} = v(t)$ satisfies the following reduced 3D *Navier-Stokes equations*

$$(5.15) \qquad \begin{cases} \bar{v}_t + \nu A_\epsilon \bar{v} + B_\epsilon(\bar{v}, \bar{v}) = M\mathbb{P}_\epsilon f \\ \bar{v}(0) = v_0 \end{cases} .$$

Let $m = (\bar{v}_1, \bar{v}_2)$, $(g_1, g_2, g_3) = M\mathbb{P}_\epsilon f$, then $\bar{v} = (m, \bar{v}_3)$ is a solution of (5.15) if and only if m is a solution of the 2D Navier–Stokes equations

$$m_t - \nu\mathbb{P}_2(\partial_1^2 + \partial_2^2)m + \mathbb{P}_2((m \cdot \nabla)m) = (g_1, g_2),$$

and $\bar{v}_3$ is the solution of the linear equation

$$\bar{v}_{3t} - \nu(\partial_1^2 + \partial_2^2)\bar{v}_3 + (\bar{v}_1\partial_1 + \bar{v}_2\partial_2)\bar{v}_3 = g_3,$$

where $\nabla = (\partial_1, \partial_2)$.

5.2.3 Useful properties. In our arguments, we shall use the following properties.

• In Lemma 3.1, we have stated the following inequality, for $w \in V_\epsilon$, $Mw = 0$, for $2 \le q \le 6$,

$$(5.16) \qquad \|w\|_{L^q(Q_3)} \le C_0\epsilon^{2/q}\|A_\epsilon^{1/2}w\|,$$

which is simply the Poincaré inequality in the case $q = 2$.

• We define the trilinear form $b_\epsilon(u^1, u^2, u^3)$ by

$$b_\epsilon(u^1, u^2, u^3) = < B_\epsilon(u^1, u^2), u^3 > = \sum_{i,j=1}^{3} \int_{Q_3} \epsilon^{-\{i\}} u_i^1 (\partial_i u_j^2) u_j^3 dx$$

where $\{1\} = \{2\} = 0$, $\{3\} = 1$. For $u^1, u^2 \in V_\epsilon$, we have the well-known equality

$$(5.17) \qquad b_\epsilon(u^1, u^2, u^2) = 0 .$$

In the case of periodic boundary conditions, if $v = (v_1, v_2, v_3) \in V_\epsilon$ depends only on the x_1 and x_2 variables, we set:

$$(5.18) \qquad \tilde{v} = (v_1, v_2, 0), \quad v^* = (0, 0, v_3) .$$

Since $\nabla_\epsilon \cdot v = 0$, we have $\nabla_\epsilon \cdot \tilde{v} = \nabla_\epsilon \cdot v^* = 0$. And $\tilde{v}$ satisfies the *conservation of enstrophy* condition

$$(5.19) \qquad b_\epsilon(\tilde{v}, \tilde{v}, A_\epsilon\tilde{v}) = b_\epsilon(v, v, A_\epsilon\tilde{v}) = 0 .$$

The use of the conservation of enstrophy leads to the possiblity of choosing larger sets $\mathcal{R}(\epsilon)$ and $\mathcal{S}(\epsilon)$.

• We shall also use the following inequalities. The inequality (5.20) is very classical (see for instance Temam (1977)). The inequalities (5.21) are a consequence of (5.16) and the definition of $b_\epsilon(\cdot,\cdot,\cdot)$. If $v^1, v^2, v^3 \in R(M)$, one has:

$$(5.20) \qquad |b_\epsilon(v^1, v^2, v^3)| \le C_1 \|v^1\|^{1/2} \|A_\epsilon^{1/2} v^1\|^{1/2} \|A_\epsilon^{1/2} v^2\|^{1/2} \|A_\epsilon v^2\|^{1/2} \|v^3\| .$$

If $v \in R(M)$ and if $Mw^1 = Mw^2 = Mw = 0$, one has

$$(5.21) \qquad \begin{cases} |b_\epsilon(w^1, w^2, u)| \le C_2 \epsilon^{1/2} \|A_\epsilon^{1/2} w^1\| \, \|A_\epsilon^{1/2} w^2\|^{1/2} \|A_\epsilon w^2\|^{1/2} \|u\| , \\[2mm] |b_\epsilon(w, u^2, u^3)| \le C_3 \epsilon^{5/32} \|A_\epsilon^{1/2} w\|^{15/32} \|A_\epsilon w\|^{17/32} \|A_\epsilon^{1/2} u^2\| \, \|u^3\| , \\[2mm] |b_\epsilon(v, w, u)| \le C_4 \epsilon^{1/4} \|A_\epsilon^{1/2} v\| \, \|A_\epsilon^{1/2} w\|^{1/2} \|A_\epsilon w\|^{1/2} \|u\| . \end{cases}$$

5.2.3 The short time argument. We choose initial data $u_0 = v_0 + w_0$ and a forcing term f satisfying

$$(5.22) \qquad \begin{cases} \|A_\epsilon^{1/2} v_0\|^2 \le \epsilon^{q_1} \rho_1^2, \quad \|M\mathbb{P}_\epsilon f\|_\infty^2 \le \epsilon^{q_2} \rho_2^2 \\[2mm] \|A_\epsilon^{1/2} w_0\|^2 \le \epsilon^p \rho_3^2, \quad \|(I-M)\mathbb{P}_\epsilon f\|_\infty^2 \le \epsilon^r \rho_4^2 \end{cases}$$

where $\rho_1, \rho_2, \rho_3, \rho_4$ are positive ("large") constants larger than 1 (one can also suppose that $\rho_i = \rho_i(\epsilon)$ satisfies $\rho_i^0 \le \rho_i(\epsilon) \le |\log \epsilon|^{\alpha_i}$, for $\alpha_i > 0$). We denote by λ_1 the first eigenvalue of A_ϵ.

We shall say that *Hypothesis* H1 is satisfied if one has:

$$(1) \qquad -2 < r < 0, \; -1 - \frac{5}{8} < p < 0, \; -1 < q_1 < 0, \; -2 < q_2 < 0,$$

$$(2) \qquad 2 + q_1 + q_2 > 0, \; 4 + q_1 + 2p > 0, \; 4 + q_2 + 2p > 0 .$$

The objective of the short time argument is to show that the w-component of the strong solution u of (5.12) *rapidly* becomes *small* and that we have a good control on the w-component.

LEMMA 5.3. — *Assume that (5.22) and the Hypothesis H1 hold. Then there are positive constants k_1, k_2 and ϵ_1 such that, for $0 < \epsilon \le \epsilon_1$, there exists a time $T_1 = T_1(\epsilon)$ so that $u(t) \in D(A_\epsilon^{1/2})$ for $0 \le t \le T_1$ and*

$$(5.23) \qquad \begin{cases} \|A_\epsilon^{1/2} v(T_1)\|^2 \le 4\epsilon^{q_1} \rho_1^2 + k_1^2 \epsilon^{2+2p} \rho_3^4 \\[2mm] \|A_\epsilon^{1/2} w(T_1)\|^2 \le k_2^2 \epsilon^{2+r} \rho_4^2 . \end{cases}$$

Sketch of the proof : We set:

$$R_1^2 = \epsilon^{q_1}\rho_1^2 + \epsilon^{2+2p}\rho_3^4, \quad R_3^2 = \epsilon^p\rho_3^2$$

and let $N > 1$ be an appropriate fixed real number. There is a time $T^N > 0$ such that

(5.24)
$$\begin{cases} ||A_\epsilon^{1/2}v(t)||^2 < NR_1^2 \\ ||A_\epsilon^{1/2}w(t)||^2 < NR_3^2 \end{cases}$$

holds for all t, $0 \le t < T^N$. We can choose T^N so that $[0, T^N)$ denotes the maximal time interval for which (5.24) is valid. If $T^N < +\infty$, we must have

(5.25)
$$\begin{cases} \text{either} \quad ||A_\epsilon^{1/2}v(T^N)||^2 = NR_1^2 \\ \text{or} \quad ||A_\epsilon^{1/2}w(T^N)||^2 = NR_3^2 \ . \end{cases}$$

Taking the inner product of the w-equation in (5.14) with $A_\epsilon w$ and applying (5.21) as well as the Poincaré inequality, we obtain:

(5.26)
$$\frac{d}{dt}||A_\epsilon^{1/2}w||^2 + (\nu - \frac{D_1}{2}(\epsilon^{5/8}||A_\epsilon^{1/2}v|| + \epsilon||A_\epsilon^{1/2}w||))||A_\epsilon w||^2$$
$$\le \frac{1}{\nu}||(I - M)\mathbb{P}_\epsilon f||_\infty^2 \ .$$

From Hypothesis H1, we see that, for $0 \le t \le T^N$, one has

(5.27)
$$\frac{D_1}{2}(\epsilon^{5/8}||A_\epsilon^{1/2}v|| + \epsilon||A_\epsilon^{1/2}w||) \le \frac{D_1}{2}(\epsilon^{5/8}N^{1/2}R_1 + \epsilon N^{1/2}R_3) \le \frac{\nu}{2},$$

for $\epsilon > 0$ small enough.

From (5.26) and (5.27) we deduce that, for $\epsilon > 0$ small enough,

(5.28)(i)
$$\frac{d}{dt}||A_\epsilon^{1/2}w||^2 + \frac{\nu}{2}||A_\epsilon w||^2 \le \frac{1}{\nu}||(I - M)\mathbb{P}_\epsilon f||_\infty^2$$

and, with the Poincaré inequality,

(5.28)(ii)
$$\frac{d}{dt}||A_\epsilon^{1/2}w||^2 + \frac{\nu C_0^{-2}\epsilon^{-2}}{2}||A_\epsilon^{1/2}w||^2 \le \frac{1}{\nu}||(I - M)\mathbb{P}_\epsilon f||_\infty^2 \ .$$

With the Gronwall inequality, we deduce from (5.28)(ii), for $0 \le t \le T^N$,

(5.29)
$$||A_\epsilon^{1/2}w(t)||^2 \le \exp(\frac{-\nu C_0^{-2}\epsilon^{-2}t}{2})||A_\epsilon^{1/2}w_0||^2 + \frac{2C_0^2\epsilon^2}{\nu^2}||(I - M)\mathbb{P}_\epsilon f||_\infty^2 \ .$$

We now introduce the time $T_1 = T_1(\epsilon)$ such that

$$\epsilon^p\rho_3^2\exp(\frac{-\nu C_0^{-2}\epsilon^{-2}T_1}{2}) = \frac{2C_0^2\epsilon^2}{\nu^2}\epsilon^r\rho_4^2 \ .$$

This time T_1 is given by

$$T_1 \equiv 2C_0^2 \epsilon^2 \nu^{-1} Q(\epsilon)$$

where

$$Q(\epsilon) = |\log(2C_0^2 \nu^{-2} \epsilon^{2+r-p} \rho_4^2 \rho_3^{-2}|\,.$$

Thus, if $T_1 < T^N$, then, for $T_1 \le t \le T^N$,

(5.30) $$\|A_\epsilon^{1/2} w(t)\|^2 \le k_2^2 \epsilon^{2+r} \rho_4^2$$

with $k_2^2 = 4C_0^2 \nu^{-2}$.

The next step is to estimate $\|A_\epsilon^{1/2} v(t)\|^2$. Taking the inner product of the v-equation in (5.14) with $A_\epsilon v$, using the estimates (5.20), (5.21) and Young inequalities, we obtain:

(5.31) $$\frac{d}{dt}\|A_\epsilon^{1/2} v\|^2 \le \left(\frac{27C_1^4}{2\nu^3}\|v\|^2\|A_\epsilon^{1/2} v\|^2\right)\|A_\epsilon^{1/2} v\|^2$$
$$+ \frac{1}{\nu}\|M\mathbb{P}_\epsilon f\|_\infty^2 + \frac{2C_2^2 \epsilon}{\nu}\|A_\epsilon^{1/2} w\|^3\|A_\epsilon w\|\,.$$

Integrating (5.31) and using the Gronwall inequality, it comes, for $0 \le t \le T^N$,

(5.32) $$\|A_\epsilon^{1/2} v(t)\|^2 \le e^{\int_0^t g(s)ds}\left(\|A_\epsilon^{1/2} v_0\|^2 + \int_0^t h(s)ds\right)$$

where

$$h(t) = \frac{1}{\nu}\|M\mathbb{P}_\epsilon f\|_\infty^2 + \frac{2C_2^2 \epsilon}{\nu}\|A_\epsilon^{1/2} w\|^3\|A_\epsilon w\|$$

and

$$g(t) = \frac{27C_1^4}{2\nu^3}\|v\|^2\|A_\epsilon^{1/2} v\|^2\,.$$

From (5.32) and some computations (involving the inequalities (5.28) and (5.29)), we can show that, for $0 \le t \le \min(T_1, T^N)$,

(5.33) $$\|A_\epsilon^{1/2} v(t)\|^2 \le e^{E_2(\epsilon)}\left(\epsilon^{q_1} \rho_1^2 + E_1(\epsilon) + \frac{7}{4}D_2 \epsilon^{2+2p} \rho_3^4\right),$$

where, due to Hypothesis H1, $E_i(\epsilon) \to 0$ as $\epsilon \to 0$, $i = 1, 2$.

Now, we set $N = 2 + \max(4, \frac{7}{2}D_2)$ and choose $\epsilon_1 > 0$ so that, for $0 < \epsilon \le \epsilon_1$,

(5.34) $$e^{E_2(\epsilon)} \le 2, \quad E_1(\epsilon) \le \epsilon^{q_1} \rho_1^2, \quad \frac{2C_0^2 \epsilon^{2+r} \rho_4^2}{\nu^2} < \epsilon^p \rho_3^2\,.$$

We claim that, for $0 < \epsilon \le \epsilon_1$, one has $T_1 < T^N$. To prove this we assume on the contrary that $T^N \le T_1 < \infty$. Then, (5.22), (5.29), (5.33) and (5.34) imply that

$$\begin{cases} \|A_\epsilon^{1/2} w(T^N)\|^2 < NR_3^2 \\ \|A_\epsilon^{1/2} v(T^N)\|^2 < NR_1^2 \end{cases}$$

which contradicts (5.25). The lemma 5.3 is proved. $\qquad\square$

5.2.4 The long time argument. For the long time argument needed in the proof of the Theorem 5.1, we begin at the point where the lemma 5.3 leaves off, i.e. we assume that

$$||A_\epsilon^{1/2} w_0||^2 \leq k_2^2 \epsilon^{2+r} \rho_4^2$$

and

$$||A_\epsilon^{1/2} v_0||^2 \leq G_1(\epsilon)$$

where

(5.35)
$$G_1(\epsilon) = 4\epsilon^{q_1} \rho_1^2 + k_1^2 \epsilon^{2+2p} \rho_3^4 .$$

We show that, for ϵ small enough, the 3D Navier–Stokes equations (5.12) have a strong solution on $[0, \infty)$ and $||A_\epsilon^{1/2} w(t)||$ stays small. We need to restrict the hypotheses. The numbers r, p, q_1, q_2 will be assumed to satisfy *Hypothesis* H2

$$-\frac{29}{24} < p < 0, \quad \frac{-5}{12} < q_1 < 0, \quad \frac{-5}{12} < q_2 < 0, \quad -2 < r < 0 .$$

LEMMA 5.4. — *Assume that Hypothesis* H2 *holds. Then, there exists* $\epsilon_0 > 0$ *such that whenever one has*

(5.36)
$$\begin{cases} ||A_\epsilon^{1/2} v_0||^2 \leq G_1(\epsilon), & ||M\mathbb{P}_\epsilon f||_\infty^2 \leq \epsilon^{q_2} \rho_2^2 \\ ||A_\epsilon^{1/2} w_0||^2 \leq k_2^2 \epsilon^{2+r} \rho_4^2, & ||(I - M)\mathbb{P}_\epsilon f||_\infty^2 \leq \epsilon^r \rho_4^2, \end{cases}$$

then the solution $u(t)$ *of* (5.12) *satisfies*

$$u(t) \in C^0([0, \infty); D(A_\epsilon^{1/2})) \cap L^\infty((0, \infty); D(A_\epsilon^{1/2}))$$

and there exists a positive constant D, *which depends only on* ν *and* λ_1, *such that*

(5.37)
$$\begin{cases} (i) & ||A_\epsilon^{1/2} v(t)||^2 \leq D(e^{-\nu\lambda_1 t} G_3(\epsilon) + G_2(\epsilon) + G_2(\epsilon)^3) \\ (ii) & ||A_\epsilon^{1/2} w(t)||^2 \leq [\exp(-\frac{\nu C_0^{-2} \epsilon^{-2} t}{2}) + \frac{1}{2}] k_2^2 \epsilon^{2+r} \rho_4^2 \end{cases}$$

for $0 \leq t < \infty$, *where*

$$G_2(\epsilon) \equiv (k_2^2 + 1)(\epsilon^{q_2} \rho_2^2 + \epsilon^{2+r} \rho_4^2 + \epsilon^{4+2r} \rho_4^4)$$

and

$$G_3(\epsilon) \equiv G_1 + (G_1 + G_2)^3 - G_2^3 .$$

Sketch of the proof: As before, we define R_0^2 by

$$R_0^2 = (G_1 + G_2) + (G_1 + G_2)^3 + k_2^2 \epsilon^{2+r} \rho_4^2$$

and choose an appropriate real number $N > 1$. We denote by $[0, T^N)$ the maximal time interval for which $\|A_\epsilon^{1/2} u(t)\|^2 < N R_0^2$ where $u(t)$ is the strong solution of (5.12). Then, if $T^N < +\infty$, one has

$$(5.38) \qquad \|A_\epsilon^{1/2} u(T^N)\|^2 = N R_0^2 .$$

Arguing as before, we obtain the inequality (5.37)(ii).

Now we write $v = \tilde{v} + v^* = (v_1, v_2, 0) + (0, 0, v_3)$. At first, one takes the scalar product of the v-equation in (5.14) with $\tilde{v}$ and one uses the identity $b_\epsilon(v, v, \tilde{v}) = b_\epsilon(\tilde{v}, \tilde{v}, \tilde{v}) = 0$, which gives us an estimate of $\|\tilde{v}\|$ and $\int_0^t \|A_\epsilon^{1/2} \tilde{v}\|^2 ds$. Then one takes the inner product of the v-equation in (5.14) with $A_\epsilon \tilde{v}$. Using the conservation of enstrophy (5.19) as well as the fact that $b_\epsilon(w, v, A_\epsilon \tilde{v}) = b_\epsilon(v, w, A_\epsilon \tilde{v}) = 0$, we obtain, after a few computations,

$$(5.39) \qquad \|A_\epsilon^{1/2} \tilde{v}(t)\|^2 \leq D_7 (e^{-\nu \lambda_1 t} G_1(\epsilon) + G_2(\epsilon)), \quad 0 \leq t < T^N .$$

Finally one estimates $\|A_\epsilon^{1/2} v^*(t)\|$. One uses the fact that the v^*-equation is linear in v^*. This implies that

$$(5.40) \qquad \begin{aligned} \|A_\epsilon^{1/2} v^*(t)\|^2 &\leq D_{10}(e^{-\nu \lambda_1 t} G_1(\epsilon) + G_2(\epsilon)) \\ &\quad + D_{11}(e^{-\nu \lambda_1 t} G_1(\epsilon) + G_2(\epsilon))^3, \quad 0 \leq t < T^N . \end{aligned}$$

By contradiction, as before one shows that $T^N = +\infty$.

5.2.5 Other regularity results. If we assume that the third component of $A_\epsilon^{1/2} v_0$ and $M \mathbb{P}_\epsilon f$ are small, we even obtain a better result of existence of strong solutions. Indeed assume that (5.22) and, in addition,

$$(5.41) \qquad \|v_0^*\|^2 \leq \kappa_1^2 \epsilon^{m_1}, \quad \|(M \mathbb{P}_\epsilon f)^*\|^2 \leq \kappa_2^2 \epsilon^{m_2},$$

hold, where κ_1, κ_2 are positive constants, and $m_1, m_2 \geq 0$. Assume also that the negative constants p, q_1, q_2, r and the nonnegative numbers m_1, m_2 satisfy *Hypothesis* H**:

(1) $\quad 0 \leq m_1 < 2, \quad m_2 \geq 0,$

(2) $\quad -2 < r < 0, \quad -\dfrac{5}{4} < q_1 < 0, \quad -\dfrac{5}{4} < q_2 < 0, \quad -\dfrac{5}{4} < p < 0,$

(3) $\quad p > \dfrac{1}{2} m_1 - 2,$

(4) $\quad 2q_i + \inf(m_1, m_2) + \dfrac{5}{4} > 0, \quad i = 1, 2,$

(5) $\quad 4q_1 + 2m_1 + \inf(m_1, m_2) + \dfrac{5}{4} > 0,$

(6) $\quad 8 + 8p + 2m_1 + \inf(m_1, m_2) + \dfrac{5}{4} > 0 .$

Example :

$$\begin{cases} m_1 = m_2 = 1, \quad p = -\frac{5}{4} + \delta_1 \\ q_1 = -\frac{17}{16} + \delta_2, \quad q_2 = -\frac{9}{8} + \delta_3, \quad r = -2 + \delta_4 \end{cases}$$

LEMMA 5.5. — *Assume that* (5.22), (5.41) *and Hypothesis* H** *hold. Then there exist positive numbers* $\epsilon_0^{**}, C_0^{**}$ *and, for* $0 < \epsilon \leq \epsilon_0^{**}$, *there is a time* $T_1^{**} = T_1^{**}(\epsilon)$, *such that, for* $0 < \epsilon \leq \epsilon_0^{**}$, (5.12) *has a (unique) strong solution* u *in* $C^0([0,\infty); V_\epsilon) \cap L^\infty([0,\infty); V_\epsilon)$ *and, for* $t \geq T_1^{**}$,

$$(5.42) \quad \begin{cases} \text{(i)} \quad \|A_\epsilon^{1/2} v(t)\|^2 \leq C^{**}(1 + \epsilon^{q_2} \rho_2^2 + \epsilon^{2q_2} \rho_2^4(\epsilon^{m_2} + \epsilon^2)) \\ \text{(ii)} \quad \|A_\epsilon^{1/2} w(t)\|^2 \leq k_2^2 \epsilon^{2+r} . \end{cases}$$

(For the proof, see Raugel and Sell (1992)).

5.2.5 Local and global attractor. We go back to the general case. We assume that the hypotheses H1 and H2 hold. We fix a forcing term f satisfying (5.22), which is time-independent. Let $S_\epsilon(t)$ be the semigroup defined by (5.12). We set

$$\widetilde{q}_1 = \min(q_1, q_2), \quad \widetilde{\rho}_1^2 = \max(8\nu^{-2}\lambda_1^{-1}\rho_2^2, \rho_1^2) .$$

We define the following bounded sets

$$\mathcal{B}_{\epsilon,0}^* \equiv \{u_0 = v_0 + w_0, \; \|A_\epsilon^{1/2} v_0\|^2 \leq \epsilon^{\widetilde{q}_1} \widetilde{\rho}_1^2, \; \|A_\epsilon^{1/2} w_0\|^2 \leq \epsilon^p \rho_3^2\},$$

$$\mathcal{B}_{\epsilon,1}^* \equiv \{u_0 = v_0 + w_0, \; \|A_\epsilon^{1/2} v_0\|^2 \leq 4\epsilon^{\widetilde{q}_1} \widetilde{\rho}_1^2 + k_1^2 \epsilon^{2+2p} \rho_3^4, \; \|A_\epsilon^{1/2} w_0\|^2 \leq k_2^2 \epsilon^{2+r} \rho_4^2\}$$

and

$$\mathcal{B}_{\epsilon,2}^* \equiv \bigcup_{t \geq 0} S_\epsilon(t)(\mathcal{B}_{\epsilon,0}^* \cup \mathcal{B}_{\epsilon,1}^*) .$$

Then, in the usual way, one can construct a unique, compact (local) attractor $\mathcal{A}_\epsilon^*$ included in $\mathcal{B}_{\epsilon,2}^*$, attracting $\mathcal{B}_{\epsilon,2}^*$ in V_ϵ (and $\mathcal{B}_{\epsilon,2}^* \cap V_\epsilon^2$ in the space V_ϵ^2). Actually this local attractor $\mathcal{A}_\epsilon^*$ is the global attractor for the weak (Leray) solutions of (5.12).

PRPOSITION 5.6. — *Let* $p < -1$, $q_1 < 0$, $q_2 < 0$, $r < 0$ *satisfy the hypotheses* H1 *and* H2. *Then there exists a positive number* ϵ_1^* *such that, if* $0 < \epsilon \leq \epsilon_1^*$, *for any* $\rho > 0$, *there exists a time* $T_\epsilon^* = T_\epsilon^*(\rho)$ *and, for any weak solution* $u(t)$ *of* (5.12) *with* $\|u(0)\| \leq \rho$, *there is a positive time* t_0, $0 < t_0 \leq T_\epsilon^*(\rho)$ *such that*

$$u(t_0) \in \mathcal{B}_{\epsilon,2}^* .$$

In particular, $u(t)$ *is a strong solution of* (5.12) *for* $t \geq t_0$ *and the local attractor* $\mathcal{A}_\epsilon^*$ *is the global attractor for the weak solutions of* (5.12).

Proof: For any weak (Leray) solution of (5.12), we deduce from (5.8) that, for almost every $t > 0$,

$$\nu \int_0^t ||A_\epsilon^{1/2} u(s)||^2 ds \leq ||u(0)||^2 + 2t\nu^{-1}(||A_\epsilon^{-1/2} M \mathbb{P}_\epsilon f||^2 + ||A_\epsilon^{-1/2}(I - M)\mathbb{P}_\epsilon f||^2) .$$

Thanks to the Poincaré inequality (5.16), we deduce from the above estimate,

$$(5.43) \qquad \frac{1}{t} \int_0^t ||A_\epsilon^{1/2} u(s)||^2 ds \leq \frac{\nu^{-1}}{t}||u(0)||^2 + 2\nu^{-2}(\lambda_1^{-1} \epsilon^{q_2} \rho_2^2 + C_0^2 \epsilon^{2+r} \rho_4^2) .$$

Since $p < -1$, $r > -2$, there exists ϵ_1^* such that, for $0 < \epsilon \leq \epsilon_1^*$,

$$(5.44) \qquad 2\nu^{-2}(\lambda_1^{-1} \epsilon^{q_2} \rho_2^2 + C_0^2 \epsilon^{2+r} \rho_4^2) \leq k \min(\epsilon^{\tilde{q}_1} \tilde{\rho}_1^2, \epsilon^p \rho_3^2) .$$

where $0 < k < 1$.

The inequalities (5.43) and (5.44) imply that

$$\frac{1}{t} \int_0^t ||A_\epsilon^{1/2} u(s)||^2 ds \leq \frac{\nu^{-1}}{t}||u(0)||^2 + k \min(\epsilon^{\tilde{q}_1} \tilde{\rho}_1^2, \epsilon^p \rho_3^2) .$$

Since $||u(0)|| \leq \rho$, there exists a positive time t_0, $0 < t_0 \leq T(\rho)$ where

$$T(\rho) = \frac{2\nu^{-1}\rho^2}{(1 - k) \min(\epsilon^{\tilde{q}_1} \tilde{\rho}_1^2, \epsilon^p \rho_3^2)}$$

such that

$$(5.45) \qquad ||A_\epsilon^{1/2} u(t_0)||^2 \leq \min(\epsilon^{\tilde{q}_1} \tilde{\rho}_1^2, \epsilon^p \rho_3^2) .$$

$\square$

REMARKS 5.2. — If $(I - M)\mathbb{P}_\epsilon f = 0$, then $\mathcal{A}_\epsilon^*$ coincides with the global attractor $\mathcal{A}_0(g)$ of the *reduced 3D Navier–Stokes equations* (5.15).

If $\lim_{n\to+\infty} ||M \mathbb{P}_{\epsilon_n} f_n - g_0|| = 0$ for some sequence ϵ_n and some function g_0 in MH_{ϵ_n}, then one can compare the attractor $\mathcal{A}_{\epsilon_n}^*$ with the global attractor $\mathcal{A}_0(g_0)$ of the reduced 3DNS equations (5.15). More precisely, the attractors $\mathcal{A}_{\epsilon_n}^*$ are upper-semicontinuous at $\epsilon_n = 0$ (see Raugel and Sell (1992), (1993a) and (1993b)).

5.3 Remarks about the Euler equations on a thin domain.

Here we still consider a thin three–dimensional domain Q_ϵ. In order to simplify, we set $F = 0$ in (5.1). When ν is taken equal to 0 in (5.1), we obtain the Euler equations,

$$(5.46) \qquad \begin{cases} U_t + (U \cdot \nabla_3)U + \nabla_3 P = 0 \\ \nabla_3 \cdot U = 0, \end{cases}$$

with appropriate boundary conditions. (We added the subscript 3 to emphasize that we are working in a three-dimensional domain). Since the Euler equations are conservative, we cannot hope to show global existence of strong solutions of (5.46) in Q_ϵ, with the methods developped in this course. But we can obtain an improved time of existence result in the case where Q_ϵ is thin and compare the solutions of (5.46) with those of a limit problem in a two–dimensional domain. The local existence results of strong solutions, as given, for instance, by Temam (1975), show that we need to work here in higher order Sobolev spaces $(W^{m,p}(Q_\epsilon))^3$ with $m > 1 + \frac{3}{p}$. Since we want to stay in the frame of Hilbert spaces, this leads us to work in the space $(H^3(Q_\epsilon))^3$.

In (Marsden, Ratiu and Raugel (1994)), two types of thin domains have been considered. Here we only briefly describe the results and refer to this paper for more details. The simplest domain considered is a thin cylinder $Q_\epsilon = Q_2 \times (0, \epsilon)$, where, as in Section 5.1, Q_2 is a smooth bounded domain in $\mathbb{R}^2$ or $Q_2 = (0, \ell_1) \times (0, \ell_2)$. In the case where Q_2 is a smooth bounded domain in $\mathbb{R}^2$, we are interested in the following boundary conditions,

(5.47)
$$\begin{cases} \text{(i)} & U(y + \epsilon e_3, t) = U(y, t) \\ \text{(ii)} & U \cdot \nu_\epsilon = 0 \text{ in } \partial Q_2 \times (0, \epsilon), \end{cases}$$

where ν_ϵ is the outward normal to ∂Q_ϵ. When $Q_2 = (0, \ell_1) \times (0, \ell_2)$, (5.46) is supplemented with the periodic conditions (5.2). As in the previous sections, we will drop the superscripts and replace $(H^3(Q_3))^3$ or $(H^3(Q_2))^2$ by $H^3(Q_3)$ or $H^3(Q_2)$, for example. If the cylinder Q_ϵ is thin enough or, what is essentially the same thing, if the initial data $U_{|t=0}$ are nearly two–dimensional, we show that the limit equations are simply the usual Euler equations in Q_2,

(5.48)
$$\begin{cases} V_t^0 + (V^0 \cdot \nabla_2)V^0 + \nabla_2 P^0 = 0 \\ \nabla_2 \cdot V^0 = 0, \end{cases}$$

with the corresponding boundary conditions. If

$$V_{|t=0}^0 \equiv M_\epsilon U_{|t=0} \equiv \frac{1}{\epsilon} \int_0^\epsilon U(y_1, y_2, y_3)_{|t=0} dy_3$$

belongs to the space $H^3(Q_2) \cap V$, then the equation (5.48) has a global strong solution $V^0(t)$ in $C^0([0, \infty); H^3(Q_2) \cap V)$. Assuming that $\|(V^0(t) - U(t))_{|t=0}\|_{H^2(Q_\epsilon)}$ is small enough and that the solution $U(t)$ of (5.46) is uniformly bounded with respect to ϵ in $H^3(Q_3)$ on a time interval $[0, T_1(\epsilon)]$, we give an estimate of $\|V^0(t) - U(t)\|_{H^2(Q_\epsilon)}$, for $t \in [0, T_1(\epsilon)]$. Finally, using estimates like in (Beale, Kato and Majda (1984)), we show that $U(t)$ is actually uniformly bounded with respect to ϵ in $H^3(Q_\epsilon)$ if the time $T_1(\epsilon)$ is chosen to be of order $C \log \log(D(\log E |\log \epsilon^\alpha|)^{1/2})$ where C, D, E are some positive constants. This gives us a lower bound of the time existence of $U(t)$. As in the previous sections, we can also work in the scaled domain Q_3. If we work in the scaled domain Q_3, we replace the norms $||\cdot||_{H^2(Q_\epsilon)}$ and $||\cdot||_{H^3(Q_\epsilon)}$ by the norms $||\cdot||_{H^2_\epsilon(Q_3)}$ and $||\cdot||_{H^3_\epsilon(Q_3)}$, where

$$||u||_{H^1_\epsilon(Q_3)} = (||u||^2_{L^2(Q_3)} + \sum_{j=1}^{2} ||\frac{\partial u}{\partial x_j}||^2_{L^2(Q_3)} + \frac{1}{\epsilon^2}||\frac{\partial u}{\partial x_3}||^2_{L^2(Q_3)})^{1/2}$$

and, for $m \geq 2$,

$$||u||_{H^m_\epsilon(Q_3)} = (||u||^2_{H^{m-1}_\epsilon(Q_3)} + \sum_{j=1}^{2} ||\frac{\partial u}{\partial x_j}||^2_{H^{m-1}_\epsilon(Q_3)} + \frac{1}{\epsilon^2}||\frac{\partial u}{\partial x_3}||^2_{H^{m-1}_\epsilon(Q_3)})^{1/2}.$$

Remark that we are obliged, unlike in the case of the Navier–Stokes equations, to assume that $V^0_{|t=0} - U_{|t=0}$ is small enough. This is due to the lack of smoothing properties of the Euler equations. Finally, let us point out that, even if we want to obtain estimates of $U(t)$ in $H^3(Q_\epsilon)$ or to compare $U(t)$ with $V^0(t)$ in $H^2(Q_\epsilon)$, we only need to take into account the limit equation (5.48). We do not need any auxiliary limit equation. The situation is quite different if we replace the thin product domain $Q_2 \times (0, \epsilon)$ by a more complicated thin domain. Let us describe such a case, where the geometry of the domain plays a bigger role.

In (Marsden, Ratiu and Raugel (1994)), we have also considered the case where Q_ϵ is the thin domain delimited by two spheres in $\mathbb{R}^3$ of radii 1 and $1+\epsilon$. The curved geometry and its interaction with the boundary conditions make the problem here much more intricate. In the case of a thin cylinder, we could embed the two-dimensional problem into the three–dimensional problem, in an obvious way. For spherical shells, because of the curvature and the boundary conditions, this cannot be done in such a straightforward way.

When Q_ϵ is the thin domain between the spheres of radii 1 and $1 + \epsilon$, the Euler flow may be decomposed as follows. At a point in Q_ϵ at a radial distance $1 + r$ with $0 < r < \epsilon$. we decompose the three–dimensional vector field U into a radial component $W(r, \cdot)$, regarded as a scalar function on the unit sphere S^2, and a component V orthogonal to the radial direction. We regard V as a time and r dependent vector field on the sphere S^2. Then the Euler equations in Q_ϵ can be written as follows:

(5.49)
$$\begin{cases} W_t + W\frac{\partial W}{\partial r} + V[W] - (1+r)k(V,V) = -\frac{\partial P}{\partial r} \\ V_t + W\frac{\partial V}{\partial r} + \nabla_V V + \frac{2}{1+r}WV = -\frac{1}{(1+r)^2}\nabla_2 P \\ \frac{1}{(1+r)^2}\frac{\partial}{\partial r}((1+r)^2 W) + \text{div}_2 V = 0 \\ W_{|r=0} = W_{|r=\epsilon} = 0, \end{cases}$$

where P is the pressure. $k(V, V)$ is the second fondamental form on S^2, ∇_2 is the gradient operator on S^2. $\nabla_V V$ is the covariant derivative of V along V using the geometry of S^2 and div_2 is the divergence on the sphere.

Let us now consider an expansion of V, W, P in r (or in $r = \epsilon s$, if one wants to work on the reference domain Q, which is the domain delimited by the spheres of radii 1 and 2):

$$(5.50) \quad \begin{cases} \text{(i) } V = V_0 + rV_1 + r^2V_2 + r^3V_3 + O(r^4) \\ \text{(ii) } W = r^2W_2 + r^3W_3 + O(r^4) \\ \text{(iii) } P = P_0 + rP_1 + r^2P_2 + r^3P_3 + O(r^4) \,. \end{cases}$$

At each order in r, one finds evolution equations for these quantities. At first, V_0 and P_0 are solutions of the standard Euler equations on S^2:

$$(5.51) \quad \begin{cases} V_{0t} + \nabla_{V_0} V_0 = -\nabla_2 P_0 \\ \operatorname{div}_2 V_0 = 0 \,. \end{cases}$$

Then, P_1 is given by

$$(5.52) \qquad P_1 = k(V_0, V_0) \,.$$

Next, V_1 is a solution of a linearized Euler equation with a known right hand side and no divergence condition

$$(5.53) \qquad V_{1t} + \nabla_{V_0} V_1 + \nabla_{V_1} V_0 = -\nabla_2 P_1 + 2\nabla_2 P_0 \,.$$

And W_2 and P_2 are given by

$$(5.54) \qquad W_2 = -\frac{1}{2}\operatorname{div}_2 V_1 \,, \qquad P_2 = k(V_0, V_1) + \frac{1}{2}k(V_0, V_0) \,.$$

Next V_2 evolves according to a linearized Euler equation with a known right hand side and no divergence condition

$$(5.55) \quad \begin{aligned} V_{2t} + \nabla_{V_0} V_2 + \nabla_{V_2} V_0 = &- \nabla_2 P_2 + 2\nabla_2 P_1 - 3\nabla_2 P_0 \\ &- \nabla_{V_1} V_1 \\ &- W_2(V_1 + 2V_0) \,, \end{aligned}$$

whereas W_3 and P_3 are given by

$$(5.56) \quad \begin{cases} -3W_3 = 2W_2 + \operatorname{div}_2 V_2 \,, \\ -3P_3 = W_{2t} + V_0[W_2] - 2k(V_0, V_2) \\ \qquad\quad - k(V_1, V_1) - 2k(V_0, V_1) \,. \end{cases}$$

In general the equation for V_n has the form

$$(5.57) \qquad V_{nt} + \nabla_{V_0} V_n + \nabla_{V_n} V_0 = \text{ known quantities}$$

where the "known quantities" were determined in the previous steps, and then W_{n+1} and P_{n+1} are given in terms of the previously computed quantities.

In the case of a thin cylinder, an expansion at the order 0 in y was sufficient to solve our question. Here, due to the more complex geometry and its interaction with the boundary conditions, we need an expansion up to order two, in order to be able to give a lower bound of the time existence of $U = (V, W)$ in $H^3(Q_\epsilon)$ (or in $H^3_\epsilon(Q)$, where Q is the corresponding scaled domain). Assuming that $(U - V^0)_{|t=0}$, $(\frac{\partial U}{\partial r} - V^1)_{|t=0}$ and $(\frac{\partial^2 U}{\partial r^2} - V^2)_{|t=0}$ are small in the norms of $H^2(Q_\epsilon)$, $H^1(Q_\epsilon)$ and $L^2(Q_\epsilon)$ and that the solution (V, W) of (5.49) is uniformly bounded with respect to ϵ in $H^3(Q_\epsilon)$ on a time interval $[0, T_2(\epsilon)]$, we give estimates of $\|(U - V^0)(t)\|_{H^2(Q_\epsilon)}$, $\|(\frac{\partial U}{\partial r} - V^1)(t)\|_{H^1(Q_\epsilon)}$ and $\|(\frac{\partial^2 U}{\partial r^2} - V^2)(t)\|_{L^2(Q_\epsilon)}$ and, as in the thin cylinder case, we show that $U(t)$ is actually uniformly bounded with respect to ϵ in $H^3(Q_\epsilon)$, if the time $T_2(\epsilon)$ is chosen to be of order $C\log\log(D(\log E|\log\epsilon^\alpha|)^{1/2})$. In these proofs, like in the sections 5.1 and 5.2, we use a mean value operator M, which here is the mean value with respect to the radial direction.

References

Abounouh M. (1994), The Cahn–Hilliard equation on a thin domain, submitted to *Applicable Analysis*.

Adams R.A. (1975), *Sobolev Spaces*, Academic Press.

Angenent S.B. (1986), The Morse–Smale property for a semilinear parabolic equation, *J. Differential Equations* **62**, 427-442.

Aganovic I. and Tutek Z. (1986), A justification of the one–dimensional model of an elastic beam, *Math. Methods in Applied Sci.* **8**, 1-14.

Arrieta J. (1994), Rates of eigenvalues on a dumbbell domain, simple eigenvalue case, Preprint.

Arrieta J., Carvalho A. and Hale J.K. (1992), A damped hyperbolic equation with critical exponent, *Comm. PDE* **17**, 841-866.

Aulbach B. (1983), Approach to hyperbolic manifolds of stationary solutions, *Lecture Notes in Math.* **1017**, 56-66, Springer–Verlag.

Babin, A.V. and Vishik, M.I. (1983), Regular attractors of semigroups of evolutionary equations, *J. Math. Pures Appl.* **62**, 441-491.

Babin, A.V. and Vishik, M.I. (1986), Unstable invariant sets of semigroups of nonlinear operators and their perturbations, *Russian Math. Surveys* **41**, 1-41.

Babin, A.V. and Vishik, M.I. (1989a), *Attractors of evolutionary equations*, Nauka (in Russian), North–Holland (in English).

Babin, A.V. and Vishik, M.I. (1989b), Uniform finite-parameter asymptotics of solutions of nonlinear evolutionary equations, *J. Math. Pures Appl.* **68**, 399-455.

Bardos C. and Tartar L. (1973), Sur l'unicité rétrograde des équations paraboliques et quelques équations voisines, *Arch. Rational Mech. Anal.* **50**, 10-25.

Bayada G. and Chambat M. (1986), The transition between the Stokes equation and the Reynolds equation: a mathematical proof, *Appl. Math. Opt.* , 73-93.

Beale J.T., Kato T. and Majda A. (1984), Remarks on the breakdown of smooth solutions for the 3–D Euler equations, *Commun. Math. Phys.* **94**, 61-66.

Besson O. and Laydi M.R. (1992), Some estimates for the anisotropic Navier–Stokes equations and for the hydrostatic approximation, *Math. Modelling and Numerical Analysis, M2AN* **26**, 855-865.

Besson O., Laydi M.R. and Touzani (1990), Un modèle asymptotique en océanographie, *C. R. Acad. Sci. Paris, Série 1* **310**, 661-665.

Bourquin F. and Ciarlet P.G. (1989), Modeling and justification of eigenvalue problems for junctions between elastic structures, *J. Funct. Anal.* **87**, 392-427.

Brunovsky P. and Chow S.-N. (1984), Generic properties of stationary solutions of reaction–diffusion equations, *J. Diff. Equat.* **53**, 1-23.

Caillerie D. (1981), Homogénéisation des équations de la diffusion stationnaire dans les domaines cylindriques aplatis, *RAIRO, Analyse Numérique* **15**, 295-319.

Camassa R. and Holm D. (1992), Dispersive barotropic equations for stratified mesoscale ocean dynamics, *Physica D* **60**, 1-15.

Casten R.C. and Holland C.J. (1978), Instability results for reaction–diffusion equations with Neumann boundary conditions, *J. Diff. Equat.* **27**, 266-273.

Ciarlet P.-G. (1988), *Mathematical Elasticity, Vol. I: Three–dimensional Elasticity*, North–Holland, Amsterdam.

Ciarlet P.-G. (1990), Plates and junctions in elastic multi-structures: an asymptotic analysis. RMA **14**, Masson.

Ciarlet P.-G. and Destuynder P. (1979), A justification of the two–dimensional plate model, *J. Mécanique* **18**. 315-344.

Ciarlet P.-G., Le Dret H. and Nzengwa R. (1987), Modélisation de la jonction entre un corps élastique tridimensionnel et une plaque, *C.R. Acad. Sci. Paris, Série I* **305**, 55-58.

Ciarlet P.-G., Le Dret H. and Nzengwa R. (1989), Junctions between three–dimensional and two–dimensional linearly elastic structures, *J. Math. Pures Appl.* **68**, 261-295.

Cimetière A., Geymonat G.. Le Dret H., Raoult A. and Tutek Z. (1988), Asymptotic theory and analysis for displacements and stress distribution in nonlinear elastic straight slender rods. *J. Elasticity* **19**, 111-161.

Cioranescu D. and Saint Jean Paulin J. (1986), Reinforced and honey–comb structures, *J. Math. Pures et Appl.* **65**, 403-422.

Ciuperca I.S. (1994), Reaction–diffusion equations on thin domains with varying order of thinness, to appear in J. Diff. Equat.

Constantin P. and Foias C. (1988), *Navier–Stokes equations*. Univ. Chicago Press, Chicago.

Conway E., Hoff D. and Smoller J. (1978), Large time behavior of solutions of systems of reaction diffusion equations. *SIAM J. Appl. Math.* **35**. 1-16.

Crouzeix M. and Rappaz J. (1990), *On Numerical Approximation in Bifurcation Theory*, Collection RMA 13. Masson.

Dancer E.N. (1988), The effect of domain shape on the number of positive solutions of certain nonlinear equations. *J. Diff. Equat.* **74**, 120-156.

Dancer E.N. (1990), The effect of domain shape on the number of positive solutions of certain nonlinear equations. II, *J. Diff. Equat.* **87**, 316-339.

Destuynder P. (1981), Comparaison entre les modèles tridimensionnels et bidimensionnels de plaques en élasticité, *RAIRO Analyse Numérique* **15**, 331-369.

Dridi H. (1982), Comportement asymptotique des équations de Navier–Stokes dans les domaines "aplatis", *Bull. Sc. Math. 2e Série* **106**, 369-385.

Feireisl E. and Zuazua E. (1992), Global attractors for semilinear wave equations with locally distributed nonlinear damping and critical exponent, Preprint.

Foias C. and Temam R. (1979), Some analytic and geometric properties of the solutions of the Navier–Stokes equations, *J. Math. Pures Appl.* **58**. 339-368.

Foias C. and Temam R. (1987), *The connection between the Navier–Stokes equation, dynamical systems, and turbulence theory*, in *Directions in Partial Differential Equations*, Academic Press, New-York, 55-73.

Garabedian P.R. and M. Schiffer (1952), Convexity of domain functionals, *J. d'Analyse Math.* **2**, 281-368.

Ghidaglia J.M. (1986), Some backward uniqueness results, *Nonlinear Analysis TMA* **10**, 777-720.

Ghidaglia J.M. and Héron B. (1987), Dimension of the attractors associated to the Ginzburg–Landau partial differential equation, *Physica D* **28**, 282-304.

Ghidaglia J.M. and Temam R. (1987), Attractors for damped nonlinear hyperbolic equations, *J. Math. Pures Appl.* **66**, 273-319.

Grisvard P. (1985), Elliptic Problems in Nonsmooth Domains, Monographs and Studies in Math. **24**, Pitman.

Hadamard J. (1907), *Mémoire sur le problème d'analyse relatif à l'équilibre des plaques élastiques encastrées*, in *Oeuvres de J. Hadamard 2*, Editions du CNRS, Paris, 1968.

Hale J.K. (1985), *Asymptotic behaviour and dynamics in infinite dimensions*, in *Research Notes in Mathematics 132*, Pitman, Boston, 1-41.

Hale J.K. (1986), Large diffusivity and asymptotic behavior in parabolic systems, *J. Math. Anal. Appl.* **118**, 455-466.

Hale J.K. (1988), Asymptotic Behaviour of Dissipative Systems, Math. Surveys and Monographs **25**, Amer. Math. Soc..

Hale J.K., Magalhães L. and Oliva W. (1984), An Introduction to Infinite Dimensional Dynamical Systems, Applied Math. Sciences **47**, Springer–Verlag.

Hale J.K. and Massat P. (1982), *Asymptotic behaviour of gradient–like systems*, in *Dynamical Systems II*, Academic Press, 85-101.

Hale J.K. and Raugel G. (1989), Lower semicontinuity of attractors of gradient systems and applications, *Annali di Mat. Pura Appl.* **(IV) (CLIV)**, 281-326.

Hale J.K. and Raugel G. (1991), *Partial differential equations on thin domains*, in *Differential Equations and Mathematical Physics: Proc. Int. Conf. in Alabama 1990*, Academic Press, 63-97.

Hale J.K. and Raugel G. (1992a), Reaction–diffusion equation on thin domains, *J. Math. Pures Appl.* **71**, 33-95.

Hale J.K. and Raugel G. (1992b), Convergence in gradient–like systems and applications, *ZAMP* **43**, 63-124.

Hale J.K. and Raugel G. (1992c), A damped hyperbolic equation on thin domains, *Trans. Am. Math. Soc.* **329**, 185-219.

Hale J.K. and Raugel G. (1992d), Dynamics of partial differential equations on thin domains, Preprint.

Hale J.K. and Raugel G. (1992e), *Attractors for dissipative evolutionary equations*, in *Equadiff 91, International Conference on Differential Equations, Barcelona 1991*, World Scientific, 1992.

Hale J.K. and Raugel G. (1993a), A reaction–diffusion equation on a thin L–shaped domain, Proc. Royal Soc. Edinburgh, to appear.

Hale J.K. and Raugel G. (1993b), *Attractors and convergence of PDE on thin L–shaped domains*, in *Progress in Partial Differential Equations: the Metz Surveys 2*, Longman, 149-171.

Hale J.K. and Raugel G. (1993c), Limits of semigroups depending on parameters, *Resenhas IME-USP* **1**, 1-45.

Hale J.K. and Sakamoto K. (1989), Shadow systems and attractors in reaction diffusion equations, *Applicable Anal.* **32**, 287-303.

Haraux A. (1985), *Two remarks on dissipative hyperbolic problems*, in *Séminaire du Collège de France 7*, Research Notes in Math.**122**, Pitman, 161-179.

Haraux A. (1986), Nonlinear Vibrations and the Wave Equation, Texto de Métodos Matematicos **20**, Rio de Janeiro.

Haraux A. (1987), Semi-linear Hyperbolic Equations in Bounded Domains, Mathematical Reports **3**, Harwood Academic Publishers, Gordon and Breach.

Haraux A. and Poláčik P. (1992), Convergence to a positive equilibrium for some nonlinear evolution equations in a ball, *Acta Math. Univ. Comenianae* **LXI**, 129-141.

Henry D. (1981), Geometric theory of semilinear parabolic equations, Lecture Notes In Math. **840**, Springer–Verlag.

Henry D. (1985a), Some infinite dimensional Morse–Smale systems defined by parabolic differential equations, *J. Diff. Equat.* **59**, 165-205.

Henry D. (1985b), Perturbations of the boundary for boundary value problems of partial differential equations, Sem. Brazileiro Anal. **22**, ATS.

Henry D. (1987), *Generic properties of equilibrium solutions by perturbations of the boundary*, in *Dynamics of Infinite Dimensional Systems, NATO ASI Series* **F 37**, Springer–Verlag, 129-139.

Henry D. (1988), Lecture notes on domain variation, Preprint.

Iosif'yan G.A., Oleinik O.A. and Shamaev A.S. (1989), On the limiting behavior of the spectrum of a sequence of operators defined on different Hilbert spaces, *Russian Math. Surveys* **44**, 195-196.

Jimbo S. (1988), Singular perturbation of domains and semilinear elliptic equation, *J. Fac. Sci. Univ. Tokyo* **35**, 27-76.

Jimbo S. (1988), Singular perturbation of domains and the semilinear elliptic equation, II, *J. Diff. Equat.* **75**, 264-289.

Jimbo S. (1989), The singulary perturbed domain and the characterization for the eigenfunctions with Neumann boundary conditions, *J. Diff. Equat.* **77**, 322-350.

Jimbo S. and Y. Morita (1992), Stability of non–constant steady state solutions to a Ginzburg-Landau equation in higher space dimensions, Preprint.

Kato T. (1966), *Perturbation theory for linear operators*, Springer–Verlag, Berlin and New-York.

Kishimoto K. and Weinberger H.F. (1985), The spatial homogeneity of stable equilibria of some reaction diffusion equations in convex domains, *J. Diff. Equat.* **57**, 15-21.

Kirchgässner K. and Raugel G. (1994a), Global stability of fronts in nonlinear parabolic systems, in preparation.

Kirchgässner K. and Raugel G. (1994b), Bifurcations of solitary or travelling waves, in preparation.

Kohn R.V. and Sternberg P. (1989), Local minimizers and singular perturbations, *Proc.Roy. Soc. Edinburgh* **111A**, 69-84.

Kohn R.V. and Vogelius (1984), A new model for thin plates with rapidly varying thickness, I, *Internat. J. Engrg. Sci.* **20**, 333-350.

Kohn R.V. and Vogelius (1985), A new model for thin plates with rapidly varying thickness, II. A convergence proof, *Quart. Appl. Math.* **43**, 1-21.

Kohn R.V. and Vogelius (1986), A new model for thin plates with rapidly varying thickness, III: Comparison of different scalings, *Quart. Appl. Math.* **44**, 35-48.

Kostin I.N. (1994a), Lower semicontinuity of a non–hyperbolic attractor, Preprint.

Kostin I.N. (1994b), Non–hyperbolic attractor for the Chafee–Infante problem, Preprint.

Kwak M. (1992a), Finite dimensional description of convective reaction diffusion equations, *J. Dyn. Diff. Eq.* **4**, 515-543.

Kwak M. (1992b), Finite dimensional inertial forms for the 2D Navier–Stokes equations, *Indiana J. Math.* **41**, 927-981.

Kurata K., Kisimoto K. and Yanagida E. (1989), The asymptotic transectional circumferential homogeneity of the solutions of reaction diffusion systems in cylinder–like domains, *J. Math. Biol.* **27**, 485-490.

Kwak M., Sell G. and Shao Z. (1994), Finite dimensional structures for Navier–Stokes equationq on thin 3D domains, Preprint.

Ladyzhenskaya O.A. (1969), *The Mathematical Theory of Viscous Incompressible Flow*, Gordon and Breach, New York.

Ladyzhenskaya O.A. (1972), On the dynamical system generated by the Navier–Stokes equations, *J. Soviet Math.* **3**, 458-479.

Lagnese J.E. and Lions J.L. (1988), *Modelling, Analysis and Control of Thin Plates*, Masson, RMA 6.

Le Dret H. (1989a), Modeling of the junction between two rods, *J. Math. Pures et Appl.* **68**, 365-397.

Le Dret H. (1989b), Folded plates revisited, *Comput. Mech.* **5**, 345-365.

Le Dret H. (1991), *Problèmes variationnels dans les multi–domaines – Modélisation des jonctions et applications*, Masson, RMA 19.

Lions J.L. (1969), *Quelques méthodes de résolution des problèmes aux limites non linéaires*, Gauthier Villars, Paris.

Lions J.L. (1973), Perturbations Singulières dans les Problèmes aux Limites et en Contrôle Optimal, Lecture Notes in Math. **323**, Springer–Verlag, Berlin.

Lions J.L. (1988), *Contrôlabilité exacte, perturbations et stabilisation de systèmes distribués,tome1*, Masson, RMA 8.

Lions J.L. (1988), Exact controllability, stabilization and perturbations for distributed domains, *SIAM Review* **30**, 1-68.

Lions J.L. and Prodi G. (1959), Un théorème d'existence et d'unicité dans les équations de Navier–Stokes en dimension 2, *C. R. Acad. Sci. Paris* **248**, 3519-3521.

Lions J.L. and Strauss W.A. (1965), Some non-linear evolution equations, *Bull. Soc. Math. France* **93**, 43-96.

Lions P.L. (1984), Structure of the set of steady–state solutions and asymptotic behaviour of semilinear heat equations, *J. Diff. Equat.* **53**, 362-386.

Lopes O. and Ceron S. (1984), Existence of forced periodic solutions of dissipative semilinear hyperbolic equations and systems, Preprint from University of Campinas, Sao Paulo, Brazil.

Malkin I. G. (1952), *Theory of Stability of Motion*, (in Russian), Moscow.

Mallet-Paret J. (1976), Negatively invariant sets of compact maps and an extension of a theorem of Cartwright, *J. Diff. Equat.* **22**, 331-348.

Mallet-Paret J. and Sell G. (1988), Inertial manifolds for reaction diffusion equations, *J. Amer. Math. Soc.* **1**, 805-866.

Marsden J., Ratiu T. and Raugel G. (1994), The Euler equations on thin domains, in preparation.

Matano H. (1978), Convergence of solutions of one–dimensional semilinear parabolic equations, *J. Math. Kyoto Univ.* **18**, 221-227.

Matano H. (1979), Asymptotic behaviour and stability of solutions of semilinear diffusion equations, *Publ. Rech. Inst. Math. Sci.* **15**, 401-458.

Matano H. (1982), Nonincrease of the lap number of a solution for a one dimensional semilinear parabolic equation, *J. Fac. Sci. Univ. Tokyo* **29**, 401-441.

Matano H. and Mimura M.(1983), Pattern formation in competition diffusion systems in non–convex domains, *Publ. RIMS Kyoto Univ.* **19**, 1049-1079.

Micheletti A.M. (1973), Perturbazione dello spettro di un operatore ellittico di tipo variazionale, in relazione ad una variazione del campo, *Annali Mat. Pura Appl.* **XCVII**, 261-281.

Mischaikow K. and Morita Y. (1994), Dynamics on the global attractor of a gradient flow arising from the Ginzburg–Landau equation, *Japan J. of Indus. and Appl. Math.* **11**, 185-202.

Mischaikow K. and Raugel G. (1994), The use of Conley index in thin domain problems, in preparation.

Morita Y. (1990), Reaction diffusion systems in nonconvex domains: Invariant manifold and reduced form, *J. Dyn. Diff. Eq.* **2**, 69-115.

Morita Y. and Jimbo S. (1992), Ordinary differential equations on inertial manifolds for reaction diffusion systems in a singularly perturbed domain with several channels, *J. Dyn. Diff. Eq.* **4**, 65-93.

Nickel K. (1962), Gestaltaussagen über Lösungen parabolischer Differentialgleichungen, *J. Reine Angew. Math.* **211**, 78-94.

Nishiura Y. (1982), Global structure of bifurcating solutions for reaction diffusion systems, *SIAM J. Math. Anal.* **13**, 555-593.

Palis J. and de Melo W. (1982), *Geometric Theory of Dynamical Systems*, Springer-Verlag, Berlin.

Pereira A.L. (1989), Auto valores do Laplaciano im regiões simétricas, Ph. D. Thesis, Univ. São Paulo.

Pliss V.A. and Sell G. (1993), *Approximations of the long-time dynamics of the Navier–Stokes equations*, in *Differential Equations and Geometric Dynamics: Control Science and Dynamical Systems*, Marcel Dekker Publ., New York.

Poláčik P. and Rybakowski K. (1994), Non convergent bounded trajectories in semilinear heat equations, Preprint.

Ramm A.G. (1985), Limit of the spectra of the interior Neumann problems when a solid shrinks to a plane one, *J. Math. Anal. Appl.* **108**, 107-112.

Raoult A. (1988), Analyse mathématique de quelques modèles de plaques et de poutres élastiques ou élasto–plastiques, Thèse de Doctorat d'Etat, Univ. Pierre et Marie Curie, Paris.

Raugel G. (1989), Continuity of attractors, *Math. Modelling and Numerical Analysis, M2AN* **23**, 519-533.

Raugel G. (1993), Persistence of Morse–Smale properties under some approximations and singular perturbations, Preprint.

Raugel G. and Sell G. (1989), Equations de Navier–Stokes dans des domaines minces en dimension trois: régularité globale, *C. R. Acad. Sci. Paris* **309**, 299-303.

Raugel G. and Sell G. (1992), *Navier–Stokes equations on thin 3D domains. II: Global regularity of spatially periodic solutions*, in *Collège de France Proceedings, Pitman Res. Notes Math. Ser.*, Longman, to appear.

Raugel G. and Sell G. (1993a), Navier–Stokes equations on thin 3D domains. I: Global attractors and global regularity of solutions, *J. Amer. Math. Soc.* **6**, 503-568.

Raugel G. and Sell G. (1993b), *Navier–Stokes equations on thin 3D domains. III: Existence of a global attractor*, in *Turbulence in Fluid Flows, The IMA Vol. in Math. and its Appl.* 55, Springer–Verlag, 137-163.

Rocha C. (1985), Generic properties of equilibria of reaction–diffusion equations with variable diffusion, *Prc. Roy. Soc. Edinburgh* **101 A**, 45-56.

Ruiz A., Unique continuation for weak solutions of the wave equation plus a potential, Preprint.

Saut J.C. and Temam R. (1979), Generic properties of nonlinear boundary value problems, *Comm. in PDE* **4**, 293-319.

Segal I. (1963), Nonlinear semigroups, *Ann. Math.* **78**, 339-364.

Sell G.R. and Taboada M. (1992), Local dissipativity and attractors for the Kuramoto–Sivashinsky equation in thin 2D domains, *Nonlinear Analysis, TMA* **18**, 671-687.

Simon J. (1981), Differentiation with respect to the domain in boundary value problems, *Num. Funct. Anal. Appl.* **2**, .

Simon J. (1983), Asymptotics for a class of nonlinear evolution equations, with applications to geometric problems, *Ann. Math.* **118**, 525-571.

Smoller J. (1983), *Shock Waves and Reaction Diffusion Equations*, Springer–Verlag.

Smoller J. and Wasserman A. (1984), Generic properties of steady state solutions, *J. Diff. Equat.* **52**, 423-438.

Stummel F. (1976), Perturbation of Domains in Elliptic Boundary Value Problems, Lecture Notes in Math. **503**, Springer–Verlag.

Tataru D. (1992), Uniform decay rates and attractors for evolution PDE with boundary conditions, Preprint.

Teman R. (1975), On the Euler equations of incompressible perfect fluids, *J. of Funct. Anal.* **20**, 32-43.

Temam R. (1977), *Navier–Stokes Equations*, North–Holland, Amsterdam.

Teman R. (1982), Behaviour at time $t = 0$ of the solutions of semilinear evolution equations, *J. Diff. Equ.* **43**, 73-92.

Teman R. (1983), Navier–Stokes Equations and Nonlinear Functional Analysis, CBMS Regional Conference Series **41**, SIAM, Philadelphia.

Temam R. (1988), *Infinite dimensional dynamical systems in mechanics and physics*, Springer–Verlag, New York.

Uhlenbeck K. (1976), Generic properties of eigenfunctions, *Am. J. Math.* **98**, 1059-1078.

Vegas J.M. (1983), Bifurcation caused by perturbing the domain of an elliptic equation, *J. Diff. Equat.* **48**, 189-226.

Vegas J.M. (1990), A functional analytic framework for the study of elliptic equations on variable domains, *Proc. Roy. Soc. Edinburgh* **116**, 367-380.

Vishik M.I. (1992), *Asymptotic Behaviour of Solutions of Evolutionary Equations*, Lezioni Lincee, Cambridge University Press.

Wells (1976), Invariant manifolds of nonlinear operators, *Pacific J. Math.* **62**, 285-293.

Yanagida E. (1982), Stability of stationary distribution in a space–dependent population growth process, *J. Math. Biol.* **15**, 37-50.

Yanagida E. (1990), Existence of stable stationary solutions of scalar reaction diffusion equations in thin tubular domains, *Applicable Anal.* **36**, 171-188.

Zelenyak T.J. (1968), Stabilization of solutions of boundary value problems for a second order parabolic equation with one space variable, *Diff. Equat.* **4**, 17-22.

Zuazua (1990), Exponential decay for the semilinear wave equation with locally distributed dumping, *Comm. PDE* **15**, 205-235.

Geneviève RAUGEL
Laboratoire d'Analyse Numérique, Bât. 425
Université de Paris–Sud (et CNRS), 91405 ORSAY Cedex (France)

C.I.M.E. Session on "Dynamical Systems"

List of Participants

A. CAPIETTO, Dip. Mat. Univ., Via Carlo Alberto 10, 10123 Torino

M. CECCHI, Dip. Ing. Elettr., Via S. Marta 3, 50139 Firenze

I.S. CIUPERCA, Lab. Anal. Num., Bât. 425, 91405 Orsay Cedex, France

E. D'AMBROGIO, Dip. Mat. Univ., P.le Europa 1, 34127 Trieste

M. FURI, Dip. di Mat. Appl., Via S. Marta 3, 50139 Firenze

M.C. GIURIN, Via Sestio Cavino 60, 00174 Roma

H. HANSSMANN, Rijksuniversiteit Groningen, Vakgroep Wiskunde, Postbus 800,
        9700 AV Groningen, The Netherlands

T. KACZYNSKI, Dép. Mat.-Info., Univ. de Sherbrooke, Québec, Canada J1K 2RI

S. LUZZATTO, SISSA, Via Beirut 4, 34014 Trieste

A. LYASHENKO, Ist. di Anal. Globale e Appl., Via S. Marta 13/1, 50139 Firenze

L. MALAGUTI, Dip. di Mat., Via Campi 213/b, 41100 Modena

M. MARINI, Dip. di Ing. Elettr., Via S. Marta 3, 50139 Firenze

C. MASCIA, Via della Farnesina 212, 00194 Roma

T. MESTL, Dept. of Math. Sciences, Agricultural Univ. of Norway, N-1432 Aas, Norway

A. MORO, Dip. Statistico, Viale Morgagni 59, 50134 Firenze

F. NARDINI, Dip. di Mat., Piazza di Porta S. Donato 5, 40127 Bologna

P. NISTRI, Dip. di Sistemi e Informatica, Via S. Marta 3, 50139 Firenze

P. PERA, Dip. di Mat. Appl., Via S. Marta 3, 50139 Firenze

M. PLANK, Westbahnstrasse 28/2/2/4, 1070 Wien, Austria

C. QUARANTA VOGLIOTTI, Dip; di Scienze dell'Inform., Via Comelico 39, 20135 Milano

G. RASTELLI, Via Casalone 3, 13030 Lignana (Vicenza)

L. SBANO, SISSA, Via Beirut 2/4, 34014 Trieste

M. SPADINI, Dip. di Mat., Viale Morgagni 67/A, 50134 Firenze

D. STOFFER-MANNALE, ETH-Zentrum, Mathematik, CH-8092 Zurich, Switzerland

I. TERESCHAK, Fac. of Math. and Phys., Comenius Univ., Mlynska dolina,
        842 15 Bratislava, Slovakia

D. TOGNOLA, ETH-Zentrum, Mathematik, CH-8092 Zurich, Switzerland

M. YEBDRI, Math. Inst., Theresienstr. 39, D-8000 Munchen 2, Germany

P. ZECCA, Dip. di Sistemi e Informatica, Via S. Marta 3, 50139 Firenze

P. ZGLICZYNSKI, Instytut Matematyki UJ, ul. Reymonta 4, 30-059 Krakow, Poland

FONDAZIONE C.I.M.E.
CENTRO INTERNAZIONALE MATEMATICO ESTIVO
INTERNATIONAL MATHEMATICAL SUMMER CENTER

"Probabilistic Models for Nonlinear PDE's and Numerical Applications"

is the subject of the First 1995 C.I.M.E. Session.

The Session, sponsored by the Consiglio Nazionale delle Ricerche (C.N.R.), the Ministero dell'Università e della Ricerca Scientifica e Tecnologica (M.U.R.S.T.) and the Azienda di Promozione Turistica Montecatini Terme/Val di Nievole, will take place, under the scientific direction of Professor DENIS TALAY (I.N.R.I.A., Sophia Antipolis) and Professor LUCIANO TUBARO (Università di Trento), at Montecatini Terme (Pistoia), **from 22 to 30 May, 1995.**

Courses

a) **Limit theorems for solutions of stochastic equations.** (6 lectures in English)
Prof. Tom KURTZ (University of Wisconsin, Madison)

1. Limit for stochastic differential equations.
This lecture would cover the basic material in my paper with Protter.

2. Models with lower dimensional limitis (one or two lectures).
The lecture or lectures would cover work of Shon Katzenberger and my joint work with Federico Marchetti in which the limiting SDEs are on a lower dimensional manifold. Katzenberger's work has application in genetics and queueing and the work with Marchetti is concerned with mechanical systems.

3. Limit theorems for infinite dimensional SDEs.
This lecture would probably cover extensions of the material of Lecture I to SDEs driven by martingale measures. (I have a student completing a dissertation in this area.) It could, however, consider models of particle systems in which the number of particles goes to infinity.

4. Stochastic embedding equations (two lectures).
This material is concerned with representations of stochastic processes in terms of Poisson and other spatial point processes. Applications include queueing models, population models, and spatial epidemic models. Limit theorems are obtained by invoking the law of large numbers and the CLT for the underlying Poisson process.

b) **Asymptotic behaviour of some interacting particle systems; McKean-Vlasov and Boltzmann models.**
(6 lectures in English)
Prof. Sylvie MÉLÉARD (Université Paris 6)

Introduction
Two examples:
1) The McKean-Vlasov equation and the approximating random mean field particle system.
2) The Boltzmann equation: the physical model and some mollified Boltzmann equations.

Propagation of chaos for exchangeable mean field systems of diffusions with jumps

1) Description of the model.

2) Definition of the propagation of chaos.

3) Relation between the propagation of chaos and the convergence of the empirical measures as probability measures on the path space.

4) Equivalence between the tightness of the laws of the empirical measures and of the laws of the diffusions.

5) A general criterion of tightness for the laws of cadlag semimartingales.

6) Characterisation of the limit values: non linear martingale problem.

7) Uniqueness of the non linear limit martingale problem. The underlying nonlinear partial differential equation.

Fluctuations for the McKean-Vlasov model

1) Convergence of the finite dimensional marginals.

2) Convergence in a Sobolev space.

A pathwise approach for some pure jumps systems with shared resources

Examples: some networks models, mollified (non spatially homogeneous) Boltzmann equations.

1) A result of convergence in variation norm; the rate of convergence.

2) A pathwise representation of the past of a fixed number of particles by using random interaction graphs.

3) Coupling and interaction chains.

4) The limit tree.

Algorithms for the Boltzmann equation

One uses the previous section to construct some random algorithms which simulate the approximating Boltzmann particle systems.

c)   **Weak convergence of stochastic integrals.** (2 lectures in English)
       Prof. Philips PROTTER (Purdue University)

Description:

   We will begin with a brief description of convergence in the Skorohod topology, followed by notions of weak convergence. We will then recall the definition of a semimartingale as a good integrator and discuss the notion of uniform tightness. We will then give necessary and sufficient conditions for stochastic integrals to converge, under appropriate assumption on the integrands (eg. the must be cadlag processes so that they are in the Skorohod space). Applications to Stochastic Differential Equations will also be given.

d)   **Kinetic limits for stochastic particle systems.** (6 lectures in English)
       Prof. Mario PULVIRENTI (Università di Roma "La Sapienza")

- Boltzmann equation for hard spheres. Hard sphere dynamics, BBKGY hierarchy. Heuristic derivation of the Boltzmann equation. (1 lecture)
- Stochastic particle methods for the Boltzmann equation. Convergence of some particle schemes. Explicit error estimate. (2 lectures)
- Stochastic particle systems with strictly local interaction. Kinetic limits for model equations. Law of large numbers. Propagation of chaos. Cluster expansion for small coupling constant. (3 lectures)

References

- C. Cercignani, R. Illner, M. Pulvirenti, The Mathematical Theory of Dilute Gases, Applied Math. Sciences, Springer-Verlag n. 106 (1994).

e)   **Branching processes and non-linear PDE.** (2 lectures in English)
      Prof. Alain ROUAULT (Université de Versailles)

Outline

Convergence of Sherman-Peskin scheme, probabilistic representation of the solutions to nonlinear PDE and superprocesses.

References

- B. Chauvin, A. Rouault, A stochastic simulation for solving scalar reaction-diffusion equations, Adv. in Appl. Probab. 22 (1990), 88-100.
- G. Ben Arous, A. Rouault, Laplace asymptotics for reaction-diffusion equations, Prob. Theory and Rel. Fields 97 (1993), 259-285.
- Course by D. Dawson: Lecture Notes n. 1541 Ecole d'Eté de Probabilités de Saint Flour XXI (1991).

f)   **Elements on probabilistic numerical methods for partial differential equations.** (6 lectures in English)
      Prof. Denis TALAY (I.N.R.I.A., Sophia Antipolis)

Abstract

The objective of the lectures is to present some recent results on probabilistic numerical methods for some deterministic Partial Differential Equations, with a particular emphasis on the construction of the methods (for which adequate probabilistic interpretations of the PDE's are necessary), and the error analysis.

Contents:

I - MonteCarlo methods for parabolic PDE's
      Principle, advantages and disadvantages of the method
      The Euler and Milshtein schemes for stochastic differential equations
      Expansion of the error
      The stationary case
      Lepingle's reflected Euler scheme for Neumann boundary conditions
      Newton's variance reduction technique

II - Introduction to Stochastic Particle Methods
      Numerical solving of the Fokker-Planck equation
      Convection-reaction-diffusion equations: Chorin-Puckett's method, Sherman-Peskin's method
      One-dimensional McKean-Vlasov equations
      The Burgers equation
      Chorin's random vortex method for the 2-D inviscid Navier-Stokes equation

References

- P. Bernard, D. Talay, L. Tubaro. Rate of convergence of a stochastic particle method for the Kolmogorov equation with variable coefficients. *Mathematics of Computation,* 63, 1994.
- M. Bossy. *Thèse d'Université.* PhD thesis, Université de Provence, 1995.
- M. Bossy, D. Talay. Convergence rate for the approximation of the limit law of weakly interacting particles: application to the Burgers equation. Rapport de Recherche 2410, Inria, novembre 1994. Submitted for publication.
  M. Bossy, D. Talay. A stochastic particles method for nonlinear PDE's of Burgers type. Submitted for publication (a part of the paper has appeared as a Rapport de Recherche INRIA (number 2180), under the title "Convergence rate for the approximation of the limit law of weakly interacting particles. 1: Smooth interacting kernesls"). 1994.
- A.L. Chorin. Vortex methods and Vortex Statistics - Lectures for Les Houches Summer School of Theoretical Physics. *Lawrence Berkeley Laboratory Prepublications,* 1993.
- A.J. Chorin, J.E. Marsden. *A Mathematical Introduction to Fluid Mechanics.* Springer Verlag, 1993.
- O.H. Hald. Convergence of random methods for a reaction diffusion equation. *SIAM J Sci. Stat. Comput.,* 2:85-94, 1981.
- O.H. Hald. Convergence of a random method with creation of vorticity. *SIAM J. Sci. Stat. Comput.,* 7:1373-1386, 1986.
- D.G. Long. Convergence of the random vortex method in two dimensions. *Journal of the American Mathematical Society,* 1(4), 1988.
- E.G. Puckett. A study of the vortex sheet method and its rate of convergence. *SIAM J. Sci. Stat. Comput.,* 10(2): 298-327, 1989.
- E.G. Puckett. Convergence of a random particle method to solutions of the Kolmogorov equation. *Mathematics of Computation.* 52(186):615-645, 1989.
- S. Roberts. Convergence of a random walk method for the Burgers equation. *Mathematics of Computation,* 52(186):647-673, 1989.
- A.S. Sznitman. Topics in propagation of chaos. In P.L. Hennequin, editor, *Ecole d'Eté de Probabilités de Saint Flour XIX,* volume 1464 of *Lecture Notes in Mathematics,* Berlin, Heidelberg, New York, 1989. Springer Verlag.

FONDAZIONE C.I.M.E.
CENTRO INTERNAZIONALE MATEMATICO ESTIVO
INTERNATIONAL MATHEMATICAL SUMMER CENTER

"Viscosity Solutions and Applications"

is the subject of the Second 1995 C.I.M.E. Session.

The Session, sponsored by the Consiglio Nazionale delle Ricerche (C.N.R.), the Ministero dell'Università e della Ricerca Scientifica e Tecnologica (M.U.R.S.T.) and the Azienda di Promozione Turistica Montecatini Terme/Val di Nievole, will take place, under the scientific direction of Professor ITALO CAPUZZO DOLCETTA (Università di Roma, La Sapienza) and Professor PIERRE LOUIS LIONS (Université Paris-Dauphine) at Montecatini Terme (Pistoia), **from 12 to 20 June, 1995.**

Courses

a) **Deterministic optimal control and differential games.** (6 lectures in English)
   Prof. M. BARDI (Università di Padova)

The dynamic programming method and Hamilton-Jacobi-Bellman equations
Necessary and sufficient conditions of optimality, synthesis of multivalued optimal feedbacks
Discontinuous viscosity solutions
Convergence of numerical schemes for Hamilton-Jacobi equations, synthesis of approximate optimal feedbacks
Two-players zero-sum differential games, Isaac's equations

References

- L.C. Evans and P.E. Souganidis, Differential games and representation formulas for solutions of Hamilton-Jacobi equations. Indiana Univ. Math. J., 33 (1984).
- G. Barles and B. Perthame, Discontinuous solutions of deterministic optimal stopping time problems, RAIRO Model. Math. Anal. Num., 21 (1987).
- M.G. Crandall, L.C. Evans and P.L. Lions, Some properties of viscosity solutions of Hamilton-Jacobi equations. Trans. Amer. Math. Soc., 282 (1984).
- M. Bardi, M. Falcone and P. Soravia. Fully discrete schemes for the value function of pursuit-evasion games, in Advances in Dynamic Games and Applications, T. Basar and A. Haurie eds., Birkhauser (1994).
- M. Bardi and I. Capuzzo Dolcetta, Optimal Control and Viscosity Solutions of Hamilton-Jacobi-Bellman Equations, Birkhauser, Boston, to appear.

b) **General theory of viscosity solutions.** (6 lectures in English)
   Prof. M. G. CRANDALL (University of California, Santa Barbara)

Scalar fully nonlinear pde's of first and second order: examples, nonexistence of classical solutions, nonuniqueness of strong solutions
The notion of viscosity solutions, uniqueness in the first order case, examples of existence and uniqueness theorems for viscosity solutions
Uniqueness for second order equations
Closure under limit operations: Perron's method and existence
Generalized boundary conditions
Extensions to infinite dimensions

References

- M.G. Crandall, H. Ishii and P.L. Lions, User's guide to viscosity solutions of second order partial differential equations, Bull. Amer. Math. Soc., 27 (1992).
- A. Pazy, Semigroups of linear operators and applications to partial differential equations, Springer-Verlag, New York (1983).
- L.C. Evans, R.F. Gariepy, Measure theory and fine properties of functions, Studies in Advanced Math., CRC Press, Ann Arbor (1992).

c) **Fully nonlinear equations and motion by mean curvature.** (6 lectures in English)
    Prof. L. C. EVANS (University of California, Berkeley)

Fully nonlinear elliptic PDE's
Regularity for convex nonlinearities
Viscosity solutions for general nonlinearities
Motion by mean curvature I: introduction
Motion by mean curvature II: level set method
Motion by mean curvature III: properties of the generalized flow

References

- D. Gilbarg and N. Trudinger, Elliptic Partial Differential Equations of the Second Order (chapter 17), Springer (1983).
- Y. Chen, Y. Giga and S. Goto, Uniqueness and existence of viscosity solutions of generalized mean curvature flow equations. J. Diff. Geom. 33 (1991).
- L.C. Evans and J. Spruck, Motion of level sets by mean curvature. J. Diff. Geom., 33 (1991).
- R. Jensen, P.L. Lions and P. Souganidis, A uniqueness result for viscosity solutions of second order, fully nonlinear PDE's, Proc. Am. Math. Soc., 102 (1988).
- M.G. Crandall, H. Ishii and P.L. Lions, User's guide to viscosity solutions to 2nd order partial differential equations, Bull. Amer. Math. Soc., 27 (1992).

d) **Optimal control and mathematical finance.** (6 lectures in English)
    Prof. M. H. SONER (Carnegie Mellon University)

Optimal control of Markov processes
Dynamic programming equations for controlled Markov diffusion processes
Viscosity solutions of Hamilton-Jacobi-Bellman equations
Investment-consumption and option pricing problems
Scheduling problems in manufacturing

References

- M.G. Crandall, H. Ishii and P.L. Lions, User's guide to viscosity solutions to 2nd order partial differential equations, Bull. Amer. Math. Soc., 27 (1992).
- W.H. Fleming and M.H. Soner, Controlled Markov Processes and Viscosity Solutions, Springer-Verlag (1993).
- F. Black and M. Scholes, The pricing of options and corporate liabilities. J. Political Economy, 81 (1973).
- M.H.A. Davis, V. Panas and T. Zariphopoulou, European option pricing with transactions costs. SIAM J. Control Optim. 31 (1993).
- I. Karatzas, J. Lehoczky, S. Sethi and S. Shreve, Explicit solution of a general consumption-investment problem. Math. Oper. Res., 11 (1986).

e) **Asymptotic problems in front propagation.** (6 lectures in English)
    Prof. P. E. SOUGANIDIS (University of Wisconsin)

General discussion and examples from phase transitions
Phase-field theory and asymptotics of reaction-diffusion equations
Macroscopic limits of stochastic Ising models with long range interactions
A general theory about moving fronts and asymptotic limits
Large scale front dynamics for turbulent reactions-diffusion equations

References

- G. Barles, H.M. Soner and P.E. Souganidis, Front propagation and phase field theory, SIAM J. Control and Optimization, 31 (1993).
- L.C. Evans, H.M. Soner and P.E. Souganidis, Phase transitions and generalized motion by mean curvature. Comm. Pure Appl. Math. XLV (1992).
- M.A. Katsoulakis and P.E. Souganidis, Interacting particle systems and generalized evolution of fronts. Arch. Rat. Mech. Anal., in press.
- M.A. Katsoulakis and P.E. Souganidis, Generalized motion by mean cruvature as a macroscopic limit for stochastic Ising models with long range interactions and Glauber dynamics. Comm. Math. Physics, to appear.
- A. Majda and P.E. Souganidis, Large scale front dynamics for turbulent reaction-diffusion equations with separated velocity scales. Nonlinearity, 7 (1994).

FONDAZIONE C.I.M.E.
CENTRO INTERNAZIONALE MATEMATICO ESTIVO
INTERNATIONAL MATHEMATICAL SUMMER CENTER

"Vector Bundles on curves. New Directions"

is the subject of the Third 1995 C.I.M.E. Session.

The Session, sponsored by the Consiglio Nazionale delle Ricerche (C.N.R.), and the Ministero dell'Università e della Ricerca Scientifica e Tecnologica (M.U.R.S.T.) will take place, under the scientific direction of Professor M. S. NARASIMHAN at Grand Hotel San Michele, Cetraro (Cosenza), **from 19 to 27 June, 1995.**

Courses

a) **Kac-Moody Groups, Their Flag Varieties and Moduli Spaces of G-bundles** (8 lectures in English)
Prof. Shrawan KUMAR (University of North-Carolina)

Outline

The basic theory of Kac-Moody groups and their flag varieties will be developed, including their constructions, elementary geometric properties, central extension, homogeneous line bundles, Picard group of the flag varieties and the analog of Borel-Weil-Bott theorem for Kac-Moody groups.

Further, the affine Kac-Moody flag variety will be realized as a parameter space for algebraic G-bundles on a smooth complex projective curve C of any given genus (where G is finite dimensional complex simple simply connected algebraic group). It will be shown that this family plays an important role in connecting the moduli space M of semistable G-bundles on C and the affine flag variety. This connection will be exploited to calculate the Picard group of the moduli space M.

b) **Drinfeld Stukas.** (8 lectures in English)
Prof. G. LAUMON (Université de Paris-Sud, Orsay)

Outline

In the seventies, Drinfeld has introduced a whole family of new mathematical objects: modules, elliptic sheaves and stukas over a function field F of characteristic p.

These objects have some rank $r \in Z_{>0}$. In the well-known analogy between function fields and number fields, rank 2 elliptic modules correspond to elliptic curves. The rank r elliptic sheaves and the rank r elliptic modules are equivalent objects. The rank r elliptic sheaves are rank r stukas of a special kind.

Using the etale cohomology of the modular variety of rank 2 stukas, Drinfeld has been able to prove the global Langlands correspondence for the group GL_2 over F in full generality. He also has suggested that the stukas of arbitrary rank r are precisely the objects which are needed to prove the global Langlands correspondence for the group GL_r over F in full generality.

More recently, Stuhler has introduced a D-version of those objects, where D is a global order of a central division algebra D over F, Rapoport, Stuhler and I have used the rank 1 D-elliptic sheaves to prove the local Langlands correspondence in positive characteristic and Lafforgue has studied the D-stukas of arbitrary rank and their modular varieties.

The purpose of these lectures is to explain what the D-stukas are and what they are good for and to give a survey of the recent progresses made on this subject.

Organization of the lectures.

* **Lecture 1**: Definition of the D-stukas, level structures.
* **Lecture 2**: Modular stacks, partial Frobenius morphisms, Hecke operators.
* **Lecture 3**: Adelic description of the D-stukas with "finite" pole and zero.

324

* **Lecture 4**: The rank one case, projectivity of the modular varieties and application to Langlands correspondence for D^x.
* **Lecture 5**: Reducible stukas and horocycles.
* **Lecture 6**: Harder-Narasimhan filtration of a D-stuka, truncations of the modular stacks.

The main references for these lectures are:

- V.G. Drinfeld. Varieties of modules of F-sheaves. Funct. Anal. and its Appl. 21 (1987), 107-122.
- V.G. Drinfeld. Proof of Peterssons's conjecture for $GL(2)$ over a global field of characteristic p. Funct. Anal. and its Appl. 22 (1988), 28-43.
- L. Lafforgue. D-stukas de Drinfeld. Thèse, Université Paris-Sud, 1994.

c) **Drinfeld modules and elliptic shaves.** (8 lectures in English)
Prof. U. STUHLER (Universität Göttingen)

Outline

The above mentioned modules were introduced around 1973 by V. Drinfeld in his fundamental paper [2]. They are an analogue of the classical concept of an elliptic curve in the situation of a function field over a finite field of constants and bear a strong resemblance with this theory. They are an indispensable tool (together with their generalisations, the stukas) to study the Langlands correspondence but have also an interesting life in their own right. Besides this analogy with elliptic curves they have a description in terms of locally free sheaves and Frobenius operations which brings them into contact with the theory of vector bundles over curves. Surprising enough additionally there are close relations with the theory of the Korteweg-de Vries and similar differential as well as difference equations. The purpose of these lectures is to introduce the participants to this beautiful theory with its many facettes. Additionally I hope to speak about some variants of this concept, the so called D-elliptic sheaves where D is a sheaf of division algebras over the curve involved.

These are useful in terms of the Langlands program for function fields but otherwise the theory is not as complete so that there are interesting open questions.

Other topics will probably concern questions of uniformisation of the moduli problems (at infinity as well as other primes), compactification questions and results concerning the cohomology and G. Andersons's t-motives.

References for these lectures are:

- P. Deligne, P. Husemoller: Survey of Drinfeld modules. In: Ribet, K.A. (ed.) Current trends in arithmetical algebraic geometry. Contemp. Math. vol. 67, pp. 25-91) Providence, R.I. Ann. Math. 1987.
- V.G. Drinfeld. Elliptic modules. Math. USSR, Sb. 23, 561-592 (1974.
- V.G. Drinfeld. Elliptic modules II. Math. USSR, Sb. 31, 155-170 (1977).
- V.G. Drinfeld. Commutative subrings of certain noncommutation rings. Funct. Anal. Appl. 11, 9-12, 1977.
- G. Laumon, Forthcoming book on Drinfeld modules and the cohomology of their moduli spaces. Cambridge University Press.
- G. Laumon, M. Rapoport, U. Stuhler: D-elliptic sheaves and the Langlands correspondence. Invent. math. 113, 217-330 (1993).

Organisation of the lectures:

* **Lecture 1,2**: Definition of Drinfeld modules, algebraic and analytic theory, level structures, rings of endomorphisms, uniformisation.
* **Lecture 3**: Description of Drinfeld modules in terms of locally free sheaves, t-motives, the noncommutative projective line. Analogies for rings of differential operators.
* **Lecture 4**: Coverings of the upper half plane and their cohomology. Rigid analytic geometry.
* **Lectures 5,6**: D-elliptic sheaves, moduli problems, good and bad fibres, applications to the Langlands program.

LIST OF C.I.M.E. SEMINARS   Publisher

1954 - 1. Analisi funzionale   C.I.M.E.
        2. Quadratura delle superficie e questioni connesse   "
        3. Equazioni differenziali non lineari   "

1955 - 4. Teorema di Riemann-Roch e questioni connesse   "
        5. Teoria dei numeri   "
        6. Topologia   "
        7. Teorie non linearizzate in elasticità, idrodinamica,aerodinamica "
        8. Geometria proiettivo-differenziale   "

1956 - 9. Equazioni alle derivate parziali a caratteristiche reali   "
       10. Propagazione delle onde elettromagnetiche   "
       11. Teoria della funzioni di più variabili complesse e delle
           funzioni automorfe   "

1957 - 12. Geometria aritmetica e algebrica (2 vol.)   "
       13. Integrali singolari e questioni connesse   "
       14. Teoria della turbolenza (2 vol.)   "

1958 - 15. Vedute e problemi attuali in relatività generale   "
       16. Problemi di geometria differenziale in grande   "
       17. Il principio di minimo e le sue applicazioni alle equazioni
           funzionali   "

1959 - 18. Induzione e statistica   "
       19. Teoria algebrica dei meccanismi automatici (2 vol.)   "
       20. Gruppi, anelli di Lie e teoria della coomologia   "

1960 - 21. Sistemi dinamici e teoremi ergodici   "
       22. Forme differenziali e loro integrali   "

1961 - 23. Geometria del calcolo delle variazioni (2 vol.)   "
       24. Teoria delle distribuzioni   "
       25. Onde superficiali   "

1962 - 26. Topologia differenziale   "
       27. Autovalori e autosoluzioni   "
       28. Magnetofluidodinamica   "

1972 - 59. Non-linear mechanics   "
        60. Finite geometric structures and their applications   "
        61. Geometric measure theory and minimal surfaces   "

1973 - 62. Complex analysis   "
        63. New variational techniques in mathematical physics   "
        64. Spectral analysis   "

1974 - 65. Stability problems   "
        66. Singularities of analytic spaces   "
        67. Eigenvalues of non linear problems   "

1975 - 68. Theoretical computer sciences   "
        69. Model theory and applications   "
        70. Differential operators and manifolds   "

1976 - 71. Statistical Mechanics   Ed Liguori, Napoli
        72. Hyperbolicity   "
        73. Differential topology   "

1977 - 74. Materials with memory   "
        75. Pseudodifferential operators with applications   "
        76. Algebraic surfaces   "

1978 - 77. Stochastic differential equations   "
        78. Dynamical systems   Ed Liguori, Napoli and Birhäuser Verlag

1979 - 79. Recursion theory and computational complexity   "
        80. Mathematics of biology   "

1980 - 81. Wave propagation   "
        82. Harmonic analysis and group representations   "
        83. Matroid theory and its applications   "

1981 - 84. Kinetic Theories and the Boltzmann Equation (LNM 1048) Springer-Verlag
        85. Algebraic Threefolds   (LNM 947)   "
        86. Nonlinear Filtering and Stochastic Control (LNM 972)   "

1982 - 87. Invariant Theory   (LNM 996)   "
        88. Thermodynamics and Constitutive Equations   (LN Physics 228)   "
        89. Fluid Dynamics   (LNM 1047)   "

1993 -	117. Integrable Systems and Quantum Groups	to appear	Springer-Verlag
	118. Algebraic Cycles and Hodge Theory	(LNM 1594)	
	119. Phase Transitions and Hysteresis	(LNM 1584)	"
1994 -	120. Recent Mathematical Methods in Nonlinear Wave Propagation	to appear	"
	121. Dynamical Systems	(LNM 1609)	"
	122. Transcendental Methods in Algebraic Geometry	to appear	"
1995 -	123. Probabilistic Models for Nonlinear PDE's and Numerical Applications	to appear	"
	124. Viscosity Solutions and Applications	to appear	"
	125. Vector Bundles on Curves. New Directions	to appear	"

Vol. 1515: E. Ballico, F. Catanese, C. Ciliberto (Eds.), Classification of Irregular Varieties. Proceedings, 1990. VII, 149 pages. 1992.

Vol. 1516: R. A. Lorentz, Multivariate Birkhoff Interpolation. IX, 192 pages. 1992.

Vol. 1517: K. Keimel, W. Roth, Ordered Cones and Approximation. VI, 134 pages. 1992.

Vol. 1518: H. Stichtenoth, M. A. Tsfasman (Eds.), Coding Theory and Algebraic Geometry. Proceedings, 1991. VIII, 223 pages. 1992.

Vol. 1519: M. W. Short, The Primitive Soluble Permutation Groups of Degree less than 256. IX, 145 pages. 1992.

Vol. 1520: Yu. G. Borisovich, Yu. E. Gliklikh (Eds.), Global Analysis – Studies and Applications V. VII, 284 pages. 1992.

Vol. 1521: S. Busenberg, B. Forte, H. K. Kuiken, Mathematical Modelling of Industrial Process. Bari, 1990. Editors: V. Capasso, A. Fasano. VII, 162 pages. 1992.

Vol. 1522: J.-M. Delort, F. B. I. Transformation. VII, 101 pages. 1992.

Vol. 1523: W. Xue, Rings with Morita Duality. X, 168 pages. 1992.

Vol. 1524: M. Coste, L. Mahé, M.-F. Roy (Eds.), Real Algebraic Geometry. Proceedings, 1991. VIII, 418 pages. 1992.

Vol. 1525: C. Casacuberta, M. Castellet (Eds.), Mathematical Research Today and Tomorrow. VII, 112 pages. 1992.

Vol. 1526: J. Azéma, P. A. Meyer, M. Yor (Eds.), Séminaire de Probabilités XXVI. X, 633 pages. 1992.

Vol. 1527: M. I. Freidlin, J.-F. Le Gall, Ecole d'Eté de Probabilités de Saint-Flour XX – 1990. Editor: P. L. Hennequin. VIII, 244 pages. 1992.

Vol. 1528: G. Isac, Complementarity Problems. VI, 297 pages. 1992.

Vol. 1529: J. van Neerven, The Adjoint of a Semigroup of Linear Operators. X, 195 pages. 1992.

Vol. 1530: J. G. Heywood, K. Masuda, R. Rautmann, S. A. Solonnikov (Eds.), The Navier-Stokes Equations II – Theory and Numerical Methods. IX, 322 pages. 1992.

Vol. 1531: M. Stoer, Design of Survivable Networks. IV, 206 pages. 1992.

Vol. 1532: J. F. Colombeau, Multiplication of Distributions. X, 184 pages. 1992.

Vol. 1533: P. Jipsen, H. Rose, Varieties of Lattices. X, 162 pages. 1992.

Vol. 1534: C. Greither, Cyclic Galois Extensions of Commutative Rings. X, 145 pages. 1992.

Vol. 1535: A. B. Evans, Orthomorphism Graphs of Groups. VIII, 114 pages. 1992.

Vol. 1536: M. K. Kwong, A. Zettl, Norm Inequalities for Derivatives and Differences. VII, 150 pages. 1992.

Vol. 1537: P. Fitzpatrick, M. Martelli, J. Mawhin, R. Nussbaum, Topological Methods for Ordinary Differential Equations. Montecatini Terme, 1991. Editors: M. Furi, P. Zecca. VII, 218 pages. 1993.

Vol. 1538: P.-A. Meyer, Quantum Probability for Probabilists. X, 287 pages. 1993.

Vol. 1539: M. Coornaert, A. Papadopoulos, Symbolic Dynamics and Hyperbolic Groups. VIII, 138 pages. 1993.

Vol. 1540: H. Komatsu (Ed.), Functional Analysis and Related Topics, 1991. Proceedings. XXI, 413 pages. 1993.

Vol. 1541: D. A. Dawson, B. Maisonneuve, J. Spencer, Ecole d´ Eté de Probabilités de Saint-Flour XXI - 1991. Editor: P. L. Hennequin. VIII, 356 pages. 1993.

Vol. 1542: J.Fröhlich, Th.Kerler, Quantum Groups, Quantum Categories and Quantum Field Theory. VII, 431 pages. 1993.

Vol. 1543: A. L. Dontchev, T. Zolezzi, Well-Posed Optimization Problems. XII, 421 pages. 1993.

Vol. 1544: M.Schürmann, White Noise on Bialgebras. VII, 146 pages. 1993.

Vol. 1545: J. Morgan, K. O'Grady, Differential Topology of Complex Surfaces. VIII, 224 pages. 1993.

Vol. 1546: V. V. Kalashnikov, V. M. Zolotarev (Eds.), Stability Problems for Stochastic Models. Proceedings, 1991. VIII, 229 pages. 1993.

Vol. 1547: P. Harmand, D. Werner, W. Werner, M-ideals in Banach Spaces and Banach Algebras. VIII, 387 pages. 1993.

Vol. 1548: T. Urabe, Dynkin Graphs and Quadrilateral Singularities. VI, 233 pages. 1993.

Vol. 1549: G. Vainikko, Multidimensional Weakly Singular Integral Equations. XI, 159 pages. 1993.

Vol. 1550: A. A. Gonchar, E. B. Saff (Eds.), Methods of Approximation Theory in Complex Analysis and Mathematical Physics IV, 222 pages, 1993.

Vol. 1551: L. Arkeryd, P. L. Lions, P.A. Markowich, S.R. S. Varadhan. Nonequilibrium Problems in Many-Particle Systems. Montecatini, 1992. Editors: C. Cercignani, M. Pulvirenti. VII, 158 pages 1993.

Vol. 1552: J. Hilgert, K.-H. Neeb, Lie Semigroups and their Applications. XII, 315 pages. 1993.

Vol. 1553: J.-L- Colliot-Thélène, J. Kato, P. Vojta. Arithmetic Algebraic Geometry. Trento, 1991. Editor: E. Ballico. VII, 223 pages. 1993.

Vol. 1554: A. K. Lenstra, H. W. Lenstra, Jr. (Eds.), The Development of the Number Field Sieve. VIII, 131 pages. 1993.

Vol. 1555: O. Liess, Conical Refraction and Higher Microlocalization. X, 389 pages. 1993.

Vol. 1556: S. B. Kuksin, Nearly Integrable Infinite-Dimensional Hamiltonian Systems. XXVII, 101 pages. 1993.

Vol. 1557: J. Azéma, P. A. Meyer, M. Yor (Eds.), Séminaire de Probabilités XXVII. VI, 327 pages. 1993.

Vol. 1558: T. J. Bridges, J. E. Furter, Singularity Theory and Equivariant Symplectic Maps. VI, 226 pages. 1993.

Vol. 1559: V. G. Sprindžuk, Classical Diophantine Equations. XII, 228 pages. 1993.

Vol. 1560: T. Bartsch, Topological Methods for Variational Problems with Symmetries. X, 152 pages. 1993.

Vol. 1561: I. S. Molchanov, Limit Theorems for Unions of Random Closed Sets. X, 157 pages. 1993.

Vol. 1562: G. Harder, Eisensteinkohomologie und die Konstruktion gemischter Motive. XX, 184 pages. 1993.

Vol. 1563: E. Fabes, M. Fukushima, L. Gross, C. Kenig, M. Röckner, D. W. Stroock, Dirichlet Forms. Varenna, 1992. Editors: G. Dell'Antonio, U. Mosco. VII, 245 pages. 1993.